Analytical Uses
of Immobilized Enzymes

GEORGE G. GUILBAULT

Laboratoire de Biologie et Technologie
  des membranes du CNRS
Université Claude Bernard, Lyon
Villeurbanne, France

Department of Chemistry
University of New Orleans
New Orleans, Louisiana

MARCEL DEKKER, INC.   New York and Basel

CHEMISTRY

7235-50/3

Library of Congress Cataloging in Publication Data

Guilbault, George G.
  Analytical uses of immobilized enzymes.

  (Modern monographs in analytical chemistry ; 2)
  Includes bibliographical references and indexes.
  1. Chemical tests and reagents. 2. Immobilized
enzymes. I. Title. II. Series.
QD77.G85 1984   543   84-4285
ISBN 0-8247-7125-7

MARCEL DEKKER, INC.
270 Madison Avenue, New York, New York 10016

Current printing (last digit):
10 9 8 7 6 5 4 3 2 1

PRINTED IN THE UNITED STATES OF AMERICA

Pour ma mère et mon père
et le petit chat rouge

Preface

Although 68 years have elapsed since Nelson and Griffin immobilized
the first enzyme, it was not until the 1950s that serious interest in
this field ensued. Since this period, an explosion in technology has
resulted, with several thousands of papers published, together with
numerous books and review articles. In 1970 when I wrote my first
book in the field of analytical enzymology, there were only 150 soluble enzymes and no immobilized enzymes commercially available. Today
427 soluble enzymes and over 50 immobilized enzymes are available
from various sources, several in different isoenzyme forms.

The advantages of the immobilized enzyme--reusability, stability,
insensitivity to inhibitors and activators, broad pH range, and
economy--when combined with the advantages of the specificity and
sensitivity provided by the enzyme itself, account for the extreme
popularity of the insolubilized biocatalyst as an analytical reagent.
Today these reagents are used in electrodes, enzyme columns, thin
layers, and other applications. Numerous commercial instruments are
available that use immobilized enzymes. It is this intense interest,
combined with the relative inaccessibility of a great diversity of
publications spread in various places in the literature, that prompted
the compilation resulting in this book.

In Chapters 1 and 2 a thorough discussion of the basic principles
of the use of enzymes as reagents and of immobilized enzymes in particular is presented. The remaining chapters are devoted to analytical

uses of immobilized enzymes: enzyme electrode probes including microbial and immunological electrodes (Chapter 3), enzyme reactors and membranes (Chapter 4), the applications of enzyme layers (solid surface, dip sticks, and other techniques), and commercial instrumentation (Chapter 5).

Finally, a listing of all commercially available soluble and immobilized enzymes is given in the appendixes. This feature was very popular in my last book, *Handbook of Enzymatic Methods of Analysis* (Dekker, 1976), and I hope it will again prove invaluable to experimentalists.

I would like to express my sincere appreciation to Mrs. Gayle Barlow, who typed the manuscript.

The dedication of this book to the three most special individuals in my life, my mother, Valerie Kothe Guilbault, my father, George Robert Guilbault, and my wife, Susan, is with deep love and appreciation for their support, encouragement, and devotion--without this, the present manuscript would never have materialized.

George G. Guilbault

Contents

PREFACE   v

1. GENERAL CONCEPTS OF USE OF ENZYME REAGENTS   1

    I.   Basic Properties of Enzymes   1
         A.   Enzymes as Catalysts   1
         B.   Enzyme Structure   2
         C.   Specificity of Enzymes   9
         D.   Enzymes as Analytical Reagents   10
         E.   Enzyme Classification   11
         F.   Definition of Enzyme Activity   14
         G.   Isoenzymes   16
    II. Properties of Enzyme-Catalyzed Reactions   17
         A.   Introduction   17
         B.   Kinetics of Uncatalyzed Reactions   17
         C.   Enzyme Kinetics of Homogeneous Systems   19
         D.   Preincubation and Lag Phase   23
    III. Factors Affecting Enzyme-Catalyzed Reactions   24
         A.   Substrate Concentration   24
         B.   Enzyme Concentration   26
         C.   Activators   27
         D.   Inhibitors   28
         E.   Temperature   31
         F.   pH   34
         G.   Buffer   35
         H.   Ionic Strength   38
    IV. Importance of Enzymatic Analysis   39
         A.   As a Diagnostic Tool in Clinical Chemistry   39
         B.   In Food Processing and Fermentation   45
         C.   In Other Fields   47
    V.   Experimental Techniques   47
         A.   Determination of Concentration of Participants
              in Enzymatic Reactions   47
         B.   Methods for Measurement of Enzymatic Reactions   56
         C.   Handling of Biochemical Reagents   69
    Suggested Readings   72
    References   73

                                                                         vii

2. PRINCIPLES OF IMMOBILIZED ENZYMES   77

      I. General Discussion   77
          A. Reusability   78
          B. Greater Stability   78
          C. Fewer Interferences   78
     II. Immobilization Methods   78
          A. Physical Entrapment   79
          B. Microencapsulation   82
          C. Adsorption   83
          D. Covalent Crosslinking by Bifunctional Reagents   83
          E. Covalent Binding to Water-Insoluble Matrices   84
    III. Properties of Immobilized Enzymes   93
          A. General Discussion   93
          B. Characteristics of Immobilized Enzymes   96
     IV. Examples of Experimental Preparation of Immobilized
          Enzymes   98
          A. Preparation of Polyacrylamide-Entrapped Enzymes   98
          B. Preparation of Covalently Bound Enzymes   98
      V. Reviews on Immobilized Enzymes   103
     VI. Analytical Applications of Immobilized Enzymes   103
          References   104

3. ENZYME ELECTRODE PROBES   112

      I. General Discussion   112
     II. Use of Electrodes to Measure Enzyme Reactions   113
          A. General Discussion   113
          B. O_2 Electrode   113
          C. Gas Membrane Electrodes   115
          D. Solid-Membrane Electrodes   117
          E. Oxidation-Reduction Methods   119
          F. Conductimetric Methods   121
    III. Enzyme Electrode Probes   121
          A. General Discussion   121
          B. Construction of Enzyme Electrodes   125
          C. Performance Characteristics of Electrodes   137
          D. Effect of Interferences   157
          E. Assay of Inorganic Ions   164
          F. Assay of Organic Species   167
          G. Review   204
     IV. Probes for Analysis of Enzymes   206
      V. Electrode Probes Utilizing Whole Cells: Microbial
          or Tissue Enzymes   211
          A. General Discussion   211
          B. Applications and Types   217
     VI. Immunoelectrode Probes   226
          A. General Discussion   226
          B. Linked Antibodies   226
          C. Bound Antigens   230
          References   230

Contents

4. ENZYME REACTORS AND MEMBRANES 245

    I. General Discussion 245
    II. Analytical Uses of Immobilized Enzyme Reactors and
        Nylon Tube Bound Enzymes 248
        A. Electrochemical Detectors 248
        B. UV and Visible Detectors 265
        C. Thermal Detectors 285
        D. Luminescence Detectors 298
    References 304

5. APPLICATIONS OF ENZYME LAYERS AND OTHER TECHNIQUES.
   COMMERCIAL INSTRUMENTATION 311

    I. Applications of Enzyme Layers 311
        A. Semisolid Surface Fluorescence 311
        B. The Immobilized Enzyme Stirrer 327
        C. Visual Color Tests Using Immobilized Enzymes 331
        D. Mass Spectrometry 332
        E. Other Techniques 333
    II. Commercial Instruments 333
        A. General Discussion 333
        B. Yellow Springs Instrument Company Glucose
           Analyzer 334
        C. Leeds and Northrup Analyzer 337
        D. Kimble Division of Owens-Illinois 338
        E. Midwest Research (CAM and IEM) 341
        F. Setric and Roche Analyzers 347
        G. Tacussel 347
        H. Enzyme Coils 348
        I. Universal Sensors 350
    References 350

Appendix 1 NUMBERING AND CLASSIFICATION OF ENZYMES 353

Appendix 2 IMMOBILIZED (INSOLUBLE) ENZYMES 361

Appendix 3 PURIFIED ENZYMES (SOLUBLE) 366

Appendix 4 SUPPLIERS OF ENZYMES 417

AUTHOR INDEX 421

SUBJECT INDEX 441

Analytical Uses
of Immobilized Enzymes

1

General Concepts of Use of Enzyme Reagents

I. BASIC PROPERTIES OF ENZYMES

A. Enzymes as Catalysts

Enzymes are special biochemical catalysts synthesized by all living organisms. They accelerate the magnitude of metabolic reactions at ordinary temperatures. As catalysts enzymes affect the rate at which equilibrium is reached yet themselves undergo no permanent change. Denaturation of enzymes may be indicated by loss of enzymatic activity long before other physical or chemical evidence of denaturation is demonstrated.

In living organisms enzymes are rapidly degraded and their supply is replenished by new synthesis. Enzyme preparations are generally unstable and must therefore be handled with special care.

The existence of enzymes has been known for about 140 years, the earliest studies having been performed by the Swedish chemist Berzelius [1], who first called them catalysts. In 1926 Summer isolated the first enzyme, urease (urea amidohydrolase, EC 3.5.1.5), a discovery that won him the Nobel Prize in 1947.

Today more than 2000 enzymes are known. Several hundred have been isolated, many in pure or crystalline form. Over 400 are available commercially, some of these from as many as 10 different sources (Appendix 3). Some enzymes are relatively small molecules with molecular weights in the order of 10,000, whereas others are very large

molecules with molecular weights of 150,000 to 2 million.   Some
enzymes are albumins, and others have the properties of globulins.

B.   Enzyme Structure

Enzymes are members of a class of substances called proteins, being
high molecular weight compounds which catalyze key reactions on which
depend the existence of life as we know it.   They are distinguishable
from each other by a geometric area called the active site, into which
will fit only one or a very limited number of compounds called substrates.

    The protein structure of the enzyme is composed of a sequencing
of amino acids (Table 1) which have the basic zwitterion structure:

$$H - \overset{\overset{R}{|}}{\underset{\underset{COO^{\ominus}}{|}}{C}} - NH_3^+$$

which exists at the isoelectric point (a pH value indicated by the
pI value in Table 1).   In the protein structure these net positive

TABLE 1   Structures of Key Amino Acids in Proteins

Amino acids	Structure	pK values[a]	pI[b]
Alanine	$\overset{CH_3}{\underset{H_2N-CH-COOH}{|}}$	2.35, 9.69	6.02
Arginine	$\overset{\qquad\qquad\overset{NH}{\|}}{(CH_2)_3-NH-C-NH_2}$ $NH_2-CH-COOH$	2.17 9.04 (α-amino) 12.48 (guanidino)	10.76
Aspartic acid	$\overset{COOH}{\underset{\overset{CH_2}{\underset{NH_2-CH-COOH}{\|}}}{\|}}$	2.09 (α-carboxyl) 3.86 (β-carboxyl) 9.82	2.97
Asparagine	$\overset{CONH_2}{\underset{\overset{CH_2}{\underset{H_2N-CH-COOH}{\|}}}{\|}}$	2.02 8.8	5.41

TABLE 1 (Continued)

Amino acids	Structure	pK values[a]	pI[b]
Cysteine	CH_2-SH NH_2-CH-COOH	1.71 8.33 (sulfhydryl) 10.78 (α-amino)	5.02
Glutamic acid	COOH CH_2 CH_2 H_2N-CH-COOH	2.19 (α-carboxyl) 4.25 (γ-carboxyl) 9.67	3.22
Glutamine	$CONH_2$ CH_2 CH_2 H_2N-CH-COOH	2.17 9.13	5.65
Glycine	H_2N-CH_2-COOH	2.34, 9.6	5.97
Histidine	HC$\diagup$N C$\diagdown$N CH_2 H_2N-CH-COOH	1.82 6.0 (imidazole) 9.17	7.58
Isoleucine	CH_3 CH_2 CH-CH_3 H_2N-CH-COOH	2.36 9.68	6.02
Leucine	CH_3 $\diagdown$ $\diagup$ CH_3 CH CH_2 H_2N-CH-COOH	2.36 9.60	5.98
Lysine	$(CH_2)_4$-NH_2 H_2N-CH-COOH	2.18 8.95 (α-amino) 10.53 (ϵ-amino)	9.74

TABLE 1 (Continued)

Amino acids	Structure	pK values[a]	pI[b]
Methionine	$(CH_2)_2$-S-CH_3 H_2N-CH-COOH	2.28 9.21	5.75
Phenylalanine	(benzene ring) CH_2 H_2N-CH-COOH	1.83 9.13	5.98
Proline	(pyrrolidine ring) N—COOH H	1.99 10.60	6.10
Serine	CH_2-OH H_2N-CH-COOH	2.21 9.15	5.68
Threonine	CH_3 HO-CH H_2N-CH-COOH	2.63 10.43	6.53
Tyrosine	OH (benzene ring) CH_2 H_2N-CH-COOH	2.20 9.11 (α-amino) 10.07 (phenolic hydroxyl)	5.65
Tryptophan	(indole ring) NH CH_2 H_2N-CH-COOH	2.38 9.39	5.88
Valine	CH_3 CH_3 CH H_2N-CH-COOH	2.32 9.62	5.97

[a] pK_a values of the amino acids.
[b] pIsoelectric point.

and negative charges due to the amino and carboxyl groups are eliminated due to interaction of these groups in forming the peptide (-C-N-)
bond. The side chains of the amino acids (e.g., the CH_2-⬡ of   O
phenylalanine) are nonpolar, positive-charged polar, negative-charged
polar, or uncharged polar; this polarity greatly influences the structure. The size, shape, and function of the protein are affected by
the number of amino acids, their specific position, and their interaction in forming disulfide (-S-S-) or hydrogen bonds which link the
enzyme into a helix coil configuration.

The primary structure of the enzyme (protein) is due to the amino
acid sequence; the secondary structure is the relationship of the amino
acids via their close proximity to one another in the primary structure;
the tertiary structure is the interrelation of the parts of the protein
too far from one another in the primary structure; and the quaternary
structure is the spatial relationship of the peptide in the multichain
protein.

Two types of secondary structures have been proposed: the α helix
by Pauling et al. [2] and the β-pleated sheet. In the α helix each
amino group of the peptide backbone is hydrogen-bonded to the carboxyl
group of the amino acid residues behind it. The right-handed α helix
appears most stable and is most frequently observed in protein structure. In the β-pleated sheet the hydrogen bonding takes place between
parallel sections of the peptide chains rather than between close
neighbors.

The actual three-dimensional shape of the protein is the result
of competition and cooperation of various factors: the hydrogen bonding
of the secondary and tertiary structure, disulfide bonds of secondary
and quaternary structure, the coulombic forces and hydrophobic bonding
in the tertiary structure, as well as, of course, the nature of the
amino acid sequence.

In 1860 Emil Fischer postulated the mechanism by which the active
site takes part in the reaction of an enzyme and its substrate. In
fact, the specificity of the enzymatic reaction was ascribed to a
"lock-and-key" fit of enzyme and substrate.

Many enzymes require the presence of a metal ion, or a coenzyme
(a cofactor), in order to be active catalysts.

$$\text{Apoenzyme} \quad + \quad \text{cofactor} \longrightarrow \text{holoenzymes} \qquad (1)$$

(protein fraction)   (organic part)   (conjugated-active
catalyst)

This apoenzyme is frequently without enzymatic activity (e.g., D-amino acid oxidase without FAD); then, when the cofactor is added, activity is maximum. The cofactor is a catalyst also, in the sense that it is regenerated or recycled, usually in a metabolic pathway. Table 2 gives a listing of some of the common coenzymes or cofactors that participate in enzymatic reactions.

TABLE 2   Structure of Coenzymes

Coenzyme	Structure

1. Adenosine phosphates
    AMP: $n = 1$
    ADP: $n = 2$
    ATP: $n = 3$	

2. Coenzyme A	

3. Cytidine phosphates
    $n = 1$ monophosphate
    $n = 2$ diphosphate
    $n = 3$ triphosphate	

TABLE 2 (Continued)

Coenzyme	Structure

4. Flavine nucleotides (FMN, FAD)

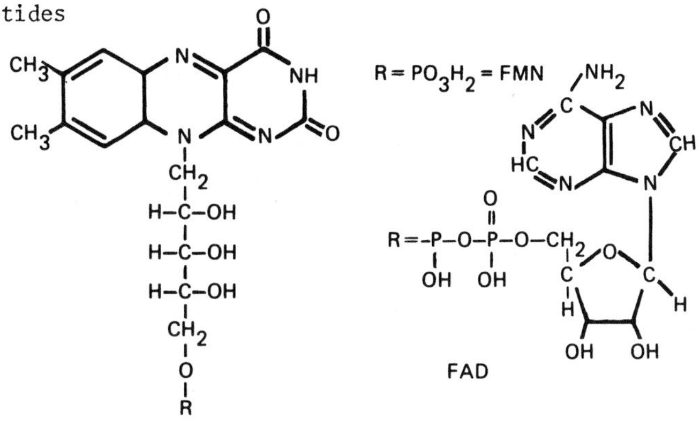

$R = PO_3H_2 = FMN$

$R = -P-O-P-O-CH_2 \ldots$

FAD

5. Guanosine and inosine

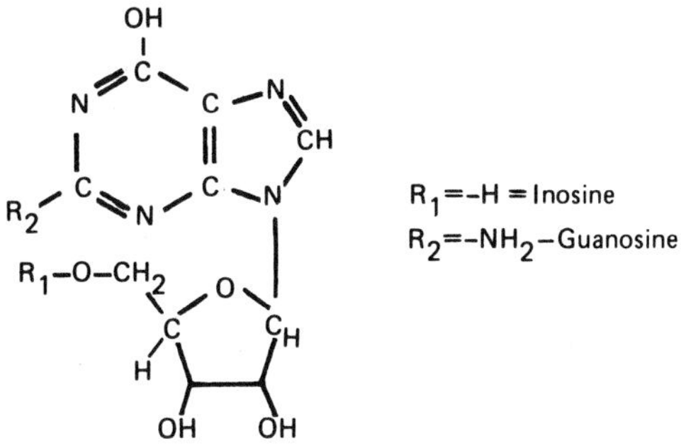

$R_1 = -H = $ Inosine

$R_2 = -NH_2 - $ Guanosine

6. Nicotinamide adenine dinucleotide (NAD, NADH)

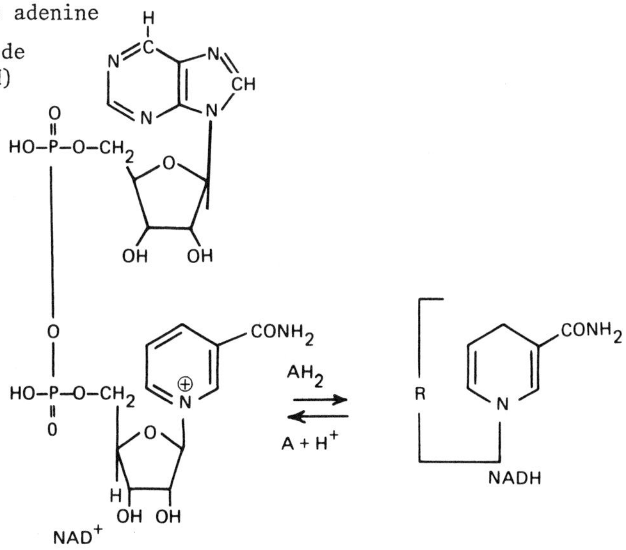

NAD^+

AH_2

$A + H^+$

$CONH_2$

NADH

TABLE 2 (Continued)

Coenzyme	Structure
7. Nicotinamide adenine dinucleotide phosphate (NADP, NADPH)	
8. Pyridoxyl phosphate	
9. Thiamine pyrophosphate	

10. Uridine phosphates
    n = 1 uridine monophosphate
    n = 2 uridine diphosphate
    n = 3 uridine triphosphate	

There are three classes of cofactors: (1) tightly bound cofactors, such as pyridoxyl phosphate held at the active site of decarboxylative enzymes, like tyrosine decarboxylases, or to transaminases, or flavine mononucleotide (FMN) in glucose oxidase and flavine adenosine dinucleotide (FAD) with D-amino acid oxidase; (2) loosely bound cofactors, which frequently must be added in nonstoichiometric amounts, such as thiamine phosphate in dehydrogenase or decarboxylase enzyme systems; and (3) the substrate cofactor, which participates stoichiometrically in the enzymatic reaction.

The best known of this third type of cofactor are NAD (nicotinamide adenine dinucleotide) and NADP (nicotinamide adenine dinucleotide phosphate), which function as electron acceptors. They are reduced by the addition of two equivalents of hydrogen ions from the substrate, as catalyzed by the dehydrogenase enzyme. For example, alcohol dehydrogenase (ADH) catalyzes the oxidation of ethanol by NAD with addition of two hydrogen atoms to NAD.

$$\text{EtOH + NAD} \xrightarrow[\text{H}^+]{\text{ADH}} \text{acetaldehyde + NADH}_2 \tag{2}$$

The coenzyme ATP (adenosine-5'-triphosphate), is involved in phosphoryl transfer reactions, such as the hexokinase-catalyzed phosphorylation of glucose:

$$\text{Glucose + ATP} \xrightarrow{\text{hexokinase}} \text{ADP + glucose phosphate} \tag{3}$$

Coenzyme A serves as an acyl group carrier in acyl transfer reactions; the functional part of the molecule is the sulfhydryl group. More than 60 enzymes utilize CoA as a coenzyme.

Various other cofactors utilized are listed in Table 2.

C. Specificity of Enzymes

Because enzymes work in complex living systems, one of their properties is specificity. Some enzymes have nearly absolute specificity for a given substrate and do not attack even closely related substrate molecules; others are far less specific and act on an entire class of substrate molecules.

Among the relatively specific enzymes is aspartase (EC 4.3.1.1),

which catalyzes the reversible addition of ammonia to the double bond
of fumarate to form L-aspartate. Phosphatases, peptidases, and amino
acid oxidases, on the other hand, show a relatively broad specificity.
This specificity of enzymes, and their ability to catalyze reactions
at extremely low concentrations, is of significant use in biochemical
analysis.

Enzymes also exhibit selectivity with respect to a particular
reaction. If one attempted to determine glucose by oxidation in an
uncatalyzed way, such as by heating a glucose solution with an oxidizing agent such as ceric perchlorate, many side reactions would
occur to yield products in addition to gluconic acid. With the enzyme
glucose oxidase, however, catalysis is so effective at room temperature
and pH 7 that the rates of the other thermodynamically feasible reactions are negligible.

D. Enzymes as Analytical Reagents

Enzyme-catalyzed reactions have been used for analytical purposes for
many years, in the determination of substrates, activators, inhibitors,
as well as enzymes themselves. Osann [3] used peroxidase for an assay
of peroxide in 1845, and enzymatic methods were accepted techniques
for the analysis of carbohydrates in the nineteenth century. In the
early 1940s, procedures based on the photometric measurement of the
reduced coenzymes NAD and NADP were described by Warburg [4]. This
and the developments of immobilized enzymes with electrochemical,
luminescence, and thermometric methods of analysis have raised enzymatic procedures to the forefront of accepted analytical techniques
in such areas as clinical chemistry.

Because of the relatively recent availability of large numbers
of highly purified, very active enzyme preparations at reasonable
costs, it is now possible to use enzymes as laboratory reagents for
routine analytical work. Many biological compounds can be determined
because of the substrate specificity of enzymes. One important application is the use of glucose oxidase (β-D-glucose: oxygen-1-oxidore-
ductase, EC 1.1.3.4) for the assay of blood glucose, the single most
common assay in the clinical laboratory. This enzyme is highly specific, catalyzing only the oxidation of β-D-glucose. 6-Deoxy-fluoro-

D-glucose and 2-deoxy-D-glucose are oxidized much more slowly by
molecular oxygen with glucose oxidase. Similarly, the use of galactose oxidase permits the selective assay of β-D-galactose in the
presence of β-D-glucose and most other reducing sugars. Many other
important constituents of serum can likewise be accurately analyzed
using enzymes.

There are two possibilities for determining the concentration
of a substrate participating in an enzyme reaction. The first is to
measure the total change that occurs in the end product or in the
unreacted starting material by chemical or physical analysis. Large
amounts of enzyme and small amounts of substrate are used to ensure
a complete reaction. In the second method, which is a kinetic method,
the initial rate of reaction is measured. This is done by following
the production of products or the disappearance of substrate with
time. In this method, the rate of reaction is a function of the
concentration of substrate, enzyme, inhibitor, and activators.

Because enzymes are catalysts, affecting the rate and not the
equilibrium or end point of a reaction, their concentration and
activity must be measured by a rate or kinetic method. The rate
method is faster than the total change method because it enables the
measurement of the initial rate without waiting for equilibrium to
be established. The accuracy and precision of both methods are comparable.

E. Enzyme Classification

A systematic nomenclature for the identification of enzymes according
to the type of reaction they catalyze was originally recommended by
the Enzyme Commission of the International Union of Biochemistry in
1961 and was updated in 1975 [5]. This system was designed to eliminate any ambiguity in identification of any enzyme. The system in-
volves subdivision of all enzymes into six major groups according to
the type of reaction catalyzed. The number of enzymes in each group
shown is taken from the list of enzymes produced in 1972. A complete
listing of the numbering and classification of enzymes can be found
in Appendix 1.

1. OXIDOREDUCTASES (about 570 enzymes)

In this class are all enzymes catalyzing oxidation-reduction reactions. The oxidized substrate is the hydrogen donor. The recommended name is dehydrogenase, but reductase can also be used. If O_2 is the acceptor, then oxidase can be used. The systematic name is donor:acceptor oxidoreductase.

The second figure in the code indicates the group that undergoes oxidation (e.g., 1 denotes a CH-OH group, 2 an aldehyde or keto group, and so on, as listed in Appendix 2).

The third figure denotes the type of acceptor (e.g., 1 denotes NAD(P), 2 a cytochrome, 3 molecular O_2, and so on).

The final digit indicates the particular enzyme (e.g., alcohol dehydrogenase is 1.1.1.1).

Examples:

Subclass 1.1.1 $S + NAD \rightleftharpoons P + NADH$ (4)

1.1.2 $S + cytochrome(ox) \rightleftharpoons P + cytochrome(red)$ (5)

1.1.3 $S + O_2 \rightleftharpoons P + H_2O_2$ (6)

2. TRANSFERASES (about 490 enzymes)

Enzymes which catalyze the transfer of a group, such as a methyl, carboxyl, or hydroxyl, from one compound (the donor) to another (the acceptor) are called transferases. The systematic name is donor:acceptor group transferase, and the recommended names are formed according to the acceptor or donor group. In many cases the donor is a coenzyme (cofactor).

The second digit in the code indicates the group transferred: 2.1 is a one-carbon group, 2.2 an aldehydic or ketonic group, 2.3 a glycosyl group, and so on (see Appendix 2).

The third figure indicates the group transferred: 1 a methyl group, 2 a hydroxymethyl, 3 a carboxyl, and so on. In subclass 2.7, the third figure indicates the nature of the acceptor group (1 an alcohol group, 2 a carboxyl, and so on).

Examples:

2.1.1.1, nicotinamide methyltransferase:

S-adenosyl-L-methionine + *nicotinamide* $\rightleftharpoons$ S-adenosyl-L-

    homocysteine + 1-methyl*nicotinamide*   (7)

2.7.1.1, hexokinase:

ATP + D-hexose $\rightleftharpoons$ ADP + D-hexose-6-*phosphate*   (8)

3. HYDROLASES (about 560 enzymes)

In this group are contained those enzymes which catalyze the
hydrolytic cleavage of C-O, C-N, C-C, and some other bonds. The
systematic name always includes hydrolase; the recommended name is
formed by adding -ase to the name of the substrate (e.g., cholinesterase is an enzyme that catalyzes the hydrolysis of cholinesters).

The second figure in the code indicates the nature of the bond
hydrolyzed (e.g., 3.1 are the *esterases*, 3.2 the *glycosid*ases, etc.).

The third figure denotes the nature of the substrate (e.g.,
3.1.1 a *carboxylic* ester hydrolase, 3.1.2 a *thiol* ester hydrolase,
etc.) (see Appendix 2).

Examples:

Subclass 3.1   ester + H_2O $\rightleftharpoons$ acid + alcohol   (9)

Subclass 3.5   L-asparagine + H_2O $\rightleftharpoons$ L-aspartate + NH_3 (10)

4. LYASES (about 240 enzymes)

Enzymes that cleave C-C, C-O, C-N, and other bonds by elimination, leaving double bonds or adding groups to double bonds, are
called lyases. It is important to add a hyphen (-) to avoid confusion with other enzymes (e.g., hydro-lyase, not hydrolyase). Recommended names are decarboxylase, aldolase, deamines, and dehydratases.

The second digit indicates the bond broken (e.g., 4.1 is a
carbon-carbon lyase, 4.2 carbon oxygen lyase, and so on) (see Appendix
2).

The third figure gives information on the group eliminated (e.g.,
CO_2 in 4.1.1, H_2O in 4.2.1).

Examples:

Subclass 4.1.1 $S \rightleftharpoons P + CO_2$ (11)

Subclass 4.3.1 $S \rightleftharpoons P + NH_3$ (12)

5. *ISOMERASES (about 85 enzymes)*

Those enzymes which catalyze structural or geometric changes within one molecule are called isomerases. Depending on the type of isomerism, they may be called racemases, epimerases, isomerases, tantomerases, cis-trans isomerases, mutases, or cycloisomerases.

The subclasses are formed according to the type of isomerism, the subsub classes to the types of substrates.

Examples:

Subclass 5.1.1   L-amino acid $\rightleftharpoons$ D-amino acid   (13)

Subclass 5.3.1   Aldolase $\rightleftharpoons$ ketose   (14)

6. *LIGASES (SYNTHETASES) (about 80 enzymes)*

Ligases are enzymes which catalyze the joining together of two molecules coupled with the hydrolysis of a pyrophosphate bond in ATP or a similar triphosphate. The bonds formed are frequently high-energy bonds. The systematic names are formed according to the system X:Y ligase (ADP-forming). The term *synthetase* can be used in the recommended nomenclature if no other short term, such as carboxylase, is available.

The second figure in the code indicates the bond formed: 6.1 for C-O bonds (enzymes acylating t-RNA), 6.2 for C-S bonds (acyl CoA derivatives), etc. Subsub classes are only in use in the C-N ligases.

Examples:

Subclass 6.2   ATP + acid + CoA $\rightleftharpoons$ AMP + PO_4^{3-} + acyl CoA   (15)

Subclass 6.3   ATP + NH_3 + acid $\rightleftharpoons$ AMP + PO_4^{3-} + amine   (16)

F. Definition of Enzyme Activity

The amount of enzyme measured cannot be expressed in classical terms, such as milligrams per deciliter, because only a very small amount of catalytically active enzyme may be present in a given weight of this substance. The result of enzyme measurements are hence reported in enzyme activity units. There is considerable confusion in such reporting, however, since several different enzyme units have come into usage, mainly through their origin in the technique of a certain investigator. For example, urease has been reported in Sumner units and glutamic-oxaloacetic transaminase (GOT) in Karmen units; there are also Bodansky, Bucher, Wroblewski, Rosalki, and Babson units and many others. Such reportage of results has made it extremely difficult to compare results on the same enzyme. Hence, the Fifth International Congress of Biochemistry acted to solve this problem by adopting the recommendations of the IUPAC and the International Union of Biochemistry Commissions on Enzymes for definition of an enzyme unit [5].

One unit (U) of any enzyme is defined as that amount which will catalyze the transformation of 1 micromole of substrate per minute or, where more than one bond of each substrate is attached, 1 microequivalent of the group concerned per minute under defined conditions. The temperature should be stated and 25°C has been recommended for use wherever possible. The other conditions, including pH and substrate concentration, should be optimal whenever practical.

It was also recommended that enzyme assays be based, wherever possible, on measurements of initial rates of reaction in order to avoid complications due to the formation of inhibitory products or to the reversibility of reactions. The substrate concentration used should be, wherever possible, sufficient for saturation of the enzyme so that the kinetics in the standard assay approach zero order in substrate. If a suboptimal concentration of substrate must be used, it is recommended that the Michaelis constant be determined so that the observed rate can be converted into that expected on saturation with substrate.

The concentration of an enzyme in solution should be expressed as units per milliliter (U/ml) or international units per liter (IU/liter).

The specific activity of an enzyme is expressed as units of enzyme per milligram of protein (U/mg protein). The molecular activity is defined as the units per micromole of enzyme at optimal substrate activity, that is, as the number of molecules of substrate transformed per minute per molecule of enzyme.

A recent proposal has been to accept the katal as the unit of enzyme activity. The katal would refer to the conversion of 1 mole of substrate per second and would equal 6×10^7 µmole/min of substrate or 6×10^7 U. One unit would equal 16.67 nkatals.

G. Isoenzymes

Enzymes are found in the cells of many different tissues in the body. An enzyme which is found in a variety of tissues possessing very similar or identical activity may not always have an identical structure but may have multiple structures. Many enzymes exist in such multiple or isoenzyme forms, which have the same structure at the active site or center but different overall configuration. However, despite this observation, very few isoenzyme measurements have become routinely used in clinical analysis. The best known isoenzymes are those of lactic dehydrogenase (LDH), creatine kinase (CK), and acid and alkaline phosphatase, and it is likely that all enzymes exist in such multiple forms.

Isoenzymes can be separated by electrophoresis and by physical and chemical differences, such as heat stability and the effect of inhibitors, and their affinity for their respective substrates and coenzymes can be demonstrated.

In 1961, Wroblewski and Gregory [6] first suggested calling the different enzymatic forms of one enzyme with similar activity "isoenzymes." Five major isoenzymes of LDH have been isolated, each composed of four subunits or monomers. A pure tetramer of four similar subunits was demonstrated in heart (H_4) while a different tetramer of four similar subunits was found in muscle (M_4). In addition, there are three tetramers of mixed subunits (H_3M, H_2M_2, and HM_3). Tissues with heart-type isoenzymes are primarily concerned with the conversion of lactate to pyruvate, whereas those with muscle-type subunits are active in the conversion of pyruvate to lactate.

In 1965 Dawson et al. [7] demonstrated the existence of three creatine phosphokinase (CK) isoenzymes, designated BB (brain), MM (skeletal muscle), and MB (brain muscle hybrid). Takahashi et al. [8] showed the hybrid form to be of cardiac origin.

II. PROPERTIES OF ENZYME-CATALYZED REACTIONS

A. Introduction

Because of their specificity and sensitivity, enzymes have a great potential usefulness in analytical chemistry. For many years enzyme kinetic methods have been used for the determination of substrates, enzymes, activators, and inhibitors. It has been only a short time since the kinetic theory of enzyme systems has been mastered such that experiments are possible with real enzyme systems. Knowledge of enzyme kinetics is important because it is necessary for the understanding of the analytical aspects of enzyme behavior. It is also a powerful tool for the understanding of properties of enzymes and the mechanism of enzyme action.

B. Kinetics of Uncatalyzed Reactions

Before the properties of enzyme-catalyzed reactions are examined, a short discussion of chemical kinetics is in order. Chemical reactions can be classified according to the number of molecules of reactants that combine to form products. Accordingly, reactions are classified as monomolecular, bimolecular, or trimolecular if the number is one, two, or three, respectively. They can also be classified according to how the concentrations of the reactants influence the rate. Reactions may be classified as first order, second order, or third order.

The simplest example is for the first order reaction:

$$A \longrightarrow P$$

in which one reactant (A) is irreversibly converted to product (P) according to the law of mass action:

$$v = \frac{-dA}{dt} = \frac{dP}{dt} = kA \tag{17}$$

where the velocity v of the reaction is the rate of disappearance of A or appearance of P, and k is the rate constant. The rate is first order because the velocity is directly proportional to the first power of the concentration of A.

If the mechanistic equation were

$$2A \longrightarrow P \tag{18}$$

the velocity expression would then be

$$v = -\frac{1}{2}\left(\frac{dA}{dt}\right) = \frac{dP}{dt} = kA^2 \tag{19}$$

and the reaction would be second order. A bimolecular reaction could involve two different compounds. For the mechanistic reaction

$$A + B \longrightarrow P \tag{20}$$

the velocity is given by

$$v = \frac{-dA}{dt} = \frac{-dB}{dt} = \frac{dP}{dt} = kAB \tag{21}$$

In this case the reaction is second order. If the concentration of A were very high and that of B very low, the concentration of A would remain virtually constant. The A concentration term of the rate equation could then be combined with the rate constant and the reaction would be pseudo first order overall. This would be the case, for example, when A is the solvent. This is especially important in analytical methods because it is desired that the initial velocity of a reaction be dependent on the concentration of only one substrate.

It is also possible to have zero order reactions. In this case the velocity does not depend on the reactant concentrations. Examples of third order reactions are less frequent in the literature. Lastly, the reaction order need not be a whole number; it may be fractional.

Most mechanisms consist of combinations of unimolecular and bi-molecular steps. Trimolecular reactions are rare because simultaneous three-way collisions must occur. For mechanisms involving more than one reaction, separate velocity equations must be written for each step.

C. Enzyme Kinetics of Homogeneous Systems

   1. *THE BASIC MODEL*

   The basic equations for enzyme kinetics were developed by
Michaelis and Menten [9]. According to the mechanism:

$$E + S \underset{k_{-1}}{\overset{k_1}{\rightleftharpoons}} ES \overset{k_2}{\longrightarrow} E + P \tag{22}$$

substrate S combines with enzyme E to form an enzyme-substrate complex ES which breaks down into product P and liberates the enzyme.
Figure 1 shows the time course of the concentration of constituents
involved for the case $k_1 = k_{-1} = k_2$ with the total concentration of
enzyme small in comparison with the substrate concentration. After
a transient initial period concentrations of E and ES vary only
slightly with time until the end of the reaction.

   Using the steady-state assumption that $dE/dt = 0$ and $d(ES)/dt =$
0, along with the kinetic equations:

$$\frac{dS}{dt} = k_{-1}(ES) - k_1(E)(S) \tag{23}$$

$$\frac{dE}{dt} = (k_{-1} + k_2)(ES) - k_1(E)(S) \tag{24}$$

$$\frac{d(ES)}{dt} = k_1(E)(S) - (k_{-1} + k_2)(ES) \tag{25}$$

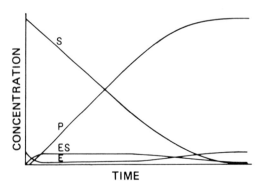

FIG. 1   Time course for $E + S \underset{k_{-1}}{\overset{k_1}{\rightleftharpoons}} ES \overset{k_2}{\longrightarrow} E + P$ with $k_1 = k_{-1} = k_2$.

$$\frac{dP}{dt} = k_2 (ES) \qquad\qquad (26)$$

and from the conservation equation for enzyme:

$$E_t = E + ES \qquad\qquad (27)$$

the following equation can be derived:

$$v = \frac{k_2 E_t S}{(k_{-1} + k_2)/(k_1) + S} \qquad\qquad (28)$$

This equation reduces to the Michaelis-Menten equation:

$$v = \frac{VS}{K_m + S} \qquad\qquad (29)$$

where $V = k_2 E_t$, $K_m = (k_{-1} + k_2)/k_1$, and S is the initial substrate concentration. V is known as the maximum velocity (V_{max}) and K_m as the Michaelis constant.

The values of the kinetic constants K_m and V may be obtained from a plot of v versus S as shown in Fig. 2. From the Michaelis-Menten equation it is seen that K_m is equal to the value of S when v equals V/2. However, other forms of the Michaelis-Menten equation are more useful in plotting experimental data and obtaining values

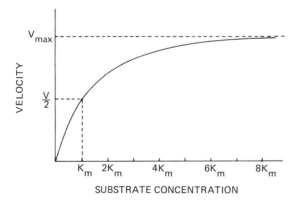

FIG. 2 Effect of substrate concentration on the initial rate of enzyme reactions.

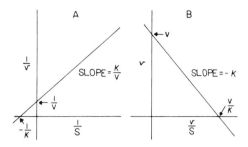

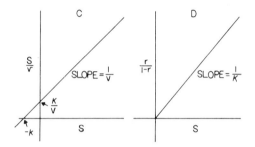

FIG. 3 Plots of linear transformations of the Michaelis-Menten
equation: (A) Lineweaver-Burk; (B) Hofstee; (C) Hanes; (D) Reiner.

for the kinetic constants K_m and V_{max}. Several of these methods are
shown in Fig. 3.

The most widely used linear transformation of the Michaelis-
Menten equation is the Lineweaver-Burk equation [10]:

$$\frac{1}{v} = \frac{K_m}{V} \cdot \frac{1}{S} + \frac{1}{v} \tag{30}$$

A plot of $1/v$ versus $1/S$ is a straight line with a slope of K_m/V and
an intercept of $1/V$ on the $1/v$ axis and $-1/K$ on the $1/S$ axis as shown
in Fig. 3A. Other widely used linear transformations and plots are
the Hofstee (Fig. 3B):

$$v = V - \left(\frac{v}{S}\right)K_m \tag{31}$$

the Hanes (Fig. 3C):

$$\frac{S}{v} = \frac{K_m}{V} + \frac{1}{V}S \tag{32}$$

and the Reiner (Fig. 3D):

$$\frac{r}{1 - r} = \frac{S}{K_m} \tag{33}$$

where $r = v/V$.

2. A MORE REALISTIC MODEL

In the development of the Michaelis-Menten equation and in the basic reaction mechanism, Eq. (21), no provision was made for any reverse reaction of the second step. A more accurate mechanistic model would include the backward reaction with k_{-2} as shown below:

$$E + S \underset{k_{-1}}{\overset{k_1}{\rightleftarrows}} ES \underset{k_{-2}}{\overset{k_2}{\rightleftarrows}} E + P \tag{34}$$

Assuming again that E_t is small in comparison with S and the steady-state assumption $d(ES)/dt = dE/dt = 0$, new kinetic equations can be solved to give

$$v = \frac{(k_1 k_2 S - k_{-1} k_{-2} P) E_t}{(k_{-1} + k_2) + k_1 S + k_{-2} P} \tag{35}$$

where

$$V_f = k_2 E_t \tag{36}$$

$$V_b = k_{-1} E_t \tag{37}$$

$$K_s = \frac{k_{-1} + k_2}{k_1} \tag{38}$$

$$K_p = \frac{k_{-1} + k_2}{k_{-2}} \tag{39}$$

v_f and v_b define the maximum velocity for the forward and backward reactions, respectively, and K_s and K_p define the Michaelis constants for the substrate (i.e., K_m) and product, respectively. Several points are of interest here. Note that if $P = 0$, Eq. (35) reduces to the basic Michaelis-Menten equation. Also, the numerator terms in Eq. (35) represent the thermodynamic driving forces due to substrate and product concentrations and correspond to the reaction in

the forward and reverse directions, respectively. The denominator
terms represent the distribution of the enzyme into its complexed
and uncomplexed forms. In many cases even the mechanism given in
Eq. (34) is too restrictive in that the substrate and product form
the same complex with the enzyme. A more complete description of
the mechanism would allow for two or more complexes between complexation with substrate and release of product. The mechanism:

$$E + S \xrightleftharpoons[k_{-1}]{k_1} ES \xrightleftharpoons[k_{-2}]{k_2} EP \xrightleftharpoons[k_{-3}]{k_3} E + P \qquad (40)$$

can be solved by using the rate equations and steady-state approxima-
tion to give an equation with the same form as Eq. (35). In this
case new kinetic constants are defined. The general rule can be made
that the addition of any extra intermediate does not change the characteristic form of the rate equation; this holds whenever we introduce unimolecular steps involving isomerizations between intermediate
enzyme-reactant complexes [11].

D. Preincubation and Lag Phase

In the case of a coupled enzyme system, for example, the assay of
GOT using the MDH (malate dehydrogenase) coupled reaction:

$$\text{Aspartate} + \alpha\text{-ketoglutarate} \xrightarrow{\text{GOT}} \text{oxaloacetate} + \text{glutamate} \qquad (41)$$

$$\text{Oxaloacetate} + \text{NADH} + \text{H}^+ \xrightarrow{\text{MDH}} \text{malate} + \text{NAD} \qquad (42)$$

the indicator reaction should not limit the first reaction by pro-
ceeding too slowly. Initially, there is a period of time in which
reaction (42) must wait for reaction (41) to generate enough oxaloacetate, so that the maximum rate of NADH conversion to NAD can be
measured. This period of time required to reach the maximum rate is
called the lag phase and is illustrated in Fig. 4. To decrease this
lag time, a 100-fold excess of the secondary enzyme should be added.

    Lag time should not be confused with preincubation. Preincuba-
tion is used to allow any interferants in the reaction mixture that
may cause the conversion of NADH to NAD to react. If this is not

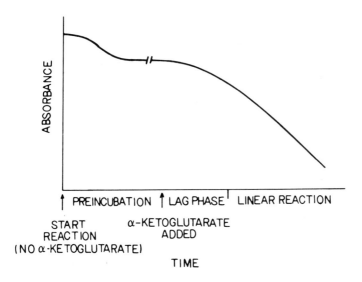

FIG. 4 Lag phase and preincubation.

done, a false high value for GOT is obtained. In preincubation all
the components of the reaction, except one, in this case α-ketoglutarate, would be placed in the sample cuvette, and the interfering
reaction would be allowed to proceed to completion before the final
assay reaction was started (by addition of α-ketoglutarate in this
case). Some assays, such as CK, have a considerable lag time; others,
such as α-hydroxybutyrate dehydrogenase (HBDH), proceed immediately.

III. FACTORS AFFECTING ENZYME-CATALYZED REACTIONS

The main factors that affect the kinetics of enzyme-catalyzed reactions are substrate concentration, enzyme concentration, activators,
inhibitors, temperature, pH, and ionic strength. Each of these is
briefly discussed below.

A. Substrate Concentration

A plot of substrate concentration versus initial velocity has the
shape of a rectangular hyperbola as shown in Fig. 2. At low substrate concentrations such that $S \leq 0.1K$, the Michaelis-Menten

equation reduces to

$$v = \left(\frac{V_{max}}{K_m}\right) S \tag{43}$$

The initial velocity is directly proportional to the substrate concentration and first order kinetics applies. When $S \geq 10K_m$, the
Michaelis-Menten equation reduces to the form:

$$v = V_{max} \tag{44}$$

The initial velocity is no longer dependent on substrate concentration and zero order kinetics applies at constant enzyme concentration.
In the region $0.1K \leq S \leq 10K$ the complete equation is required and the
reaction order is mixed. The equations previously derived were for
reactions involving only one substrate. In these cases any substrate
may be determined as above if the concentrations of the others are in
a nonlimiting excess such that there is pseudo first order kinetics.

Frequently, a decrease in the rate of an enzyme reaction is observed at high substrate concentration. This substrate inhibition is
not predictable from the Michaelis expression and may be due to a number of different causes. For example, in the hypoxanthine-xanthine
oxidase reaction, monitored by the reduction of the highly colored
methylene blue, the inhibition by substrate is due to competition
with methylene blue. In other cases this inhibition is due to the
formation of ineffective complexes with two or more substrate molecules combined with one active site [12]. This is observed with many
enzymes which have two or more groups, each combining with a particular part of the substrate molecules. In the "effective" complex one
substrate molecule is combined with all these groups. If some of
these groups are blocked with other molecules, an "ineffective" complex can be formed in which a substrate combines with only one group
on the enzyme. The chance of an ineffective complex forming increases
at high substrate concentration, where the substrate molecules tend to
crowd onto the enzymes. Methods for assay of substrates are described
in Chapter 4.

B.   Enzyme Concentration

As predicted from the Michaelis equation, the initial rate of an
enzymatic reaction is proportional to the initial enzyme concentration $[E]_0$. This dependence is illustrated in Fig. 5.

$$V_0 = \frac{k_3 [E]_0 [S]_0}{K_m + [S]_0} \qquad\qquad (45)$$

Theoretically, an increase in rate should be observed with an
increase in enzyme concentration ad infinitum. However, sometimes
it is found that there is a falling off from linearity at very high
enzyme concentrations. This does not indicate a true decrease in the
activity of the enzyme but represents a limitation in the technique
of measurement. Thus, from a measurement of the initial rate and a
calibration plot of the rate versus enzyme concentration, one can
easily calculate the concentration of this biochemical catalyst.

Some enzymes have been observed to give a nonlinear plot of rate
versus enzyme concentration, usually showing a curvature toward the
horizontal axis. For example, proteinase acting on protein fits the
equation $V_0 = k[E]^{1/2}$, called the Schütz law [13,14]. A number of

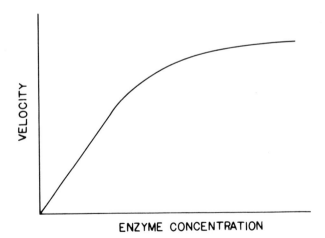

FIG. 5.   Effect of enzyme concentration on the initial rate of an
enzymatic reaction.

proteinase preparations acting on hemoglobin or gluten fit the equation $V_0 = k[E]^{2/3}$ and some authors suggest the use of this equation in the determination of proteolytic activity [15]. Roy [16] found that the velocity of the aryl sulfatase A reaction followed the form $V_0 = k[E]^{3/2}$ and offered a theoretical explanation in terms of dimerization of the enzyme. In all these cases, however, deviations from linearity are probably due in some measure to the presence of activators or inhibitors in the enzyme preparation. In the vast majority of cases an exact proportionality between initial velocity and enzyme concentration has been found, and in most kinetic studies this proportionality is assumed.

C. Activators

An activator is either a substance which is required for an enzyme to be an active catalyst or one which increases the efficiency of an active enzyme (i.e., converts the inactive apoenzyme to the active holoenzyme). The effect of activator concentration on the initial rate of an enzyme reaction is similar to that of substrate concentration. At low concentrations of activator there is a first order dependence of the rate on activator concentration (Fig. 6). At higher

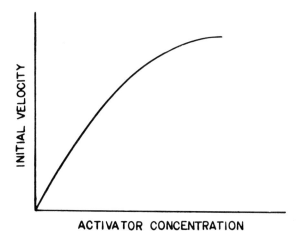

FIG. 6 Variation of the initial velocity of an enzyme reaction with increasing concentration of an activator.

concentrations the enzyme is maximally activated and the rate is
independent of activator concentration. Usually activation is not
specific and several different substances may activate the same
enzyme.

Many workers have described the activation of the enzyme iso-
citric dehydrogenase (EC 1.1.1.42) by metal ions, such as Mn^{2+} and
Mg^{2+} [17-19]. As little as 5 ppb of Mn^{2+} can be determined based
on this activation.

For the analysis of substrates in enzyme-activated systems, an
excess non-rate-limiting concentration of activator is used. Under
these conditions the rate becomes pseudo first order with respect to
substrate concentration.

Methods for assay of cofactors are described in Chapter 4.

D. Inhibitors

An inhibitor is a substance which causes a decrease in the rate of a
catalytic reaction, either by reacting with the catalyst to form a
catalyst-inhibitor complex or by reacting with one of the reactants.
Enzymatic inhibitors are either reversible or irreversible. In reversible inhibition, the enzyme can recover its activity when the
inhibitor is removed; in the case of an irreversible inhibitor it
does not. With an irreversible inhibitor the inhibition increases
progressively with increasing inhibitor concentration and becomes
complete if enough inhibitor is present to combine with all the
enzyme. With a reversible inhibitor, inhibition is progressive but
quickly reaches an equilibrium value which depends on the inhibitor
concentration. The action of the "nerve gases" sarin and tabum on
cholinesterase is an example of irreversible inhibition; eserine,
however, is a reversible inhibitor of cholinesterase.

1. REVERSIBLE INHIBITION

Both substrate and inhibitor compete equally for the active site
of the enzyme in competitive inhibition. A fully competitive type of
reversible inhibition can be represented by the following reactions:

$$E + S \xrightleftharpoons[k_{-1}]{k_1} [ES] \tag{46}$$

$$E + I \xrightleftharpoons[k_4]{k_3} [EI] \tag{47}$$

$$[ES] \xrightleftharpoons[k_{-2}]{k_2} E + P \tag{48}$$

Applying steady-state kinetics and solving for V_0, one obtains:

$$V_0 = \frac{k_2[E]}{1 + \dfrac{k_{-2} + k_2}{k_1[S]}\left(1 + \dfrac{k_3[I]}{k_4}\right)} \tag{49}$$

Writing K_m for $(k_{-1} + k_2)/k_1$ and k_i for k_4/k_3, we obtain:

$$V_0 = \frac{V_{max}}{1 + \dfrac{K_m}{[S]}\left(1 + \dfrac{[I]}{K_i}\right)} \tag{50}$$

The effect of the competitive inhibitor is to produce an apparent increase in K_m by the factor $1 + [I]/K_i$. Thus the apparent K_m increases without limit as $[I]$ increases. At any inhibitor concentration, the limiting velocity with excess of substrate is always equal to V_{max}, the maximum velocity of the uninhibited reaction. Typical plots of V_0 versus $[S]$ and $1/V_0$ versus $1/[S]$ for competitive inhibition are indicated in Fig. 7.

2. *IRREVERSIBLE INHIBITION*

Equations for noncompetitive, irreversible inhibition are as follows:

$$E + S \rightleftharpoons [ES] \tag{51}$$
$$E + I \rightleftharpoons [EI] \tag{52}$$
$$[EI] + I \longrightarrow [EIS] \tag{53}$$
$$[ES] + I \longrightarrow [EIS] \tag{54}$$
$$[ES] \longrightarrow E + P \tag{55}$$

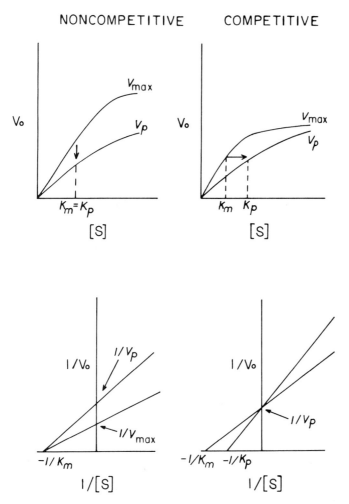

FIG. 7 Plots of V_0 versus [S] and $1/V_0$ versus [S] for noncompetitive and competitive inhibition. V_p is the maximum velocity in excess substrate. K_p is the substrate concentration giving half the maximum velocity.

Assuming the complex [EIS] does not break down and the velocity is entirely that of the breakdown of [ES], an expression for the initial velocity V_0 is

$$V_0 = \left(\frac{V_{max}}{1 + ([I]/K_i)} \right) \left(1 + \frac{K_m}{[S]} \right) \tag{56}$$

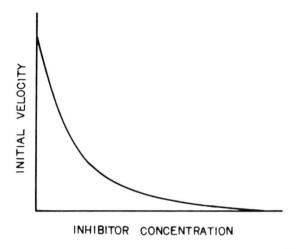

INHIBITOR CONCENTRATION

FIG. 8 Variation of the initial velocity of an enzyme reaction with increasing concentration of an inhibitor.

The initial rate of an enzyme reaction, then, will decrease with increasing inhibitor concentration, linearly at low inhibitor concentration and then asymptotically when reaching zero (Fig. 8). Generally, kinetic methods are extremely sensitive for determining substances which are catalytic inhibitors. For example, as little as 10^{-10} g/ml of organophosphorus compounds can be determined by their inhibition of the enzyme cholinesterase, which catalyzes the hydrolysis of choline esters [20]. Moreover, a great deal of specificity is built into inhibitors, thus providing an additional advantage in analysis. A specific method for the determination of nanogram quantities of fluoride in the presence of phosphate was described by Linde [21] and by McGaughey and Stowell [22]. Fluoride inhibits the enzyme liver esterase, but phosphate is a weak inhibitor. This provides one of the few direct methods for fluoride in the presence of large amounts of phosphate.

E. Temperature

The overall enzyme reaction consists of three successive stages: the formation of enzyme-substrate complex, conversion of this to an enzyme-product complex, and dissociation to products and free enzyme.

The total effect of temperature on the reaction will be the resultant of the separate effects of these individual steps. The heat, free energy, and entropy of the process for each of these stages will contribute to the overall thermodynamic parameters observed. A classical derivation yields:

$$\frac{d \log k}{dt} = \frac{\Delta H + RT}{2.303RT^2} = \frac{E^*}{2.303RT^2} \tag{57}$$

A plot of log k versus 1/T will give a straight line with a slope of $-(\Delta H + RT)/2.303R$. It should be noted that this plot gives E^* and not ΔH.

Thus enzyme functions by lowering the amount of energy required (activation energy) for the reaction it catalyzes to proceed (Fig. 9). The E^* for enzyme-catalyzed reactions is of the order of 5-10 kcal/mole, compared with 13-20 kcal/mole for uncatalyzed reactions.

The effect of temperature on enzyme reactions is usually given in terms of the temperature coefficient Q_{10}, which is the factor by which the rate of reaction is increased by raising the temperature 10°C:

$$E = \frac{2.303RT^2 \log Q^{10}}{10} \tag{58}$$

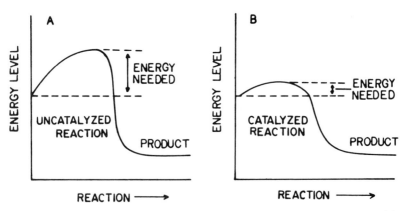

FIG. 9 Activation energy for uncatalyzed (A) and catalyzed (B) reactions.

Generally, the temperature coefficient of an enzymatic reaction lies between 1 and 2; hence, an increase in temperature increases the rate of the reaction, and a decrease in temperature decreases the rate. Since a 10°C rise in temperature approximately doubles the reaction rate in many cases, a strict control to within ±0.1°C is generally sufficient to obtain good reproducible results.

Heat denaturation is a continuous process and the rate of denaturation increases with temperature. As a result, bell-shaped curves, such as the one in Fig. 10, are obtained for enzyme systems.

The temperature to be used for following an enzymatic reaction is a subject of considerable controversy. The original IUPAC-IUB [5] recommendation was 25°C. This was objected to by many countries whose warmer climates suggested that 30°C would be more reasonable. Other countries stated that the optimum temperature for all biochemical reactions, 37°C (98.6°F), should be used. For this reason, most instruments described operate at many selectable temperatures, and procedures and calibration factors have been designed for various temperatures.

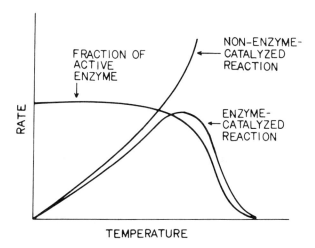

FIG. 10 Effect of temperature on the rate of reactions.

F. pH

Most enzymes are active over only a limited range of pH and in most
cases a definite optimum pH is observed. This pH optimum might be
due to a number of effects: (1) an effect of pH on the stability of
the enzyme, which may become irreversibly destroyed on one or both
sides of the optimum pH; (2) an effect on the V_{max} itself; (3) an
effect on the affinity of the enzyme, the fall on either side of the
optimum being due to a decreased saturation of the enzyme with substrate, caused by decreased affinity; and (4) an effect of pH on the
indicator reaction, if one is used to monitor the progress of the
enzymatic reaction by a coupled reaction sequence. These effects may
occur in combination and can easily be distinguished experimentally.

In the development of an analytical procedure, the pH dependence
of the total enzyme system, including the indicator reaction, is determined experimentally, and the optimum pH is used for analysis. In
many cases a compromise must be made if the optimum pH of each reaction in a coupled sequence is different. For example, in the determination of sucrose below, a three-step enzymatic reaction sequence
is used. In the first step sucrose is cleaved to glucose by the
enzyme invertase (β-fructosidase, EC 3.2.1.26) at a pH of 4.6. Glucose is then converted to peroxide by glucose oxidase (EC 1.1.3.4)
at pH 7-8:

$$\text{Sucrose} \; \underset{\text{pH 4.6}}{\overset{\text{invertase}}{\rightleftharpoons}} \; \text{glucose} \tag{59}$$

$$\text{Glucose} \; \underset{\text{pH 7-8}}{\overset{\text{glucose oxidase}}{\longrightarrow}} \; H_2O_2 \tag{60}$$

$$H_2O_2 + \text{indicator} \; \underset{\text{pH 10}}{\overset{\text{peroxidase}}{\longrightarrow}} \; \text{fluorescence} \tag{61}$$

In the final step, the peroxide produced is measured by an indicator
reaction which proceeds best at a pH 10. A compromise pH of 8.5
allows the reactions (60) and (61) to proceed but not reaction (59),
which does not go above pH 6. At this pH of 6 or below, however, the
latter reactions do not work. Therefore, one must use a pH of 4.6 to
convert all the sucrose to glucose. Then the pH is raised to pH 8.5
and the glucose produced is measured by the latter reactions.

G. Buffer

A list of some of the most widely used buffers is presented in Table 3. Frequently, the activity of an enzyme system will be noted to depend not only on the pH (see Sec. F) but also on the type of buffer. As shown in Fig. 11 for the enzyme urease, a pH optimum of 6.7 is observed in citrate buffer, 6.3 in acetate, and 7.3 in phosphate. Also the overall activity is higher in citrate buffer than in acetate or phosphate.

The buffer system can not only affect the activity and stability of an enzyme due to charge, anion activation, or surface change effects, but can also influence the thermodynamics of the reaction. This is frequently true when an intermediate product is formed first, which then leads to the desired product. For example, in the oxidation of L-amino acid oxidase, an α-peroxy keto acid is first formed, which then dissociates to peroxide and the α-keto acid:

$$\text{L-amino acids} + O_2 \longrightarrow \alpha\text{-peroxy keto acid} + NH_3 \longrightarrow$$
$$\alpha\text{-keto acid} + H_2O_2 + NH_3 \tag{62}$$

If H_2O_2 is to be measured, complete dissociation of the peroxy keto acid is necessary.

Similarly, the hydrolysis of urea is shown to proceed via an ammonium carbamate, which then yields ammonium ion and bicarbonate [23]. A buffer-mediated proton transfer is responsible for the conversion of carbamate to final products:

$$\text{Urea} + H_2O + H^+ \xrightarrow{\text{urease}} \text{ammonium carbamate} \tag{63}$$
$$\downarrow$$
$$NH_4 + HCO_3^- \tag{64}$$

so that the nature of buffer plays a key role.

Finally, the nature of the buffer will play a key role in the immobilization of enzymes. For example, a covalent attachment using a reagent such as glutaraldehyde will proceed very nicely in a phosphate buffer solution, but not in tris buffer where cross-reaction with the $-NH_2$ groups of the buffer will occur.

TABLE 3 Common Buffer Systems

Buffer	pK_a
McIlvaine (Na_2HPO_4-citric acid)	2.1, 3.1, 4.7, 6.4, 6.7
Tartrate	2.70, 4.05
Potassium biphthalate (Clark and Lubs)	2.80, 5.0
Acetate	4.76
Michaelis Universal (acetate/diethyl barbiturate)	4.76, 8.0
Citrate	3.08, 4.76, 6.40
Succinate	4.13, 5.38
MES [2-(N-morpholino)-ethane sulfonic acid]	6.15
Carbonate	6.46, 8.40
Bis-tris-[bis-(2-hydroxyethyl)imino-tris-hydroxymethyl methane]	6.50
ADA [N-(2-acetamido)-2-iminodiacetic acid]	6.60
Pyrophosphate	6.63, 9.29
Sørensen phosphate	6.7
PIPES [piperazine-N,N'-bis(2-ethane sulfonic acid)]	6.80
Bis-tris-propane [1,3-bis[tris(hydroxymethyl)methylamino] propane]	6.8, 9.0
ACES (2-[(2-amino-2-oxoethyl)amino]ethane sulfonic acid)	6.9
Imidazole	7.00
Citrulline-arsenate	7.10
BES [N,N-bis(2-hydroxyethyl)-2-aminoethane sulfonic acid]	7.15
MOPS [3-(N-morpholine)propane sulfonic acid]	7.20

Buffer	pH
Phosphate	7.21
Collidine (Gomori)	7.40
Arsenate	7.40
TES [N-2-tris(hydroxymethyl)methylaminoethane sulfonic acid]	7.50
HEPES [N-(2-hydroxyethyl)-1-piperazine-N'-ethane sulfonic acid]	7.55
Ethylglycinate	7.57
HEPPS [N-(2-hydroxyethyl)-1-piperazine-N'-3-propane sulfonic acid]	8.00
Veronal (Veronal diethylbarbiturate)	8.00
Tricine [N-tris-(hydroxymethyl)methyl glycine]	8.15
Glycinamide (Glycine amide hydrochloride)	8.20
Tris [tris(hydroxymethyl)aminomethane]	8.30
Bicine [N,N-bis(2-hydroxyethyl)glycine]	8.35
Glycylglycine	8.40
TAPS [3-tris(hydroxymethyl)methylaminopropane sulfonic acid]	8.40
2-Amino-2-methyl-1,3-propanediol	8.78
Borate-KCl-NaOH (Clark-Lubs)	9.2
AMP [2-amino-2-methyl-1-propanol]	9.30
CHES [2-(N-cyclohexylamino)ethane sulfonic acid]	9.50
Borate	9.78
Glycine	9.78
N,N-Dimethylglycine	9.95
Carbonate-bicarbonate	9.9
CAPS (Cyclohexylaminopropane sulfonic acid)	10.40

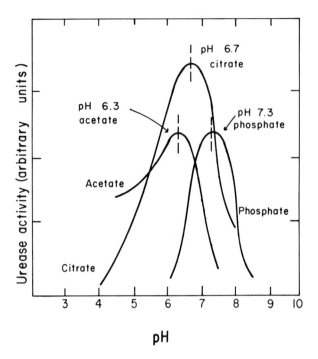

FIG. 11   pH activity curves for the enzyme urease illustrating both
the optimum and the effect of buffer on the pH optimum.   [From N.
Teitz, Fundamentals of Clinical Chemistry, 2nd ed., W. B. Saunders
Co., Philadelphia, 1977. © W. B. Saunders Company.]

H.   Ionic Strength

The presence of diverse salts can affect the rate of reaction, either
by shifting the equilibrium of formation of the activated complex or
by combining with reactants. The first, called a salt effect of the
first kind, can be calculated from the basic equation:

$$\log K = \log K_0 + Z_A Z_B \mu + (k_A + k_B - k^*)\mu \tag{65}$$

where K is the rate constant of the reaction; K_0 is the rate constant
without diverse substances; Z_A and Z_B are the charges on the reacting
substances A and B; μ is the ionic strength; and k_A, k_B, and k^* are
empirical coefficients for A, B, and activated complex, respectively.

In the salt effect of the second type, the diverse substances
may serve to tie up the effective concentration of one of the reac-

tants via the formation of complex ion or a precipitate, by shifting
the ionization equilibrium of weak acid or weak base reactants, or
some similar effect. Hence, to achieve reproducible results, one
must carefully eliminate harmful foreign ions and control the ionic
strength of the medium. In an ionic reaction, the rate will vary
with the dielectric constant of the solvent used [24,25].

IV.   IMPORTANCE OF ENZYMATIC ANALYSIS
A. As a Diagnostic Tool in Clinical Chemistry

Enzymes are valuable and extremely important aids in the diagnosis
of a wide variety of disorders (Table 4). This usefulness stems from
the fact that enzymes occur throughout the body, with varying amounts
found in such organs as the heart, liver, kidney, brain, muscle, bone,
and in body fluids and secretions, even in tears (Table 5). Some
enzymes are present in some organs in high concentrations but are
absent in others.

      Diagnostically significant enzymes can be divided into several
groups: (1) plasma-significant enzymes, those involved in blood coagu-
lation; (2) nonspecific plasma enzymes, subdivided into (a) the se-
creted enzymes, such as amylase and lipase, and (b) the intracellular
metabolic enzymes, which can be further subdivided into the organ-
specific enzymes, such as α-HBDH, and the multitissue enzymes such as
LDH, alkaline phosphatase (AP), GOT, GPT, and CK. The assay of enzymes in blood serum diagnostically serves to indicate the function
of a particular organ or tissue because damage to that species generally causes a sudden release of the enzyme associated with it into
the bloodstream. For example, La Due et al. [27] first reported
altered enzyme activity levels in serum following myocardial infarction (Fig. 12). The time of the infarct can be pinpointed by assay
of different enzymes because GOT and CK rise rapidly, reaching a
maximum in 24-36 hr, whereas LDH and HBDH reach maximal activities
much later. Enzyme tests can be used for differential diagnosis to
prove the existence of one disease rather than another. For example,
an electrocardiogram (ECG) will not differentiate between a myocardial

TABLE 4   Enzymes in Diagnosis

Enzyme	Liver			Heart (myocardial infarction)		Pancreas	Muscular dystrophy	Prostate
	Hepatitis	Cirrhosis	Carcinoma	Early	Late			
Aldolase	++	+	+		+		++	
Amylase						++		
Lipase						++		
CK				++	+		++	
Glutamate dehydrogenase	++	±						
GOT	+++	+	+	+	++		++	
GPT	+++	±			±			
ICDH	+++	±		+	++			
LDH	+	+	++	+	++			
LAP	+	+	++					
ChE	−							
Phosphatase acid								+
Phosphatase alkaline	+	+	++					

Note: − = Decrease in activity; + = increase in activity to very large increase (+++).
Source: Ref. 26.

TABLE 5 Examination of Biological Fluids

Body fluid	Enzyme	Interpretation
Serous fluid (liquid in pleural, peritoneal, and pericardial cavities)	Lactic dehydrogenase (greater than in serum)	Primary and metastatic tumors
Cerebrospinal fluid	Phosphohexose isomerase ↑	Brain tumors. Infection of CNS caused by bacteria, fungi, and viruses
	Adenosine deaminase ↑	Tuberculosis/meningitis
Synovial fluid	Alkaline phosphatase (greater than in serum)	Rheumatoid and inflammatory arthritis
Amniotic fluid	Several	Antenatal detection of familial metabolic disorders
Urine	Acid α-14-glucosidase ↓	Pompe's disease
Tears	Hexosamidase A (10 x by vol. > serum)	Tay-Sachs disease
	α-Galactosidase A absent	Fabry's disease
Blood	Acid phosphatase ↑	Prostatic carcinoma
	Alkaline phosphatase ↑	Liver disease with cholestasis. Bone disease in conditions in which bone is regenerated, e.g., rickets

* ↑ = Elevated level; ↓ = low level.
Source: Ref. 26.

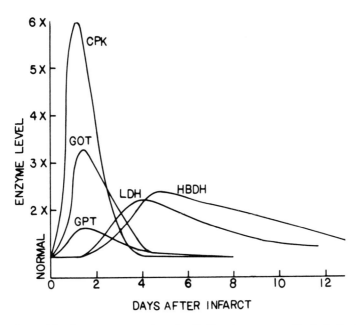

FIG. 12   Serum enzyme levels following myocardial infarct.

infarction and pulmonary embolism, but enzyme tests will distinguish
the two, as shown in Table 6.   Enzyme tests have proved to be very
useful in the diagnosis of various liver diseases [28].   The proper
selection of enzyme tests permits the differentiation between obstructive jaundice and viral hepatitis.   They also prove useful in detecting
infectious mononucleosis, cirrhosis, and cholangiolitic hepatitis.   The
latter is a disease which exhibits symptoms similar to those of obstructive jaundice.

The enzyme tests most commonly employed in the detection of liver
disorders are GPT, GOT, AP, and isocitrate dehydrogenase (CDH) (Fig.
13).

GOT and GPT levels are very high in cases of viral hepatitis.
In about 75% of the cases, GPT elevations are greater than GOT levels
[29].   Aldolase is elevated in viral hepatitis, hemolytic anemia, and
acute alcoholic psychosis.   Minimal elevations appear in cirrhosis,
obstructive jaundice, and primary hepatoma [30].

TABLE 6 Distinguishing Myocardial Infarction from
Pulmonary Embolism

Test	Myocardial infarction	Pulmonary embolism
GOT	Elevated in 92-98% cases	Normal for the first 4 days
LDH	Elevated after 48-72 hr, remains up for 8-10 days	Elevated for about 10 days
Bilirubin	Normal	Elevated
CK	Elevated in about 90% of cases	Normal
ECG	May be identical for both	

Source: Ref. 28.

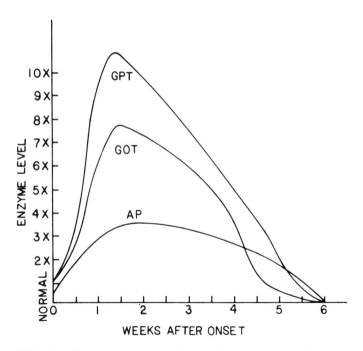

FIG. 13 Serum enzyme levels during acute hepatitis.

Serum MDH (malate dehydrogenase) activity also is elevated in acute viral hepatitis and in hepatic cirrhosis. This determination is not widely used because of special considerations of laboratory techniques [31].

Serum LDH (lactic dehydrogenase) activity is also increased in cirrhosis, hepatitis, and metastatic involvement of the liver. It is small, however, in comparison with the degree of transaminase elevation [31].

Serum ICDH activity measurement is one of the most sensitive and specific methods for the detection of liver cell injury. ICDH activity shows considerable elevation in cases of acute viral hepatitis, whereas it usually remains normal in biliary obstruction or hepatic cirrhosis. Elevations also occur in alcoholics and in about half the cases of hepatic metastases [31].

In muscle diseases [28], raised levels of serum transaminase may be found with cases of progressive muscular dystrophy. The highest incidence of this is found in patients with the Duchenne type.

Raised serum CK activity is usually found in patients with various types of muscular dystrophy. This may be due to increased cell permeability resulting from the disease. As with the transaminases, the highest and most frequent increases are found with the Duchenne type of muscular dystrophy. Levels as much as 50-100 times the upper limit have been encountered. Raised CK levels have also been found in female carriers of muscular dystrophy [32,33].

As the disease progresses, the muscle deteriorates and the fiber is replaced by fat. For this reason, CK values fall with advancing age, paralleling the progressive diminution of functioning muscle; the highest levels are usually found in infants and children. Often elevated levels are observed long before other clinical symptoms become apparent. Therefore, this test is helpful in early detection of the diseases [34-37].

In cases of muscular dystrophy of the Duchenne type, HBDH levels are also considerably elevated.

Aldolase activity also is elevated by progressive muscular dystrophy but not by other diseases of skeletal muscle. Measurement of
serum aldolase activity is done to differentiate progressive muscular
dystrophy from other types of myopathy.

Levels of serum acid phosphatase are significantly increased in
patients with metastatic carcinoma of the prostate [38]. Since acid
phosphatases from other sources, principally erythrocytes and plate-
lets, also appear in serum, it becomes necessary to differentiate
between the enzymes from these sources and those from the prostatic
source. Sodium thymolphthalein monophosphate, the substrate utilized
in some current methodology, possesses an almost complete specificity
for the prostatic isozyme [39]. Thus, it is unnecessary to utilize
additional differentiation techniques and the method is suitable for
both screening and quantitative purposes.

Raised serum alkaline phosphatase levels are found in bone disease and liver disorders. Highest levels are encountered in Paget's
disease. Hyperparathyroidism with skeletal involvement also gives
high values. Rickets is accompanied by elevations in serum alkaline
phosphatase levels which reflect the severity of the disease.

Finally, enzyme assays have been able to pinpoint inherited
disorders and those with ethnic correlation [25]. Such diagnosis
have been of great value in clinical chemistry (Table 7).

B. In Food Processing and Fermentation

Enzymes have been applied in the field of food processing for many
centuries, beginning with the use of malt diastase to produce fer-
mentable sugars and the application of crude protease to tenderize
meat. The application of enzymes in these areas today still retains
much craft, but enzyme technology has been fused with modern technology to provide a range of new and better products.

In food processing, enzymes have been used in the production of
foods by starch hydrolysis and the enzymatic manufacture of glucose,
dextran, and isomerized sugar. Proteases (papain, ficin, bromalin)
have been used in the tenderizing of meat. A test for GOT differ-

TABLE 7 Ethnicity of Disease: Simply Inherited Disorders

Ethnic group	Relatively high frequency	Relatively low frequency
Ashkenazi Jews	Abetalipoproteinemia	Phenylketonuria
	Pentosuria	
	Tay-Sachs disease	
Mediterranean peoples (Italians, Greeks, Sephardic Jews)	Thalassemia (mainly β)	Cystic fibrosis
	G6PD deficiency	
Africans	G6PD deficiency	Phenylketonuria
	Adult lactase deficiency	
Japanese (Koreans)	Acatalasia	
Chinese	α-Thalassemia	
	G6PD deficiency	
	Adult lactase deficiency	
Scots		Phenylketonuria
Armenians		G6PG deficiency
Finns		Phenylketonuria
Eskimos	E₁S (pseudocholinesterase deficiency)	

entiates fresh meat from thawed frozen meat [40,41]. Pectin is used
to clarify fruit juices; glucose oxidase is used in the treatment of
egg. Proteolytic enzymes, such as papain, are used for chill-haze
stabilization of beer, and hydrogen peroxide/catalase treatment of
milk is used in nonrefrigerated storage.

To test for the effectiveness of pasteurization, a test for
alkaline phosphatase [42] is used. Renin and lipase are used in
cheese making to give texture and flavor to cheeses.

In the beverage industry, enzymatic tests for ethanol, CO_2,
malic acid, lactate, glycerol, and carbohydrates [43] are often
conducted.

C. In Other Fields

Other fields of interest for enzymatic methods of analysis are botany
and agricultural analysis [44] and microbiology [45]. Because of
their more limited interest, these are not discussed here.

V. EXPERIMENTAL TECHNIQUES

A. Determination of Concentration of Participants
    in Enzymatic Reactions

The concentration of a material participating in an enzyme reaction
can be calculated in one of two ways: by measuring the total change
that occurs by chemical, physical, or enzymatic analysis of the product or unreacted starting material; or from the rate of the enzyme
reaction, which depends on the concentration of the substrate, coenzyme, activator, or inhibitor, as discussed above.

   1. TOTAL CHANGE OR EQUILIBRIUM METHOD

   In the first method, large amounts of enzyme and small amounts
of substrate are used to ensure a relatively rapid reaction. The
reaction is allowed to go to completion, and the amount of substrate
S in the sample can be calculated from the amount of product P formed:

$$\text{Substrate} \xrightarrow{\text{enzyme}} \text{product} \tag{66}$$

P must be in some way chemically and/or physically distinguishable
from S. For example, ethanol can be determined by the enzymatic reaction using alcohol dehydrogenase in conjunction with the coenzyme
nicotinamide adenine dinucleotide (NAD):

$$\text{Ethanol} + \text{NAD} \xrightarrow{\text{alcohol dehydrogenase}} \text{acetaldehyde} + \text{NADH} \tag{67}$$

The reduced form of NAD, NADH, is formed; it has a strong absorbance
at 340 nm, where NAD does not absorb. The total amount of NADH produced is therefore a measure of the amount of ethanol present. In
the determination of uric acid using the enzyme uricase, peroxide
is produced:

Uric acid $\xrightarrow{\text{uricase}}$ H_2O_2 (68)

Uric acid has a strong absorbance at 292 nm, where the products do not absorb. The total uric acid present can therefore be determined by noting the total change in the absorbance at 292 nm.

Alternatively, a coupled reaction can be used to indicate how much substrate has been decomposed. In the determination of glucose an enzyme reaction using glucose oxidase yields hydrogen peroxide. The extent of reaction can be determined by monitoring the uptake of oxygen using an oxygen-sensitive electrode or, more easily, with the aid of an indicator reaction which yields a colored dye from a color-less leukodye (i.e., o-dianisidine). An example of an enzyme reaction is:

Glucose + H_2O + O_2 $\xrightarrow{\text{glucose}}$ gluconic acid + H_2O_2 (69)

An example of an indicator reaction is:

H_2O_2 + leukodye $\xrightarrow{\text{peroxidase}}$ H_2O + dye (70)
   (colorless)   (colored)

The total intensity of color of the dye produced is a measure of the concentration of glucose present.

   2.   *KINETIC METHOD*

   In the second method, the kinetic method, the initial rate of reaction V_0 is measured in one of the many conventional ways by fol-lowing either the disappearance of substrate or the production of product.

$$\frac{dx}{dt} = k[A]_t$$ (71)

The rate is a function of the concentration of substrate (S), enzyme (E), inhibitor (I), and activator (A). For example, the concentra-tion of glucose can be determined by measuring the initial rate of production of the colored dye in reaction (70).

   Since the enzyme is a catalyst, and as such affects the rate and not the equilibrium of a reaction, its activity must be measured

by a kinetic (or rate) method or by a direct titration of the active
site [46]. Likewise, activators or inhibitors that affect the enzyme's catalytic ability can be measured via either a total change
or a kinetic method. The former method frees the technician from
continuous measurements in nonautomated procedures; rate methods,
however, are faster because the initial reaction can be measured
without waiting for the reaction to go to completion. The accuracy
and precision of both methods are comparable [47], and it is no
longer true that equilibrium methods are more reliable than rate
methods. The rate of reaction is affected by conditions of pH,
temperature, and ionic strength, however; all these factors must be
carefully controlled for good results. Work by Guilbault et al.
[48,49] and Pardue et al. [50] indicated that, with reasonable care,
precision and accuracies of better than 1% can be obtained. Furthermore, some of the difficulties encountered because of side reactions
are eliminated in rate methods and greater sensitivities can be obtained in many cases. With the automated equipment now available
for performing rate methods, such techniques are likely to be the
ones of choice in the future. Indeed, the majority of commercial
instruments use kinetic methods.

There are several possible methods to calculate the rate of an
enzymatic reaction: (1) the initial slope method, (2) fixed concentration or variable time method, and (3) fixed time method.

In the slope method the change in the concentration of a reactant, product, or indicator substance is plotted as a function of
time by an automatic recording of the rate curve. The initial slope
of this curve is obtained by extrapolation to time zero and is related
to the concentration of reactant or enzyme to be determined. Generally, the change in some physiochemical parameter of the reactant,
product, or inhibitor is measured (absorbance, pH, fluorescence,
etc.). The slope method is illustrated in Fig. 14, which shows a
typical curve for the hydrolysis of a substrate, such as p-nitrophenyl
phosphate (p-NPP) by alkaline phosphatase. The continuous measurement
of the slope is made by a double-beam spectrophotometer, and the most
linear slope is determined by a computer using a least-squares curve

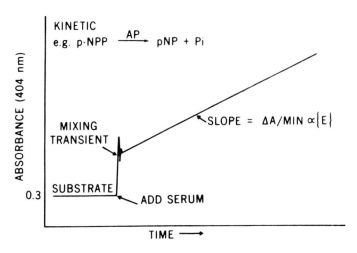

FIG. 14   Initial slope method of calculating reaction rate.   (From Ref. 3.)

fit calculation as used in the Rotochem II or in the LKB-Wallac Reaction Rate Calculator.   Alternatively, the data can be presented graphically, and the technician can calculate the activity of the enzyme or concentration of the substrate by relating the observed rate $\Delta Abs/\Delta t$ to that of a standard.   The trend in all instruments is to have a computer calculate the best slope and print out the activity or concentration directly.

In the fixed concentration (variable time) method, the time that is required for the concentration of a reactant to reach a set level is recorded.   Any property of this substance denoting its concentration (i.e., its absorbance, fluorescence, pH) could be used. Thus, the time required for a preset fluorescence or absorbance level to be reached would be inversely proportional to the concentration, and a plot of $1/t$ versus concentration would be linear.   This is predictable from basic kinetics [4-6].   Integration of Eq. (1) yields

$$X = K[A]_0 t \tag{72}$$

$$[A]_0 = K'\left(\frac{1}{t}\right) \tag{73}$$

where K' is a constant term which includes a set, specified concentration of substance at that point in the reaction at time t. Thus, one might determine the concentration of substrate A or enzyme E by noting the time it takes to reach a preset absorbance, fluorescence, or pH value.

In a typical automated procedure using the variable time method, the reaction is initiated by the injection of enzyme. After a delay time (usually about 30 sec to allow the attainment of a steady rate), a timer starts automatically. After a fixed change in absorbance or fluorescence (measured by the sensing unit) is reached, the timer automatically shuts off and a reading is taken. The concentration of substance present in the sample is inversely proportional to the time required to reach a fixed absorbance or fluorescence change. An analog-to-digital converter and printer can be used to convert the time signal to a direct printout of concentration.

In the fixed concentration procedure the time is measured between two precise points near the initiation of the reaction. One point is not zero time because many reaction curves are not linear immediately after initiation due to mixing, temperature, and stabilization effects. Generally 15-30 sec is sufficient to establish a measurable rate.

The fixed time method is similar to the fixed concentration method. In this method the reaction is allowed to proceed for a predetermined time interval and, after this time has passed, the concentration of one of the reactants (or products) is determined by some physiochemical means (absorbance, fluorescence, pH).

This method is used in many commercial instruments, in either the two-point, three-point, or multipoint methods. For example, calculation by the two-point method, illustrated in Fig. 15, is used in the Technicon AutoAnalyzer and SMAC, the Beckman DSA-560, and others. The experimental conditions are set such that A_0 or t_0 is a measure of the activity of enzyme. This technique is accurate only if the reaction progress is indeed linear between the two points.

A better method than the two-point method is the multipoint (three or more points) method, which provides more data about the

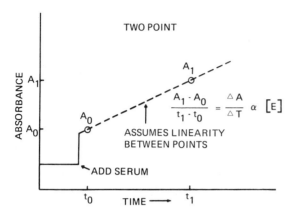

FIG. 15   The fixed time two-point assay method.   (From Ref. 3.)

kinetics of the reaction.   It is shown in Fig. 16.   By comparing
successive ΔA values it is possible to confirm the linearity of the
reaction and validate the printout of the activity based on such a
measurement.   This approach is exemplified by the three-point measurements in SMAC for GOT and GPT, or by the multiple point system of the
Vitatron AKES, which continues to take successive points, A_1, A_2, A_3,
etc., every 5 sec until it validates a linear slope.

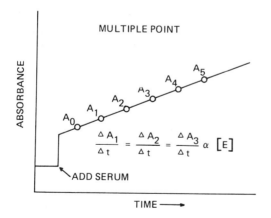

FIG. 16   Multiple-point method for calculation of slope.

3. VISUALIZATION OF UV METHODS

One of the most common kinetic assays is the continuous measurement of the disappearance of NADH in coupled reactions for glucose, urea, GOT, GPT, CK, etc., or in the enzyme reaction itself for HBDH, LDH, ethanol, pyruvate, etc. This measurement is done at 340 nm, a UV assay.

Many reports have appeared on the "visualization" of this reaction by coupling to tetrazolium dyes [51] or resazurin [52]. In the former method, colorless tetrazolium salts are used which are converted to intensely colored formazans by reduction with NADH in the presence of either phenazine methosulfate (PMS) or diaphorase:

$$\text{NADH + tetrazolium salt} \xrightarrow[\text{diaphorase}]{\text{PMS or}} \text{NAD + formazan} \qquad (74)$$

Many different tetrazolium salts have been used but the most common are INT [2-p-iodophenyl-3-(p-nitrophenol)-5-phenyltetrazolium chloride] [53], which gives a dye absorbing at 492 or 540 nm; or NBT ([2,2'-di-(p-nitrophenyl)-5,5'-diphenyl-3,3'-(3,3'-dimethoxy)-4,4'-diphenylene]ditetrazolium chloride) [54], which on reduction gives an absorbance at 495-500 nm. With resazurin as the dye, an intense fluorescence is produced (λ_{ex}540 nm, λ_{em}580 nm).

A listing of some of the dyes and electron acceptors used with dehydrogenases to yield sensitive spectrophotometric assays are given in Table 8.

4. TRAPPING REAGENTS

Many enzymatic reactions possess unfavorable equilibrium constants; hence, a quantitative analytical determination is very difficult to effect.

For example, at pH 7 the equilibrium constant for the conversion of pyruvate to lactate is 2×10^4, and an assay of lactate cannot be accomplished. Similarly, the quantification of LDH activity is best performed using lactate as a substance, rather than pyruvate, because the forward reaction requires the more unstable NADH rather than NAD.

$$\text{Pyruvate + NADH + H}^+ \xrightarrow{\text{LDH}} \text{lactate + NAD} \ (K = 2 \times 10^4) \qquad (75)$$

TABLE 8 Characteristics of Dyes and Electron Acceptors Used with Dehydrogenases

Name	Oxidized Formula	Reduced	E^0 (pH 7.0)	λ_{max} (nm)
Methylene blue	(Blue)	(Colorless)	+0.011	668
2,6-Dichloroindophenol	(Blue)	(Colorless)	+0.22	600
Triphenyltetrazolium	(Colorless)	(Red)		485

Brillant cresyl blue

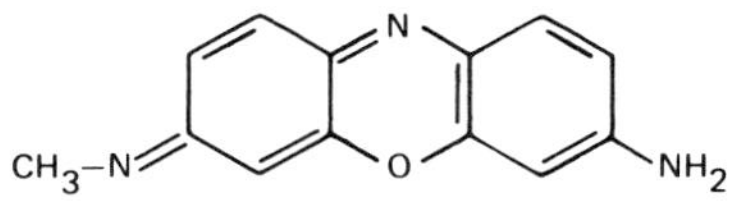

CH_3-N (Blue) NH_2 H   (Colorless) NH_2 CH_3-N H

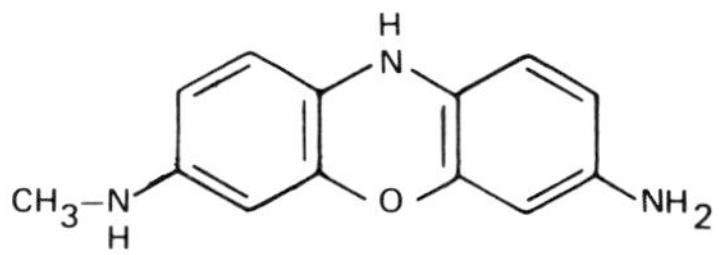

+0.045   632

Phenazine methosulfate (PMS)

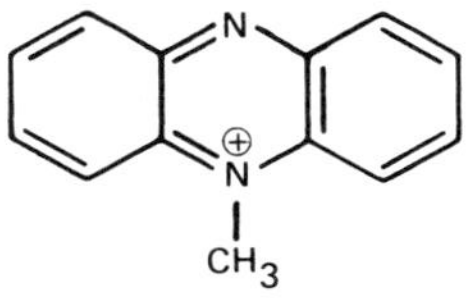

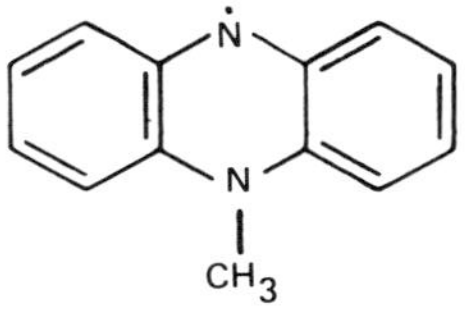

N $\oplus N-CH_3$ $N-CH_3$ (Colored)

+0.080   388

Ferricyanide

$Fe(CN)_6^{3-}$ $Fe(CN)_6^{4-}$

+0.36   410

Resazurin-resorufin

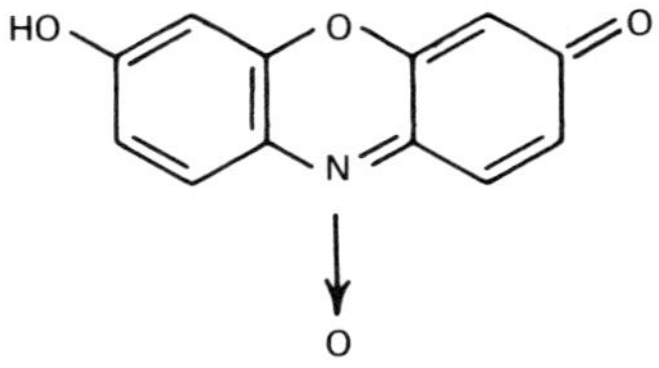

(Colorless) → O   (Red)

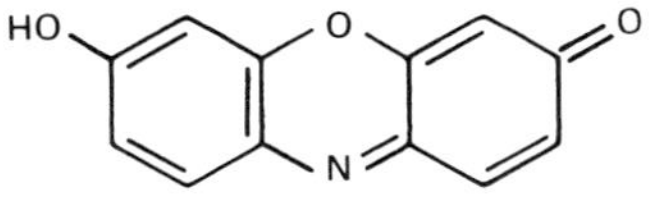

540   (λ_{flur} 580)

O-Dianisidine

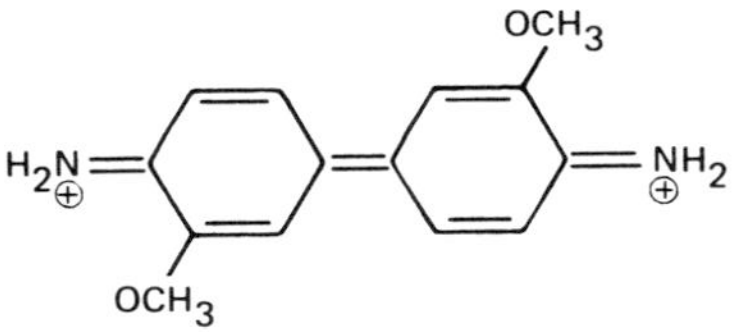

OCH_3 $\overset{\oplus}{NH_2}$ OCH_3 $H_2N\oplus$ OCH_3 (Red) OCH_3 NH_2 H_2N OCH_3 (Colorless)

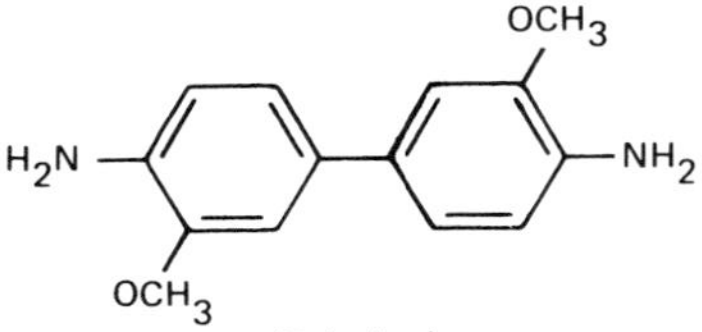

530   (ε 8.6 × 10^7)

55

If there is a high pH (~10) and a chemical compound, called a trapping agent, is added to force the equilibrium to the left (Le Chatelier's principle), an assay of lactate of LDH with NAD becomes feasible.

Aldehydes or ketones (like pyruvate) can be "trapped" by addition of hydrazine or semicarbazide, yielding hydrazones or semicarbazones. For reaction (75), for example, the equilibrium constant is changed more than 6 orders of magnitude by simply raising the pH to 9.5 and adding phenylhydrazine.

Of course, the addition of an indicator reaction [Eq. (74)] to visualize the NADH produced in the visible range is another method for trapping the reaction products. (In this case the NADH is trapped and removed from the reaction, thus forcing the equilibrium in favor of more NADH produced. Generally, electron transfer reagents, such as phenazine methosulfate or diaphorase, are added to catalyze the reaction.)

Finally, if a gas is produced in the reaction, it can be removed using either (1) heat or (2) the proper conditions of pH. For example, at high pH NH_3 is removed, at low pH CO_2 is removed.

B. Methods for Measurement of Enzymatic Reactions

1. GENERAL CONSIDERATIONS

In order to measure the progress of an enzymatic reaction, one must (1) follow the change with time of the concentration of either one of the products of the reaction or of one of its reactants (kinetic method) or (2) measure the total change in the concentration of reactant or product (end point or equilibrium method).

Usually it is more desirable to be able to continuously monitor a chemical reaction without having to draw and titrate samples. One can do this by (1) following the appearance or disappearance of some species by monitoring one of its physiochemical properties directly or (2) by using a coupled reaction sequence.

One might monitor the reaction:

$$\text{Substrate + enzyme} \longrightarrow \text{products} \tag{76}$$

by following the change in absorbance of the system, if any reactant
or product is absorbing, or the reaction can be monitored electrochemically if either a reactant or product is electroactive. The
change in pH can be recorded and equated to enzymatic activity if H^+
is one of the reactants or products.

2. MANOMETRIC METHODS

If one of the products of an enzyme reaction is a gas, the
extent of reaction can be indicated by a volumetric measurement of
the gas produced using a manometer. Such techniques, called manometric methods, were originally proposed by Bancroft [55] and were
developed by Warburg [56]. Such techniques have been extensively
used in the measurement of (1) gas-consuming reactions in which the
O_2 uptake is measured (e.g., oxidative enzyme systems, such as glucose oxidase, peroxidase, cytochrome oxidase); (2) gas-producing
reactions, in which one of the products of the enzymatic reaction
is a gas (e.g., CO_2 found in a decarboxylase enzyme system or NH_3
formed from a deaminase or lyase enzyme system); or (3) acid-forming
enzymatic reactions which are carried out in the presence of bicarbonate in equilibrium with a gas mixture containing a definite percentage of CO_2. In this latter system any acid produced in the
enzyme reaction reacts with bicarbonate to give a corresponding
amount of CO_2, which can be measured on the manometer. Since all
NAD-dependent dehydrogenase reactions yield a proton in the reduction of NAD, manometric methods have been extended to these enzyme
systems as well. A detailed discussion on manometric techniques can
be found in books by Umbreit et al. [57] and by Dixon [58]. Manometric methods have mainly been replaced by electrode methods since
good direct-reading electrodes are available for O_2, CO_2, NH_3, and
H^+.

3. SPECTROPHOTOMETRIC METHODS

If either one of the reactants or one of the products of an
enzyme reaction is absorbing in the ultraviolet, visible, or infra-
red region of the spectrum, then it is possible to monitor the

progress of such an enzyme reaction spectrophotometrically. Consider, for example, a typical reaction:

$$A \xrightarrow{\text{enzyme}} B + C \qquad\qquad (76)$$

If A has a strong absorbance in the UV, with a λ_{max} of 295 nm, yet products B and C absorb strongly in the visible, with λ_{max} of 420 and 560 nm, respectively, then one may follow the progress of this enzymatic reaction by noting the decrease in absorbance of A at 295 nm or the increase in absorbance at 420 or 560 nm as B and C are formed. Experimentally, the reaction can best be monitored at 560 nm if both B and C absorb at 420 nm, and generally it is better to follow an increase in absorbance than a decrease. Furthermore, if C has a higher molar absorptivity (absorbance per mole per unit centimeter path length) than A, then a greater sensitivity can be realized. Today there are many good spectrophotometers available that cover the entire UV-visible, near-IR region of the spectrum. Generally, an instrument that reads the change in absorbance with time automatically, such as the Beckman DB, Cary 14, or Zeiss, is preferred to a null point instrument, such as the Beckman DU, which requires a point-by-point plot of absorbance changes. Likewise, a double-beam instrument is better than a single-beam one because all measurements can be made against a cuvette containing all the reagents of an assay mixture except one (reagent blank).

In studying enzyme reactions spectrophotometrically, it is essential that the cell containing the reacting mixture be thermostatically controlled since a change in temperature of 1°C causes an approximate 10% change in the rate. The temperature of the cell must be controlled within ±0.2°C. Since some instruments are not built with such a thermostat, it is necessary to have a jacketed cell holder through which water can be circulated from an external thermostatted water bath.

It was the evolution of spectrophotometric methods three decades ago that proved a boon to enzymatic methods, starting when Warburg showed that reduced coenzymes NADH and NADPH absorb at 340 nm

TABLE 9   Spectral Characteristics of Amino Acids and Cofactors

Compound	λ_{max} (nm)	λ_{fluor} (nm)	$\epsilon\lambda_{max}$ (nm)
ADP, ATP	257	390	
FADH, FMNH	340, 365	520	
NAD, NADP	270	460	
NADH, NADPH	340,	460	6.22×10^{-6}
	365		3.4×10^{6}
Guanine	273	350	
Phenylalanine	259		2×10^{2}
Purine	271	370	
Tryptophan	279		5.2×10^{3}
Tyrosine	278		1.1×10^{3}

(or 365 nm with a Hg lamp). The oxidized coenzymes absorb at 270 nm but not at 340 nm, thus providing a method for the assay of dehydrogenase systems.

Table 9 lists the spectral characteristics (λ_{max}, λ_{fluor}, and molar absorptivities) of several amino acids and cofactors. The rather intense absorbance of NADH and NADPH accounts for its high popularity as a cofactor for studying enzymatic reactions. Not only are NAD and NADP involved as cofactors in dehydrogenase enzymes (subclasses 1.1.1, 1.4.1, 1.5, 1.7, 1.8), but also for transferases (class 2).

In many cases an appreciable absorption change is not observed in the enzyme reaction being studied. In such a case a coupled reaction sequence is used, with a second reaction used to indicate the progress of the enzyme reaction. For example, glucose is catalytically oxidized in the presence of glucose oxidase to peroxide. None of the reactants or products is absorbing, yet glucose can be easily assayed by using an indicator reaction and following the rate of formation of a colored dye. An example of an enzyme reaction is:

$$\text{Glucose} + H_2O_2 + O_2 \xrightarrow{\text{glucose oxidase}} \text{gluconic acid} + H_2O_2 \quad (78)$$

An example of an indicator reaction is:

$$H_2O_2 + leukodye \xrightarrow{\text{peroxidase}} H_2O + dye \tag{79}$$

Milano and Pardue [59] have proposed the use of silicon vidicon tube
spectrometer for the spectrophotometric assay of enzymatic reactions.
The advantages of speed, multiwavelength measurement which permits
assay of several enzymatic systems simultaneously, and fewer sample
and reagent preparation steps have made these devices attractive.

Lactate dehydrogenase and alkaline phosphatase activities in the
same medium were determined simultaneously, at 350 and 550 nm, with a
vidicon spectrometer. Substrate concentrations and pH were made optimum for the combined analysis. Such conditions result in activities
for lactate dehydrogenase that are equivalent to those found by methods
in common use and in activities for alkaline phosphatase that are about
31% below the maximum values that can be obtained with its substrate
used at the same pH and temperature in the absence of NAD and lactate.
However, activities measured by the simultaneous analysis were proportional to those obtained by other methods used in clinical laboratories, and the coefficients of variation were 2.3% for lactate dehydrogenase and 3% for alkaline phosphatase.

Many enzymatic reactions liberate NH_3 as a product of the reaction (the reductases of subclass 1.4.1, the oxidases of 1.4.3, the
deaminases of subclass 3.5, the dehydratases of subclass 4.2, the
lyases of subclass 4.3) or consume NH_3 as a reactant (the oxidases of
subclass 1.7 or the synthetases of subclass 6.3). Spectrophotometric
methods for NH_3 are the Berthelot, Nessler, and ninhydrin reactions,
and the coupling to NADH with glutarate dehydrogenase (Table 10).

Similarly there are several enzymatic systems that liberate H_2O_2
(subclasses 1.1.3, 1.4.3, 1.5) or use H_2O_2 as a substrate (subclass
1.11). Three of the highest volume clinical assay procedures are for
uric acid, cholesterol, and glucose. All liberate H_2O_2 in a enzymatic
procedure:

$$Glucose + O_2 \xrightarrow{\text{glucose}\atop\text{oxidase}} H_2O_2 \tag{80}$$

TABLE 10 Spectrophotometric Procedures for Assay of Ammonia

Procedure	λ_{max} (nm)
Berthelot reaction: $NH_3 + OCl^- + 2$ [phenol] $\longrightarrow$ indophenol	615
Nessler's reaction: $2NH_3 + 2[HgI_2 \cdot 2KI] \longrightarrow NH_2Hg_2I_2$ (yellow)	460
Ninhydrin method: $NH_3 + 2$ [ninhydrin] $\longrightarrow$ [diketohydrindylidene–diketohydrindamine]	546
Glutamate dehydrogenase: $NH_4^+ + NADH + \alpha\text{-ketoglutarate} \xrightarrow[H^+]{GDH} NAD + H_2O + glutamate$	340

$$\text{Cholesterol} + O_2 \xrightarrow[\text{oxidase}]{\text{cholesterol}} H_2O_2 \qquad\qquad (81)$$

$$\text{Uric acid} + O_2 \xrightarrow{\text{uricase}} H_2O_2 \qquad\qquad (82)$$

Peroxide is assayed by coupling the enzymatic reactions, such as Eqs. (80)-(82), with an indicator reaction which utilizes a leuko dye:

$$H_2O_2 + \text{leuko dye} \longrightarrow \text{chromogen} \qquad\qquad (83)$$

Some of the dyes most commonly used are 2,4-dichlorophenol and O-dianisidine (Table 9), malachite green, 4-methoxy-α-naphthol, and benzidine.

4. POLARIMETRIC METHODS

Because many enzymes catalyze the reaction of only one optical isomer of a substrate, yielding an optically inactive product, the reaction can be followed by the change of optical rotation. Like-wise, the production of an optically active isomer from an optically inactive substrate can be easily monitored. The activity of sugar enzymes, such as sucrase, can be nicely monitored by such a tech-nique, which requires only a commercially available polarimeter with a thermostatted tube.

In those cases where the substrate or product has too low a rotation to measure directly, the rotation may be increased by form-ing a complex. Lactic acid, for example, and other hydroxy acids form strong complexes with molybdate that have high specific rota-tions. Many dehydrogenases can be monitored by these techniques.

5. ELECTROCHEMICAL METHODS

a. Ion-Selective Electrodes

The most common electrochemical method that has been used in enzymology is that with a glass electrode in following reactions which involve the production of acid. Because changes in pH affect the activity of the enzyme and also the rate of reaction, direct readings of pH changes are generally employed in which the pH is maintained at a constant value by frequent addition of alkali. The rate at which base is added then gives the reaction velocity inde-pendent of the amount of buffer [60].

Several convenient automatic "pH stat" instruments are available, probably the most common being the Radiometer (Copenhagen) and the Metrohm (Brinkman Instrument Co., United States). Both instruments maintain a constant pH by continuous automatic additions of acid or alkali, and at the same time automatically record the amount added as a function of time.

The oxygen electrode has found increasing use in the enzymatic analysis of oxygen-consuming enzymatic systems:

Subclass 1.1.3 $S + O_2 \rightleftharpoons P + H_2O_2$

Subclass 1.4.3 $S + H_2O + O_2 \rightleftharpoons P + NH_3 + H_2O_2$

Subclass 1.5 $S + O_2 \rightleftharpoons P + H_2O_2$

Subclass 1.7 $S + O_2 + H_2O \rightleftharpoons P + H_2O_2$

Subclass 1.8 $S + O_2 + H_2O \rightleftharpoons H_2O_2 + \text{products}$

Subclass 1.10 $2S + O_2 \longrightarrow 2P + 2H_2O$

Subclass 1.99.2 $S + O_2 \rightleftharpoons P$

In cases where an O_2 electrode can be used to monitor the uptake of this reactant or where the product H_2O_2 can be monitored, it is best to follow O_2 because it is a more specific indicator of the true progress of the enzymatic reaction. Peroxide, being an oxidoreductant, can react with many coproducts and/or interferences present, resulting in a totally distorted picture of the true enzymatic procedure.

Many other ion-selective electrodes have been used to follow enzymatic reactions, such as the NH_3 gas, the NH_4^+ cation, the H^+, the CO_2 gas, and the I^- and CN^- electrodes. A review of the use of ion-selective electrodes has been prepared by Rohm and Guilbault [61], and some typical examples are given in Chapter 2.

b. Potentiometry at Small Current

Potentiometric, amperometric, and polarographic techniques have been widely used by analysts to follow enzyme activity. Guilbault et al. [62] proposed a kinetic method for enzyme reactions based on the electrochemical measurement of the rate of cleavage of a substrate by the enzyme to be assayed. Such rates are measured by recording the difference in potential between two platinum electrodes polarized with a small, constant current. Any reaction of the type

$$A \xrightarrow{\text{B}} C + D \tag{84}$$

where the substrate A undergoes enzymolysis by B to form products C and D, can be followed provided C and/or D have redox potentials different from the substrate A. For example, in the cholinesterase-catalyzed hydrolysis of thiocholine esters, a thiol is produced upon enzymatic hydrolysis which is more electroactive (has a lower oxidation potential) than the substrate. A reduction in potential results and plots of $\Delta E/\text{min}$ versus cholinesterase activity yield straight-line calibration plots:

$$R\text{-}C\text{-}O\text{-}S\text{-}R' + H_2O \xrightarrow{\text{cholinesterase}} R'\text{-}SH + R'\text{-}COOH \tag{85}$$

Pesticides, such as sarin, systox, parathion, and malathion, which inhibit cholinesterase, can be determined at 10^{-9}-g concentrations by this technique, with a deviation of 1% [63]. Other enzyme systems, such as glucose-glucose oxidase [64], xanthine oxidase [65], and peroxidase and catalase [66], can be determined by the electrochemical technique. Care must be taken to rinse the electrodes thoroughly after each use because proteins absorb on Pt and cause a decrease in sensitivity after a number of measurements.

c. Amperometry

In amperometric methods, a constant potential is applied between two electrodes immersed in a solution of the material to be analyzed. The change in the current is then recorded with change in reaction conditions (time, addition of reagent, etc.). For example, in the assay of glucose with glucose oxidase, Blaedel and Olson [67] measured the change in current that resulted upon oxidation of ferrocyanide to ferricyanide at a tubular platinum electrode:

$$\text{Glucose} + \text{glucose oxidase} \longrightarrow \text{peroxide} \tag{86}$$

$$\text{Peroxide} + \text{ferrocyanide} \longrightarrow \text{ferricyanide} \tag{87}$$

The total current, which is proportional to the relative ratio
of ferricyanide and ferrocyanide present, is measured and related to
the concentration of glucose present. Pardue [68,69] utilized a similar system for the analysis of glucose and galactose, except for the
use of iodide instead of ferrocyanide. The total current is again
proportional to the relative amounts of iodide and iodine and, hence,
to the amount of galactose or glucose present:

$$\text{Galactose} \xrightarrow{\text{galactose oxidase}} \text{peroxide} \tag{88}$$

$$\text{Peroxide} + \text{iodide} \xrightarrow{\text{Mo(VI)}} \text{iodine} \tag{89}$$

d. Coulometry

Coulometry has also found considerable use in enzymatic methods
of assay. Coulometric methods are based on the exact measurement of
the quantity of electricity that passes through a solution during the
occurrence of an electrochemical reaction. The substance to be determined may be either oxidized or reduced at one of the electrodes (primary coulometric analysis) or may react quantitatively in solution
with a single product of electrolysis (secondary coulometric analysis).

For example, Purdy et al. [70] described a method for analysis
of urea based on the urease hydrolysis of urea to form ammonia. The
resulting ammonia is titrated with coulometrically generated hypo-
bromite, using a direct amperometric end point. Simon et al. [71]
described a coulometric method for glucose in human serum. Glucose
oxidase specifically catalyzes the aerobic oxidation of glucose to
hydrogen peroxide; the peroxide reacts with iodide in the presence
of Mo(VI) catalyst to form iodine. A known excess of thiosulfate
reduces the iodine as it is produced and the excess thiosulfate is
titrated coulometrically with electrogenerated iodine.

e. Polarography

Polarographic methods (in which the change in diffusion current
is recorded with change in the potential applied) have been widely
used in enzymatic analysis. Cholinesterase has been determined by
a measurement of the change in current resulting from the production

of thiol from acetylthiocholine iodide effected by cholinesterase
[72]. Catalase [73,74] and 3-hydroxyanthranilic oxidase [75] have
been determined by similar polarographic techniques.

A thorough discussion of the advantages and disadvantages of
polarography in biochemical analysis can be found in a book by
Purdy [76]. Also discussed are interferences and problems associated with polarography and other electrochemical methods.

In enzymatic analysis, polarography is used in three types of
measurements: O_2 (at -0.85 V versus SCE), H_2O_2 (at +0.85 V versus
SCE), and NADH.

The molecule NADH does give a reproducible polarographic wave
at a carbon paste electrode, but the sensitivity is quite low; at
Pt the wave is very ill defined but the sensitivity greater. The
polarographic wave for oxidation of NADH does not occur until potentials at least 1 V higher than the formal potential (-0.320 V versus
SCE) are applied at a Hg electrode.

The use of glassy carbon by Blaedel and Jenkins [77,78] and
Thomas and Christian [79] resulted in useful analytical procedures
for lactate [77], ethanol [78], and LDH [79]. Blaedel and Mabbott
found that even though oxidation of NADH proceeds more easily on
pretreated glassy carbon than on pyrolytic carbon film, low current
sensitivities result.

Work by Guilbault and Winartasaputra [81] showed that better
results can be obtained using either ferricyanide or 2,6-dichlorindo-
phenol, followed by a measurement of the polarographic current resulting from the oxidized product. Good results are obtained from
the use of ferricyanide, for example.

6. FLUORESCENCE METHODS

Because fluorescent procedures are several orders of magnitude
more sensitive than colorimetric methods (10^{-12} M concentrations
determinable compared with 10^{-6} in spectrophotometry), they have
replaced the older absorbance methods in many cases. Fluorometric
analysis depends on the production of a fluorogenic compound as a
result of an enzyme-catalyzed reaction. The rate of production of

the fluorescent compound is related to both the enzyme concentration
and the substrate concentration. This rate can be quantitatively
measured by exciting the fluorescent compound as it is produced and
recording the quantity of fluorescence emitted per unit of time with
a fluorometer.

The instrumentation for fluorescence is very similar to that
used in spectrophotometry, differing only in a right angle measurement rather than a straight-line one and in the use of a second filter, or monochromator. In fact, almost any commercial spectrophotometer, such as the Beckman DU or DK, can be easily converted to a
fluorometer. In addition, many companies today sell both filter and
monochromator (grating or prism) fluorometers: Turner, American Instrument, Farrand, Zeiss, Baird-Atomic, and others.

Fluorescence measurements are generally several orders of magnitude more sensitive than colorimetric ones because in fluorescence
an increase in signal over a zero background is measured while in
spectrophotometry a decrease in a large standing current is measured.
This fluorescence signal is a maximum when the optimum wavelengths
for excitation and emission are used.

Because of their sensitivity and specificity fluorescence
methods have found increasing usage in enzymology. For example,
the reduced forms of nicotinamide adenine dinucleotide (NADH) and
nicotinamide adenine dinucleotide phosphate (NADPH) are highly
fluorescent. Thus all NAD- and NADP-dependent reactions involved
in enzymatic analysis can be measured fluorometrically (Table 9)
with an increase of 2-3 orders of magnitude in sensitivity over
colorimetric techniques. Fluorescence methods have also been used
extensively for the determination of hydrolytic enzymes based on the
enzyme-catalyzed hydrolysis of a nonfluorescent ester to a highly
fluorescent alcohol or amine. Guilbault and Kramer [82,83], for
example, described a rapid, simple method for the determination of
lipase based on its hydrolysis of the nonfluorescent dibutyryl ester
of fluorescein. Fluorescein is produced on enzymolysis, which is
highly fluorescent:

$$\text{Dibutyryl fluorescein} \xrightarrow{\text{lipase}} \text{fluorescein} \qquad\qquad (90)$$
$$\text{(nonfluorescent)} \qquad\qquad \text{(fluorescent)}$$

This reaction can be monitored by measurement of the rate of production of fluorescein with time ($\Delta f/min$).

One of the most serious problems associated with the use of fluorescence in enzymatic analysis is fluorescent quenching. Highly absorbing molecules (e.g., dichromate) rob energy from the molecule under study, lowering the total fluorescence observed. Other interferences are molecules that absorb or fluorescence at the same wavelengths as the substance being determined. Proteins, for example, are serious interferences in fluorescence measurements made in the ultraviolet region because they contain amino acids (i.e., tryptophan and tyrosine) that are fluorescent in this region. For this reason it is better to make fluorescence measurements in the visible region whenever possible. Hence a fluorogenic substrate that is cleaved to a red fluorescent compound is preferred to one that yields a blue fluorescent substance [84,85].

7. RADIOCHEMICAL METHODS

The activity of an enzyme can be measured using a tagged substrate which upon enzymolysis yields a radioactive product. The amount of such radioactive product formed with time is then proportional to the concentration of enzyme.

[14C]-1-Acetylcholine, for example, has been employed as a substrate for acetylcholinesterase by Reed et al. [86] and by Potter [87]. After a removal of unhydrolyzed substrate by ion exchange, the [14C]-1-acetic acid formed by enzymatic activity is measured:

$$(CH_3)_3\text{-N-}CH_2\text{-}CH_2\text{-O-}\underset{\underset{O}{\|}}{C}\text{-}C^{14}H_3 \xrightarrow{\text{ChE}} (CH_3)_3\text{-N-}CH_2\text{-}CH_2\text{-OH} + \underset{\underset{O}{\|}}{HOC}\text{-}C^{14}H_3$$

$$(91)$$

Many radioactive substrates are available commercially from such companies as New England Nuclear. Radioactivity can be measured with an instrument as simple as the Vanguard 4π paper strip counter. Since both the substrate and the product are radioactive, prior separation

of the two must be effected before measurement. This can usually be
done by distillation or chromatography.

C. Handling of Biochemical Reagents

Enzymes are relatively fragile substances which have a tendency to
undergo inactivation or denaturation if not handled properly. The
first consideration should also be given to the proper handling of
the enzyme so as to avoid inactivation. Table 11 lists some stability
data on some important enzymes at -25, 0, and +25°C.

Generally, high temperatures and acid or alkaline solutions are
to be avoided. Most enzymes are inactivated above 35-40°C and in
solutions of pH less than 5 or greater than 9. In adjusting the pH
of an enzyme solution, one must be careful not to create a zone of
destruction around a drop of reagent added, which would tend to in-
activate some of the enzyme in solution. Solutions of enzymes should
be well stirred and acids or bases added dropwise along the side of
the vessel to avoid any denaturation in adjusting pH.

The lifetime of many enzymes can be greatly prolonged by cold
storage. For most enzymes a storage in a refrigerator at 2-5°C is
sufficient for long-term stability in the dry state. Other enzymes
are unstable even at 2-5°C and must be stored in a freezer well below
0°C. Some enzymes are stabilized at high concentrations of salts and
can be kept for long periods as a suspension in ammonium sulfate.
Such solutions can often be stored in a refrigerator or freezer for
months without loss of activity, although repeated freezing and thawing is to be avoided. Lyophilized enzymes should be stored at -25 or
0°C, dry, and free of moisture.

Organic solvents, such as alcohol, acetone, or ether, denature
most enzymes at room temperature except at low concentrations (less
than 3%). Care must be taken in changing the composition of a solution from aqueous to partly nonaqueous.

In some cases care must be taken to avoid the presence of air
or oxygen in storage of enzyme solutions. The sulfhydryl group in
CoA is easily oxidized by atmospheric oxygen, for example, while

TABLE 11 Stability of Some Serum Enzymes Under Storage (≥10% change in given time)

Enzyme	Room temperature (25°C)	Refrigerator (0-4°C)	Frozen (-25°C)
Aldolase (EC 4.1.2.13)	2 Days	2-3 Days	Unstable[a] (15% loss 0-2 days; 45% loss 7-8 days)
α-HBDH (EC 1.1.1.27) (LDH isoenzyme 1)	Unstable	3 Days	Unstable[a] (35% loss 7 days)
α-Amylase (EC 3.2.1.1)	1 Month	6 Months	2 Months
Cholinesterase (EC 3.1.1.8)	1 Week	2-3 Months	1 Year
CK (EC 2.7.3.2)	4 Hr (75% loss 1 day)	Unstable (94% loss 7 days)	7 Days
ICDH (EC 1.1.1.42)	5 Hr	3 Days	3 Weeks
LDH (EC 1.1.1.27)	1 Week	2 Days	2 Days[a]
LAP (EC 3.4.11.1)	1 Week	1 Week	1 Week
MDH (EC 1.1.1.37)	8 Hr (20% loss 1 day)	1-2 Days (10% loss 1 day)	3 Days (20% loss 7 days)
Acid phosphatase (EC 3.1.3.2)	4 Hr	3 Days[b] (25% loss 1 day)	3 Days[b]
Alkaline phosphatase (EC 3.1.3.1)	2 Days[c] (30% loss 7 days)	2 Days[c] (30% loss 7 days)	1 Month
GOT (EC 2.6.1.1)	3 Days (20% loss 7 days)	1 Week (20% loss 2 weeks)	1 Month
GPT (EC 2.6.1.2)	3 Days (35% loss 7 days)	1 Week	Unstable (20% loss 2-3 days)

[a]Change caused by thawing.
[b]With added acetic acid.
[c]Change due to pH change on standing.

NADH and NADPH must be protected from light and stored in a desiccator in the cold. NADPH has about 10 times greater stability when stored in the dark than when stored in direct sunlight. Frequent difficulties with the instability of NADH and NADPH are due to moisture (50% loss in 1 day), light (20% loss in 7 weeks), or temperature (no loss at 4°C, 10% loss in 3 weeks at 33°C). CoA is easily hydrolyzed also, so that solution pH should be carefully controlled. A pH of 4-5 is optimal.

Many enzymes are denatured at surfaces; therefore, the formation of froth should be avoided. The vigorous shaking of an enzyme solution in dissolution is poor practice. Likewise, stirrers which whip up the surface should be avoided. Frequently, an enzyme may be observed to be unstable, yet this same enzyme in a carefully cleaned vessel free of detergent can be quite stable [88].

Some buffer solutions, especially phosphate, are good growth media for microorganisms. These should be carefully handled and stored in dark, sterilized bottles. To avoid contamination, each day's supply of buffer should be removed by pouring rather than pipetting.

Because traces of metal ions can cause a loss of enzymatic activity, it is necessary that the water used in preparation of reagents and the like be carefully purified, either by deionization or by a double distillation from $KMnO_4$ in an all-glass still. Fresh water should be used for preparation of solutions. All pipettes, flasks, and containers should be thoroughly cleaned. After a double rinse with distilled water, pipettes should be dried in an oven at a moderate temperature (never by rinsing with acetone or other organic solvents).

Solvents of NADH and NADPH are acid-labile; their pH must be above 7.5. Solutions of NAD and NADP are alkali-labile; their pH should be less than 7. ATP is most stable at pH 9.

SUGGESTED READINGS

The following general reference books are useful sources of information on enzymes.

General

Beckman Instruments Co., *Clinical Enzyme Primary*. Fullerton, Cal., 1973.

P. D. Boyer (ed.), *The Enzymes*, Vols. 1 and 2. Academic Press, New York, 1970. Chapters under the general headings of Structure, Control, Kinetics, and Mechanism.

Worthington Biochemical Co., *Manual of Clinical Enzyme Measurements*. Freehold, N.J., 1971.

M. Dixon and E. C. Webb, *Enzymes*. Longman, London, 1964. A very useful source of general information.

Analytical

H. U. Bergmeyer (ed.), *Methods of Enzymatic Analysis*. Springer Verlag, Berlin, 1974, 4 Vols.

G. G. Guilbault, *Handbook of Enzymatic Methods of Analysis*. Marcel Dekker, New York, 1976.

Kinetics

W. W. Cleland, in *The Enzymes,* Vol. 2 (P. D. Boyer, ed.). Academic Press, New York, 1970, pp. 1-65.

K. J. Laidler and P. S. Bunting, *The Chemical Kinetics of Enzyme Action*. Oxford University Press, New York, 1973.

K. F. Tipton, in *Companion to Biochemistry* (G. Bull, ed.). Longman, London, 1974, pp. 225-251.

Specific Enzymes

T. E. Barman, *Enzyme Handbook*. Springer Verlag, Berlin, 1969, 2 Vols., plus supplement (1974). Lists almost every enzyme and gives key reference.

M. Florkin and E. Stotz (eds.), *Enzyme Nomenclature, Recommendations* (1973). Comprehensive Biochemistry Series, Vol. 13, 3rd ed. Elsevier, New York.

P. D. Boyer (ed.), *The Enzymes,* Vol. 3 et seq. Academic Press, New York, 1971 et seq. Each volume is dedicated to a specific class of enzyme with chapters on individual enzymes of that class.

S. P. Colowick and N. O. Kaplan, *Methods in Enzymology*. Academic
Press, New York. A multivolume work giving preparation and assay
details for most enzymes.

REFERENCES

1. T. Bennet and E. Freiden, *Modern Topics in Biochemistry*. Mac-
    millan, London, 1969, p. 43.

2. L. Pauling, P. B. Corey, and H. R. Branson, *Proc. Natl. Acad.
    Sci. 37*, 205 (1951).

3. G. Osann, *Poggendorf's Ann. 67*, 372 (1845).

4. O. Warburg, *Wassenstoffubertragende Fermente*. Springer Verlag,
    Berlin, 1948.

5. M. Florkin and E. Stotz (eds.), *Enzyme Nomenclature, Recommenda-
    tions* (1973). Comprehensive Biochemistry Series, Vol. 13, 3rd
    ed. Elsevier, Amsterdam, New York.

6. F. Wroblewski and F. Gregory, *Ann. N.Y. Acad. Sci. 94*, 921 (1961).

7. D. Dawson, H. Eppenberger, and N. Kaplan, *Biochem. Biophys. Res.
    Commun. 21*, 346 (1965).

8. K. Takahashi, S. Ushikubo, M. Oimomi, and T. Shinko, *Clin. Chim.
    Acta 38*, 285 (1972).

9. L. Michaelis and M. Menten, *Biochem. Z. 49*, 333 (1913).

10. H. Lineweaver and D. J. Burk, *J. Am. Chem. Soc. 56*, 658 (1934).

11. K. M. Plowman, *Enzyme Kinetics*. McGraw-Hill, New York, 1972.

12. S. P. Colowick and N. O. Kaplan (eds.), *Methods in Enzymology*.
    Academic Press, New York, 1957.

13. E. Schütz, *Z. Physiol. Chem. 9*, 577 (1885).

14. J. Schütz, *Z. Physiol. Chem. 30*, 1 (1900).

15. B. S. Millar and J. A. Johnson, *Arch. Biochem. Biophys. 32*, 200
    (1951).

16. A. B. Roy, *Biochem. J. 57*, 465 (1954).

17. P. Baum and R. Czok, *Biochem. Z. 332*, 121 (1959).

18. E. Adler, G. Gunther, and M. Plass, *Biochem. J. 33*, 1028 (1939).

19. W. J. Blaedel and G. P. Hicks, in *Advances in Analytical Chem-
    istry and Instrumentation,* Vol. 3 (C. N. Reilley, ed.). Inter-
    Science, New York, 1964, pp. 105-140.

20. G. G. Guilbault, D. N. Kramer, and P. L. Cannon, *Anal. Chem.
    34*, 1437 (1962).

21. H. Linde, *Anal. Chem. 31*, 2092 (1959).

22. C. McGaughey and E. Stowell, *Anal. Chem. 36*, 2344 (1964).

23. N. D. Jespersen, *J. Am. Chem. Soc. 97,* 1662 (1975).

24. K. B. Yatsimirskii, *Kinetic Methods of Analysis.* Pergamon Press, Oxford, 1966.

25. G. G. Guilbault, in *Practical Fluorescence: Theory, Instrumentation and Practice* (G. G. Guilbault, ed.). Marcel Dekker, New York, 1967, pp. 306-307.

26. M. Whittaker, *Chemical Society Meeting on Enzymes.* University of Exeter, England, July 1975.

27. J. LaDue, F. Wroblewski, and A. Karmen, *Science 120,* 497 (1954).

28. Worthington Biochemical, *Manual of Clinical Enzyme Measurements.* Freehold, N.J., 1971, pp. 35, 46-49.

29. H. Zimmerman and J. Henry, *Clinical Diagnosis by Laboratory Methods* (I. Davidson and J. Henry, eds.). Saunders, Philadelphia, 1969, p. 691.

30. H. Zimmerman and J. Henry, *Clinical Diagnosis by Laboratory Methods* (I. Davidson and J. Henry, eds.). Saunders, Philadelphia, 1969, p. 694.

31. R. Seary, *Diagnostic Biochemistry.* McGraw-Hill, New York, 1969, p. 515.

32. R. Schapira, J. Dreyfus, G. Schapira, and J. Demos, *Rev. Franc. Etud. Clin. Biol. 5,* 990 (1960).

33. V. Aebi, R. Richterich, J. Colombo, and E. Rossi, *Enzymol. Biol. Clin. 1,* 61 (1962).

34. S. Ebashi, Y. Toyokura, H. Momoi, and H. Sugita, *J. Biochem. 46,* 103 (1959).

35. S. Okinaka, H. Kumagai, S. Ebashi, H. Momoi, Y. Toyokura, and Y. Fojie, *Arch. Neurol. 4,* 520 (1961).

36. V. Aebi, R. Richterich, H. Stillhart, J. Colombo, and E. Rossi, *Helv. Paediatr. Acta 16,* 543 (1961).

37. J. Dreyfus, G. Schapira, and J. Demos, *Etud. Clin. Biol. 5,* 384 (1960).

38. H. Zimmerman and J. Henry, *Clinical Diagnosis by Laboratory Methods* (I. Davidson and J. Henry, eds.). Saunders, Philadelphia, 1969, p. 720.

39. A. Roy, M. Brower, and J. Woodbridge, Paper presented at 1970 ASCP Meeting, Atlanta, Ga., 1970.

40. R. Hamm and L. Körmendy, *Fleischwirtschaft 46,* 615 (1966).

41. G. Gantner, *Fleischwirtschaft 48,* 1671 (1968).

42. W. Haab and L. Smith, *J. Dairy Sci. 39,* 1644 (1957); *40,* 546 (1959).

43. J. Schormüller, in *Methods of Enzymatic Analysis* (H. Bergmeyer, ed.). Springer Verlag, Berlin, 1974, pp. 74-77.

44. J. Schormüller, in *Methods of Enzymatic Analysis* (H. Bergmeyer, ed.). Springer Verlag, Berlin, 1974, pp. 82-85.

45. J. Schormüller, in *Methods of Enzymatic Analysis* (H. Bergmeyer, ed.). Springer Verlag, Berlin, 1974, pp. 87-91.

46. B. F. Erlanger, S. N. Burbaum, R. A. Sack, and A. G. Cooper, *Anal. Biochem. 19,* 542 (1967).

47. G. G. Guilbault, in *Practical Fluorescence: Theory, Instrumentation and Practice* (G. G. Guilbault, ed.). Marcel Dekker, New York, 1967, pp. 297-358.

48. G. G. Guilbault, P. Brignac, and M. Zimmer, *Anal. Chem. 40,* 190 (1968).

49. G. G. Guilbault, P. Brignac, and M. Juneau, *Anal. Chem. 40,* 1256 (1968).

50. H. Pardue, M. Burke, and D. O. Jones, *J. Chem. Ed. 44*(11), 684 (1967).

51. R. Kuhn and D. Jerchtel, *Ber. Dtsch. Chem. Ges. 74,* 941 (1941).

52. G. G. Guilbault, in *Carbohydrate Metabolism* (W. Wood, ed.). Methods in Enzymology, Vol. 41. Academic Press, New York, 1975, p. 53.

53. S. Fox and E. Atkinson, *J. Am. Chem. Soc. 72,* 3629 (1950).

54. K. Tsou, C. Cheng, M. Nachlas, and A. Seligman, *J. Am. Chem. Soc. 78,* 6139 (1956).

55. J. Bancroft, *J. Physiol. 37,* 12 (1908).

56. O. Warburg, *Biochem. Z. 152,* 51 (1924).

57. W. W. Umbreit, R. H. Burris, and J. F. Stauffer, *Manometric Techniques.* Burgess, Minneapolis, Minn., 1945.

58. M. Dixon, *Manometric Methods.* Cambridge University Press, Cambridge, England, 1951.

59. M. Milano and H. Pardue, *Clin. Chem. 21,* 211 (1975).

60. G. Charlton, D. Read, and J. Reed, *J. Appl. Physiol. 18,* 1247 (1963).

61. T. Rohm and G. Guilbault, *Int. J. Environ. Anal. Chem. 4,* 51 (1975).

62. G. G. Guilbault, D. N. Kramer, and P. L. Cannon, *Anal. Chem. 34,* 842 (1962).

63. G. G. Guilbault, D. N. Kramer, and P. L. Cannon, *Anal. Chem. 34,* 1437 (1962).

64. G. G. Guilbault, B. Tyson, D. N. Kramer, and P. L. Cannon, *Anal. Chem. 35,* 582 (1963).

65. G. G. Guilbault, D. N. Kramer, and P. L. Cannon, *Anal. Chem. 36,* 606 (1964).

66. G. G. Guilbault, *Anal. Biochem. 14*, 61 (1966).

67. W. J. Blaedel and C. Olson, *Anal. Chem. 36*, 343 (1964).

68. H. Pardue, *Anal. Chem. 35*, 1240 (1963).

69. H. Pardue and R. Simon, *Anal. Biochem. 9*, 204 (1964).

70. W. C. Purdy, G. D. Christian, E. C. Knoblock, Paper presented at the Northeast Section, American Association of Clinical Chemists, 16th National Meeting, Boston, Mass., August 17-20, 1964.

71. R. K. Simon, G. D. Christian, and W. C. Purdy, *Clin. Chem. 14*, 463 (1968).

72. V. Fischerova-Bergerova, *Pracovni Lekarstvi 16*(1), 8 (1964).

73. T. H. Ridgway and H. B. Mark, *Anal. Biochem. 12*, 357 (1965).

74. H. Jacob, *Z. Chem. 4*, 189 (1964).

75. M. N. Gadaleta, E. Lofrumento, C. Landriscina, and A. Alifano, *Bull. Soc. Ital. Biol. Sper. 39*(24), 1866 (1963).

76. W. C. Purdy, *Electroanalytical Methods in Biochemistry*. McGraw-Hill, New York, 1965.

77. W. J. Blaedel and R. A. Jenkins, *Anal. Chem. 47*, 1337 (1975).

78. W. J. Blaedel and R. A. Jenkins, *Anal. Chem. 48*, 1240 (1976).

79. L. C. Thomas and G. D. Christian, *Anal. Chim. Acta 78*, 271 (1975).

80. W. J. Blaedel and G. A. Mabbott, *Anal. Chem. 50*, 933 (1978).

81. G. G. Guilbault, H. Winartasaputra, and S. S. Kuan, *Anal. Chem. 54*, 1987 (1982).

82. G. G. Guilbault and D. N. Kramer, *Anal. Chem. 35*, 588 (1963).

83. G. G. Guilbault and D. N. Kramer, *Anal. Chem. 36*, 409 (1964).

84. G. G. Guilbault, *Practical Fluorescence*. Marcel Dekker, New York, 1974.

85. S. Udenfriend, *Fluorescence Assay in Biology and Medicine*. Academic Press, New York, 1963, 1967.

86. D. L. Reed, K. Goto, and C. H. Wang, *Anal. Biochem. 16*, 59 (1966).

87. L. T. Potter, *J. Pharm. Ex. Ther. 156*, 500 (1967).

88. E. Schmidt and F. Schmidt, *Enzymol. Biol. Clin. 3*, 80 (1963).

2
Principles of Immobilized Enzymes

I. GENERAL DISCUSSION

The use of enzymes for analytical purposes has been limited because of certain disadvantages, such as their instability, poor precision, and lack of availability. Moreover, aqueous solutions of enzymes often lose their catalytic ability fairly rapidly and the enzymes can be neither recovered from such solutions nor their activity regenerated. These difficulties have been removed or minimized by the development of enzyme immobilization techniques. The free enzyme is immobilized (insolubilized) by trapping it in an inert matrix, such that the immobilized enzyme retains its catalytic properties for a much longer time than the free enzyme and can be used continuously for many more analyses.

The science and technology of immobilized enzymes have experienced a phenomenal growth in the recent past. Over 50 years ago, Nelson and Griffin [1] adsorbed invertase onto charcoal and noted that the "immobilized enzyme" retained biological activity and could be reused many times, but this field lay dormant until the 1950s when a renewed interest spouted forth. A book on immobilized enzyme technology has been published [2], and many specific reviews of the application of immobilized enzymes in analytical chemistry [3-12], clinical chemistry [13], automated analysis [14,15], urinalysis [16], and the analytical aspects of immobilized enzyme columns [17] have been

published. The immobilized or insolubilized enzyme offers three
advantages over the soluble enzyme as an analytical reagent.

A. Reusability

Most enzymes are very expensive by comparison with other reagents,
and routine analysis requires large and costly amounts of these
materials. Insolubilization of the enzymes provides a sample that
can be reused many times, even up to 10,000 times in some cases,
thus representing a tremendous cost saving.

B. Greater Stability

When enzymes are immobilized they are held in an environment more
like that in which they are found naturally and are generally more
stable. Many types of immobilization impart greater stability to
the enzyme, making them more useful over wider pH ranges and at
higher temperatures.

C. Fewer Interferences

The immobilized enzyme appears to be less susceptible to the normal
activators and inhibitors that affect the soluble enzyme. Only the
strongest inhibitor will decrease its activity, only the strongest
activators will boost its catalytic power. Thus, in a complex analytical sample, such as blood or sewerage, the immobilized enzyme
is much more useful than the soluble as an analytical reagent.

II. IMMOBILIZATION METHODS

Within the last few years, a new technology based on enzyme immobilization has rapidly emerged. Five methods have been used for the
preparation of water-insoluble derivatives of enzymes: (1) microencapsulation within thin-wall spheres, (2) adsorption of inert
carriers, (3) covalent crosslinking by bifunctional reagents into
macroscopic particles, (4) physical entrapment in gel lattices, and
(5) covalent binding to water-insoluble matrices. Let us now consider some of the different methods of enzyme immobilization and the

procedures followed to immobilize enzyme used in electrodes. For
more technical details one is referred to several comprehensive
reviews published recently [2,18,19].

The two major techniques used commonly to immobilize an enzyme
are (1) the chemical modification of the molecule by the introduction
of insolubilizing groups. This technique, which results in a chemical
"tying down" of the enzyme, is in practice sometimes difficult to
achieve because the insolubilizing groups can attach across the active
site destroying the activity of the enzyme; (2) the physical entrapment of the enzyme in an inert matrix, such as starch or polyacrylamide gels.

The major difference between the entrapped and the attached
enzymes is that the former are isolated from large molecules which
cannot diffuse into the matrix. The attached enzyme may be exposed
to molecules of all sizes. Hence, the two types of immobilized
enzymes will differ in the form of the kinetics observed and in the
kinds of interferences observed. Thus for the assay of large substrates as proteins with proteolytic enzymes, an attached enzyme
must be used and not an entrapped enzyme. Either enzyme could be
used for the assay of small substrates such as urea. Since the physical entrapment techniques offer advantages of speed and ease of
preparation, these will be discussed first.

A. Physical Entrapment

Vasta and Usdin [20] showed that cholinesterase could be insolubilized by entrapment in a starch gel. The preparation of immobilized
cholinesterase for use in analytical chemistry was described by
Bauman et al. [21]. The enzyme, immobilized by the use of a starch
matrix and placed on a polyurethane foam pad, was found to be stable
and active for 12 hr under continuous use. The activity of the
enzyme was monitored electrochemically, using two platinum electrodes
and an applied current of 2 μA. This immobilized enzyme was used to
determine the substrates acetyl- and butyrylthiocholine iodide, both
in individual samples and continuously. A fluorometric system for
the assay of anticholinesterase compounds using this immobilized

cholinesterase was described by Guilbault and Kramer [22]. As long
as the enzyme is active a fluorescence is produced because of the
hydrolysis of the 2-naphthyl acetate to 2-naphthol. Upon inhibition,
the fluorescence drops to a value approaching zero. Hicks and Updike
[23] demonstrated the immobilization of enzyme activity in polyacrylamide gel. The preparation is stable and can be lyophilized and
stored conveniently. Several enzyme systems were trapped in the
gels, namely, glucose oxidase, catalase, lactic dehydrogenase, amino
acid oxidase, glutamic dehydrogenase, and enzymes active in human
serum.

The polyacrylamide enzyme gels were found to show little loss
of activity after 3 months of storage at 0-4°C. Lactic dehydrogenase
(LDH) lost about 30% of its activity in 3 months, while glucose oxidase (GO) showed no loss in activity. Hydration of a LDH gel causes
loss of all activity in 3 months at 4°C, whereas a hydrated GO gel
exhibited a loss of only 5% at 0°C in 3 months. The gels are very
resistant to flow loss and were used for the assay of the substrates
glucose and lactic acid.

To test the stability of the immobilized enzyme in comparison
with the soluble enzyme, the activity of a series of identical glucose oxidase columns and glucose oxidase solutions were compared
after heating for 10 min at temperatures of 37-70°C. About half of
the activity of GO in both gel and solution was destroyed in 10 min
at 60°C and all of the activity of both were lost at 70°C.

Updike and Hicks [24] coupled the entrapped glucose oxidase
system with an electrochemical sensor for the determination of glucose in blood. The oxygen electrode was used to monitor the oxygen
uptake:

$$\text{Glucose} + O_2 \xrightarrow{\text{glucose oxidase}} H_2O_2 + \text{gluconic acid} \tag{1}$$

The apparatus used to monitor the reaction is indicated in the article
by Updike and Hicks [25]. Immobilized glucose oxidase is placed in a
miniature chromatographic column, and samples containing glucose to be
analyzed are pumped over the column at a rate of 0.3 ml/min with a

peristaltic pump. Using an immobilized enzyme and an oxygen electrode as the sensing device, a "reagentless" analyzer was achieved.

Wieland et al. [26] also prepared insoluble enzymes in polyacrylamide gels. The enzymes alcohol dehydrogenase, trypsin, and
lactic dehydrogenase were immobilized by physical entrapment using
a procedure similar to that described above.

McLaren and Peterson [27], Barnett and Bull [28], and Nikolaev
and Mardashev [29] attempted the physical entrapment of the enzymes
asparaginase, ribonuclease, and chymotrypsin by adsorption, absorption, or ion exchange, and enzymes have been encapsulated in semipermeable microcapsules made of synthetic polymers [30]. Bernfeld
and Wan [31] used polyacrylamide gels to entrap enzyme activity.

The lattice entrapment of hexokinase, phosphoglucoisomerase,
phosphofructokinase, and aldolase in a polyacrylamide gel was described by Brown et al. [32]. Van Duijn et al. [33] discussed the
theoretical and experimental aspects of an enzyme determination in
a cytochemical model system of polyacrylamide films containing alkaline phosphatase. Lojda et al. [34] used the phosphates of the
naphthol AS series in the quantitative determination of alkaline and
acid phosphatase activities in polyacrylamide membrane model systems.

Guilbault and Das [18] conducted a thorough investigation of
various parameters affecting the immobilization of the enzymes cholinesterase and urease. The immobilization of these enzymes in
starch gel, polyacrylamide, and silicone rubber was investigated,
and the stability of the insolubilized enzymes in storage and in use
was reported.

It was found that the optimal method for immobilization of
either cholinesterase or urease appears to be by physical entrapment
in a polyacrylamide gel. The silicone rubber polymerization is too
rough on the enzyme, and about 80% of the activity is lost. The
starch gel-entrapped enzyme is too weakly held, and much of the
enzymatic activity is lost due to leeching. The stability of both
enzymes in polyacrylamide is better, both in dry and wet storage
between use, than the corresponding starch gel pads.

However, the preparation of the starch gel pad is simpler and
very little enzyme is lost during preparation. Some enzyme is lost
during the acrylamide polymerization, and the amount lost depends on
the experimental conditions of the polymerization (range 10-25% loss).
Another disadvantage in the use of polyacrylamide lies in the fact
that riboflavin and $K_2S_2O_8$ must be used as catalysts for the polymerization. Riboflavin, being fluorescent, interferes in the fluorometric monitoring of the enzyme. One has to wait for the fluorescence baseline to become steady (most of the riboflavin is washed
out of the pad) before starting the actual experiments. In cases
in which potentiometric measurements are made, $K_2S_2O_8$ should not be
used. Consideration of all factors indicates that polyacrylamide is
the best of all three gel materials tried, and starch gel should be
used only when long-term use is not desired.

Tris buffer should be used in all studies for maximum stability.
The enzyme may be dry-stored for up to 80 days with little appreciable loss of activity. The urease enzyme pads could be used for up
to 80 days of continuous use and wet storage. The cholinesterase
pads can be reproducibly used for about 40 hr when stored wet between
use.

B. Microencapsulation

Microencapsulation within thin-wall spheres, the newest approach to
enzyme immobilization, was introduced several years ago by Chang
[35-40]. The thin wall of the spheres is semipermeable such that
the enzymes are physically prevented from diffusing out of the micro-
capsule while reactants and products can readily permeate the encap-
sulating membrane.

Microencapsulation is performed by depositing polymer around
emulsified aqueous droplets; either by interfacial coactivation or
by interfacial polycondensation. Such a technique has not yet been
applied to the preparation of electrodes but can be.

The main disadvantage of the use of spheres is that many of the
interfacial polymerization procedures cause enzyme deactivation.

Rony [41] circumvented this problem by first producing hollow fibers instead of spheres and then placing the enzyme in the fiber and sealing the ends. Hollow fibers have the added advantage of (1) being easy to fill with many types of catalyst, (2) allowing recovery of the trapped enzyme, and (3) being mass produced and thus commercially available.

C. Adsorption

Adsorption of enzymes on insoluble supports results from ionic, polar, or hydrogen bonding or by hydrophobic or π-electron interactions. In early works, adsorption to inorganic carriers was used as a technique for the characterization of enzymes, and since then the characterization of the adsorption phenomenon has been the major concern of most studies. Early works on protein adsorption have been reviewed by Silman and Katchalski [42]. A few of the adsorbents that have been used are glass [43], quartz [44], charcoal [45], silica gel [46], alumina [47], ion-exchange resins [48,49], and bentonite [50].

The major advantage of this method of immobilization is its extreme simplicity. Attachment is by simple exposure and conditions are mild. But the disadvantage is a very serious one. Adsorbed enzymes are highly dependent on such factors as pH, solvent, substrate, and temperature, and can be easily desorbed by changing these factors.

D. Covalent Crosslinking by Bifunctional Reagents

Bifunctional reagents have been used to insolubilize enzymes and other proteins by intermolecular crosslinking, with the concomitant formation of macroscopic particles [51-53]. Bifunctional reagents can be divided into two classes, "homobifunctional" and "heterobifunctional," depending on whether the reagent possesses two identical or two different functional groups.

A few homobifunctional reagents are glutaraldehyde [54-57], bisdiazobenzidine-2,2'-disulfonic acid [58,59], 4,4'-difluoro-3,3'-dinitrodiphenyl sulfone [60], diphenyl-4,4'-dithiocyanate-2,2'-

disulfonic acid [61], 1,5-difluoro-2,4-dinitrobenzene [62], and
phenol-2,4-disulfonyl chloride [63].

$$
\begin{array}{c}
\text{CHO} \quad \boxed{\text{Enzyme}} \\
| \qquad\quad | \\
(\text{CH}_2)_3 + \text{NH}_2 \\
| \\
\text{CHO} \qquad \text{NH}_2 \\
\text{Glutaraldehyde} \ \boxed{\text{Albumin}}
\end{array}
\longrightarrow
\begin{array}{c}
\text{H} \\
\text{C=N-Enzyme} \\
|\\
(\text{CH}_2)_3 \\
|\\
\text{C=N-Albumin} \\
|\\
\text{H}
\end{array}
\qquad (2)
$$

A few heterobifunctional reagents are toluene-2-isocyanate-4-
isothiocyanate [64], trichloro-o-triazine [65,66], and 3-methoxydi-
phenyl methane-4,4'-diisocyanate [64]. The advantages of this method
are (1) simplicity and (2) chemical binding of the enzyme, enabling
control of the physical properties and particle size of the final
product. The major disadvantage is that many enzymes are sensitive
to the coupling reagents such that they lose activity in the process.

E. Covalent Binding to Water-Insoluble Matrices

Within the last decade, a new technology based on the immobilization
of enzymes by covalent coupling to insoluble carriers has rapidly
emerged. Because covalent coupling places enzymes in a more natural
environment, they usually function more efficiently, have increased
stability, and have the added advantage of immobilization which is
not reversed by pH, ionic strength, substrate, solvents, or temperature. It is little wonder that covalent binding of enzymes to insoluble carriers is now the most widely used method of enzyme insolubilization.

Carriers are chosen by their properties of solubility, functional
groups, mechanical stability, surface area, swelling, and hydrophobic
or hydrophilic nature. Essentially, three types of carriers have been
employed: inorganics, natural polymers, and synthetic polymers. A few
of the more common carriers which have been used are given in Table 1.
These carriers are usually activated by transformation into various
derivatives. Several of the more common methods are shown in Fig. 1.
Table 2 shows the amino acid residues which are reactive in several
common immobilizing methods.

TABLE 1   Common Carriers Used for Covalent Binding of Enzymes

Carrier	Ref.
Porous glass	67, 68
Polyacrylamide	69, 70
Polyacrylic acid derivatives	71, 72
Polyaspartic acid	73
Polyglutamic acid	71
Polystyrene	74
Nylon	75, 76
Cellulose	77, 78
Sephadex	79, 80
Ethylene-maleic anhydride copolymer	73, 81
Agarose	82, 83
Sepharose	84, 85
Carboxymethyl cellulose	73, 77

Binding is carried out by way of functional groups on the enzyme which are not essential for its catalytic activity. Even though a particular amino acid residue may be necessary for catalytic activity, it is usually present several times in nonessential positions. Random chemical reaction will denature only a portion of the molecules present. Because binding must be carried out under conditions which do not cause denaturation, it is necessary to understand the effects of chemical modification of enzymes on their activity. Several reviews have been written on this subject [86,88].

The amino acid residues suitable for covalent binding are (1) α- and ϵ-amino groups, (2) the phenol ring of tyrosine, (3) β- and γ-carboxyl groups (4) the sulfhydryl group of cysteine, (5) the hydroxyl group of serine, and (6) the imidazole group of histidine. Of these, the most widely used are the first three. There are a large number of methods of covalently coupling enzymes to water-insoluble carriers.

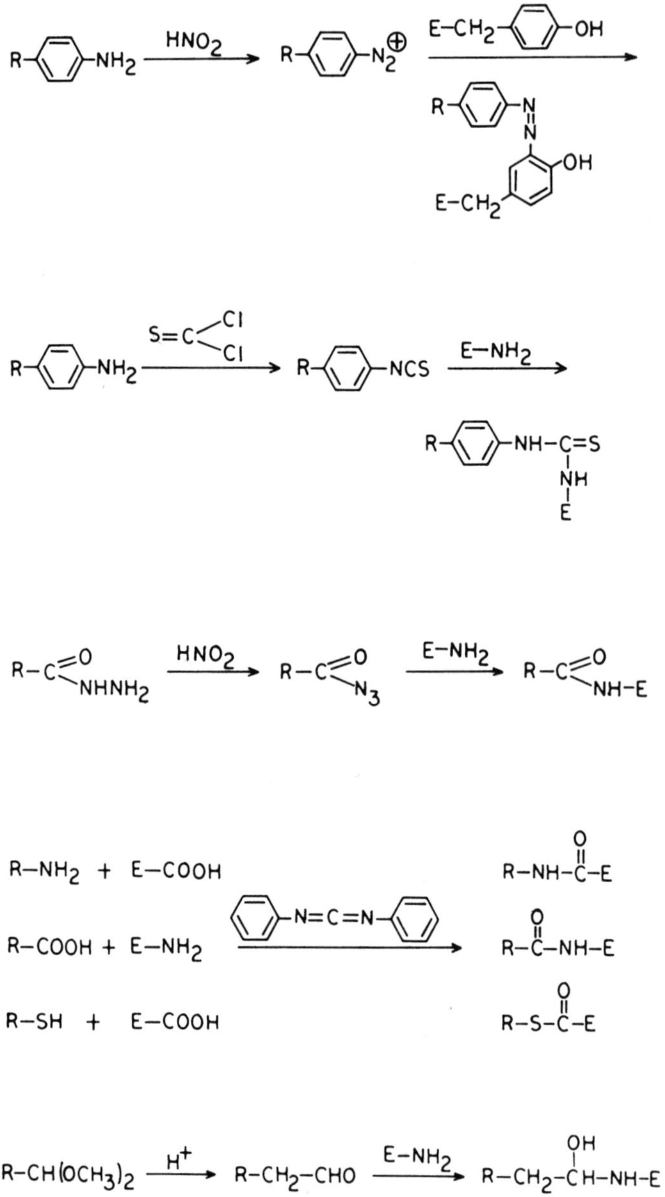

FIG. 1 Common coupling reactions.

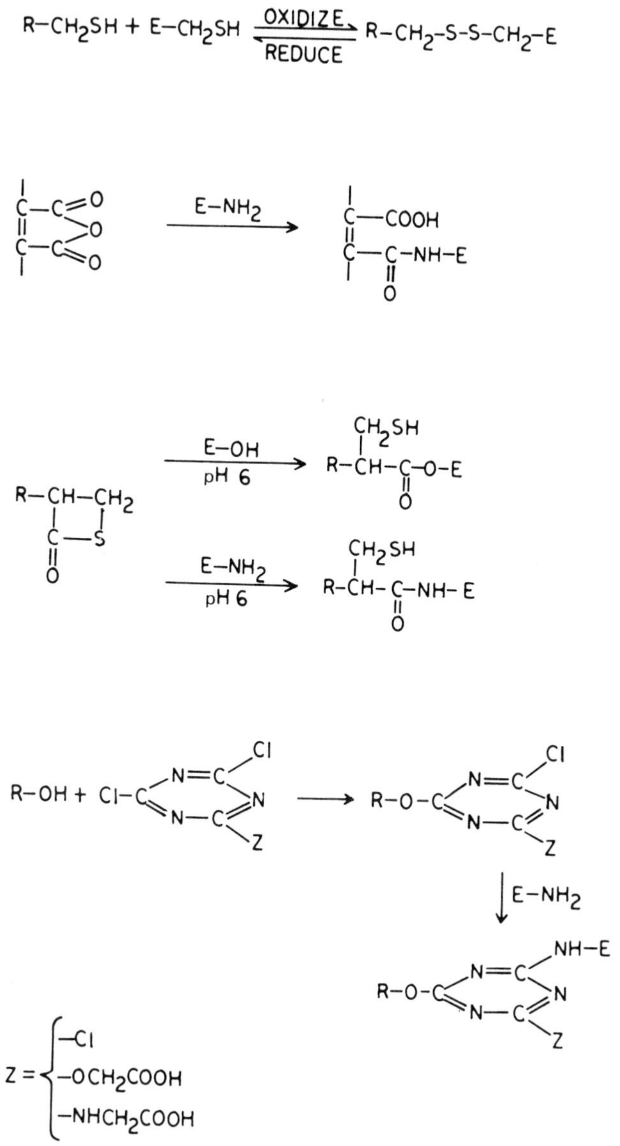

$$R-CH_2SH + E-CH_2SH \underset{\text{REDUCE}}{\overset{\text{OXIDIZE}}{\rightleftharpoons}} R-CH_2-S-S-CH_2-E$$

FIG. 1   (Continued)

TABLE 2 Reactive Amino Acid Residues in Common Insolubilizing Methods

Insolubilizing reagent	Reactive amino acid residues								Ref.
Chloro-sym-trianzinyl derivative	Lys								65, 66
Diazo derivative	Lys	His	Tyr	Arg	Cys				90
	Lys	His	Tyr						91
	Lys		Tyr	Arg					92
		His	Tyr						93–95
Isothiocyanato derivative	Lys		Tyr	Arg					96, 97
N-ethyl-5-phenylisoxazolium-3'-sulfonate	Lys					Asp	Glu		98
Diimide	Lys		Tyr		Cys	Asp	Glu		99, 100
Acid azide	Lys		Tyr		Cys			Ser	101
Maleic acid or maleic anhydride copolymer	Lys								102
Glutaraldehyde	Lys								103–107
	Lys	His	Tyr		Cys				108
Cyclic iminocarbamate	Lys								109

Source: Ref. 89.

Scientists of the Department of Biophysics of the Weismann
Institute (Rehovoth, Israel) have pioneered in the preparation of
enzyme insolubilized by covalent bonding to polymeric lattices.
These modified enzymes retain significant fractions of their native
activities while, according to initial studies, certain other properties have in fact been altered [110-114]. Three of the insolubilized covalently bound enzymes (trypsin, chymotrypsin, and papain)
are available as lyophilized powders from Miles (Elkhart, Indiana).

1.   INSOLUBILIZED TRYPSIN AND CHYMOTRYPSIN

An appropriate amount of trypsin or chymotrypsin is added to a
copolymer of maleic anhydride and ethylene, previously crosslinked
with hexamethylenediamine to decrease its water solubility.   The
reaction occurs in buffer solution overnight at 4°C.   By altering
the ratio of enzyme to carrier, derivatives of differing characteristics are produced [82].

Trypsin or chymotrypsin

2.   INSOLUBILIZED PAPAIN

The polymer is prepared by coupling native papain to a water-
soluble diazonium salt derived from a copolymer of ρ-amino-DL-
phenylalanine and L-leucine, the reaction occurring at 4°C over a
20-hr period [80,83].   The product is a stable, water-insoluble
papain derivative retaining up to 70% of the original papain activity on low molecular weight substrates and up to 30% on high molecular weight substrates.   This insolubilized papain preparation has

been used to study the structure of rabbit γ-globulin. Because this product is active in the hydrolysis of protein in the absence of added reducing agents, it is possible to differentiate protein fragments produced by proteolysis from those produced by reduction [110].

Papain

Katchalski [114] prepared water-insoluble derivaties of papain by adsorption of papain chemical derivatives on a collodion column. The acetyl-, succinyl-, poly-L-ornithyl-, poly-γ-benzyl-L-glutamyl-, water-insoluble (maleic acid-ethylene)-, and (4-aminobiphenyl-4-N'-aminoethyl) starch-papain derivatives were prepared for investigation.

3. INORGANIC MATRICES

Weetall and Hersh described procedures for the covalent coupling of enzymes to inorganic materials with the aid of an intermediate coupling agent [115-119]. Inorganic carriers are not subject to microbial attack, they do not change configuration over an extensive pH range or under various solvent conditions, and, with their greater rigidity, they immobilize enzymes to a greater degree than organic polymers.

Alkaline phosphatase [115], urease [117], trypsin, and papain [116] were covalently coupled to porous 96% silica glass with a silane coupling agent. The glass (100-mesh particle size containing pores of 790-Å diameter) was cleaned and then coupled to α-amino-propyluriethoxysilane in toluene solution. The aminoalkyl group was then converted to an aminoaryl group by coupling ρ-nitrobenzoic acid. This was reduced, diazotized, and added to a solution containing the enzyme. The product was washed and stored at 4°C.

The alkaline phosphatase immobilized on glass contained the equivalent of 0.74 mg of active enzyme/g of glass [115], the urease 1.0 mg/g of glass [117], and the trypsin 0.12-25 mg/g of glass. All products were used continuously in a column for long periods of time with no loss of activity.

In a later paper, Weetall and Hersh [118] insolubilized glucose oxidase by covalently binding the enzyme to NiO on an Ni screen through a silane coupling agent. The stability of the chemically bounded enzyme appears to have been increased over the soluble enzyme in the range of 10-40°C. No change in pH optimum and only a slight change in K_m were observed with the insolubilized glucose oxidase. This would indicate that no charge-charge interaction is involved.

In still another application of bound enzymes, Hersh et al. [119] ionically bound heparin to glass. The heparin coating resisted fluid shear stresses as high as 10^4 dynes/cm^2 at 30°C for 300 hr.

4. COLLAGEN MEMBRANES

Enzymes have been immobilized onto collagen membranes by a number of different techniques. Vieth et al. [142,143] used the macromolecular complexation technique, in which the enzyme is added to a collagen dispersion that is then cast into a thin membrane. The dried membrane is then crosslinked with glutaraldehyde.

An improved, mild method for covalent coupling of enzymes to chemically activated collagen films was developed by Coulet, Julliard, and Gautheron of the Laboratory of Biology and Membrane Technology of University Claude Bernard in Lyon, France [144-148]. Films of highly polymerized collagen were chosen for surface covalent binding of enzymes because of their insolubility, mechanical resistance, protein nature, hydrophilic properties, and their abundance in chemically activatable -COOH groups. Untanned films, previously acid-methylated, were activated by acyl azide formation. After removal of reagents by repeated washing, the coupling of enzyme was performed by immersion of the activated film in the enzyme solutions. The procedure is particularly mild because the enzymes never come into contact with

chemical reagents, thus avoiding all denaturing processes. All the
enzymes tested were successfully bound: glucose oxidase, urease,
lactate, glutamate and malate dehydrogenases, hexokinase, trypsin,
aspartate aminotransferase, glutamate pyruvate transaminase, and so
on. Excellent stabilities are obtained (no loss of activity in up to
6 months) in spite of repeated use at 30°C [144], and the membranes
show good resistance to deactivation by heat and denaturing reagents
[149].

The procedures for activation of collagen membranes and covalent
immobilization of enzymes is shown in Eqs. 3-6:

Activation

Collagen

$$\text{—COOH} \quad \xrightarrow[\substack{20\text{-}22°C, \ 3\text{-}20 \ \text{days} \\ \text{Washing in distilled water } (20°C)}]{\text{CH}_3\text{OH}/0.2 \text{ N HCl}} \quad \text{—COOCH}_3 \qquad (3)$$

$$\text{—COOCH}_3 \quad \xrightarrow[\substack{20\text{-}22°C, \ 1\text{-}15 \ \text{hr} \\ \text{Washing in distilled water } (0\text{-}4°C)}]{1\% \ \text{NH}_2\text{NH}_2} \quad \text{—CONHNH}_2 \qquad (4)$$

$$\text{—CO—NHNH}_2 \quad \xrightarrow[\substack{0\text{-}4°C, \ 3\text{-}5 \ \text{min} \\ \text{Washing in buffer } (0\text{-}4°C)}]{0.5 \text{ M NaNO}_2/0.3 \text{ N HCl}} \quad \substack{\text{—CON}_3 \\ \text{acyl azide}} \qquad (5)$$

Coupling

$$\text{—CON}_3 \quad \xrightarrow[\substack{0.4°C, \ \text{pH } 8\text{-}9, \ 2\text{-}5 \ \text{hr}}]{\text{H}_2\text{N-protein}} \quad \text{—CO-NH-protein} \qquad (6)$$

$$\text{Washing with 1 M KCl, 15-30 min } (0\text{-}4°C)$$
$$\text{Storage in buffer solution, 0-4°C}$$

The properties and behavior of these collagen films have been
extensively investigated by Coulet and his group [149-153]. Various
applications have shown that the stability enhancement is partially
due to diffusional limitations of the substrate in the collagen
matrix [151]. The kinetics were somewhat modified in the coupled
enzyme as compared with the enzyme in solution [152,153].

A patent on this immobilization technique has been issued to Coulet et al. [154]. The enzyme films are now prepared under industrial conditions by the Center Technique du Cuir, Lyon, and are incorporated into electrodes for glucose that are sold commercially by Tacussel (see also Chapters 3 and 6).

5. *CYANOGEN BROMIDE*

The activation of crosslinked dextrans, including agarose and even cellulose, is a simple and attractive method of covalently coupling proteins to water-insoluble carriers. The coupling reaction works best at pH 9.0, although many workers prefer a more neutral pH. The ε-amino lysine group is the group through which coupling to the protein generally occurs.

III. PROPERTIES OF IMMOBILIZED ENZYMES

A. General Discussion

Many properties of immobilized enzymes are different from those of their solubilized counterparts. We have already mentioned effects on the kinetics brought about by diffusion control and heterogeneous catalysis. One of the most important properties is an increased long-term stability and temperature stability. In general, stability can be increased by the proper choice of coupling procedure and insoluble carrier. Another property which has been observed is a change in reactivity. Immobilization may alter kinetic constants due to a change in activation energy (Table 3), pH profile (Fig. 2), Michaelis constant (Table 4), or even specificity. When an enzyme is immobilized, generally a decrease in the K_m is observed, although in some cases no change or an increase is observed (Table 4). All enzymes exhibit optimum activity at a certain pH (generally 98.6°F or 37°C).

If the matrix is ionic, the microenvironment of the active site may be altered by the electrostatic field. The pH within the matrix will be different from that in the external solution. If substrates are also charged, the apparent Michaelis constant may be altered, although in general no large differences are observed [120]. Covalent coupling itself may bring about these effects as well as a

TABLE 3 Comparison of Activation Energies of Some Soluble and
Immobilized Enzymes[a]

	Activation energy (kcal/gmole)	
Enzyme	Soluble	Immobilized
Papain (amide linkage)	13.8	11.0
Papain (azo linkage)	...	13.8
Glucose oxidase (azo linkage)	6.6	9.0
Glucoamylase (Shiff base)	16.3	13.8
Yeast lactase (Shiff base)	10.5	11.3
Microbial lactase (Shiff base)	10.4	6.5

[a]Enzymes immobilized on inorganic supports. Comparative values were
obtained between the same temperature ranges where possible.
Source: Ref. 120.

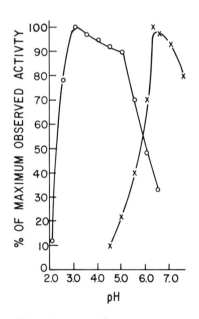

FIG. 2 pH profile of yeast lactase in solution (x-x) and covalently
attached to ZrO_2-coated control-pore glass, 550-Å pore diameter,
40/80 mesh (o-o). Substrate was 10% lactose solution. (From Ref.
164.)

TABLE 4   Comparison of K_m Values of Some Soluble and
Immobilized Enzymes[a]

Enzyme	Substrate	K_m (mM) Soluble	Immobilized
Invertase	Sucrose	0.448	0.448
Aryl sulfatase	ρ-Nitrophenylsulfate	1.85	1.57
Glucoamylase	Starch	1.22	0.30
Alkaline phosphatase	ρ-Nitrophenylphosphate	0.10	2.90
Urease	Urea	10.0	7.60
Glucose oxidase	Glucose	7.70	6.80
L-Amino acid oxidase	L-Leucine	1.00	4.00

[a]All enzymes were immobilized on ZrO_2-coated control porous 96%
silica glass particles. K_m values were determined under identical
conditions for both soluble and immobilized derivatives for comparison.
Source: Ref. 120.

change in specificity due to an alteration of the enzyme's net charge,
nearest neighbor effects on the active site region, perturbations of
intramolecular interactions, and conformational changes (Fig. 2). By
proper choice of matrix, a shift in the pH profile can be effected,
either to high or to low pH values.

Covalent binding of enzymes to electrically neutral carriers, by
way of nonactive site residues, has been shown to have no effect on
their kinetic behavior toward low molecular weight substrates [58,121,
123]. Detailed discussions of the effects of immobilization on enzyme properties can be found elsewhere [81,122,124,125].

Finally, the immobilization of an enzyme makes it much less susceptible to temperature denaturation, allowing greater operational
lifetimes at higher temperatures. Glucoamylase, for example, covalently coupled to ZrO_2-coated porous glass showed a half-life--that
point at which the activity is 50% of what it was initially--of 900
days at 40°C and 13 days at 60°C, drastically improved over the soluble enzyme [120]. Similar results have been obtained with many other

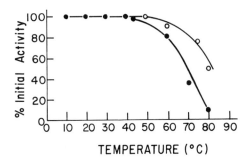

TEMPERATURE (°C)

FIG. 3 Thermal stability of an enzyme is in some cases increased by immobilization. This figure compares papain in solution (●-●) and immobilized on control-pore glass (o-o) to storage at increasing temperatures for 30-min intervals. Soluble enzyme was dissolved while immobilized derivative was suspended in water. Residual activity was measured by assay with 1% casein at 25°C. (From Ref. 165.)

enzymes. The temperature stability of papain, for example, is shown in Fig. 3.

B. Characteristics of Immobilized Enzymes

When an enzyme is immobilized--either within a matrix or on the surface of a carrier--several changes may occur in the enzyme's apparent behavior. The factors affecting this behavior are many. However, several of the observed changes occur quite commonly and will be considered here.

1. pH PROFILE

All enzymes have an optimal pH at which they show a maximum reaction rate. When the enzyme is immobilized, the optimal pH may shift, depending on the nature of the carrier. Goldstein and co-workers [111] studied this phenomenon in detail. In a nutshell, they report that if a carrier is negatively charged, then a high concentration of positively charged ions (H^+) will accumulate at the boundary layer between the carrier and the surrounding solution. This accumulation of hydrogen ions will cause the pH at the carrier surface to drop below that of the bulk solution. The enzyme, there-fore, sees a pH below that of the bulk solution. In this manner, the apparent pH of the immobilized enzyme may be increased. If the carrier is positively charged, the opposite may occur.

2. *KINETICS*

For the most part, the kinetics of immobilized enzyme reactions are studied in terms of "apparent" values. Several excellent discus-sions of the kinetics of immobilized enzymes are available [16,20,21]. When an enzyme is immobilized, one generally observes an increase in K_m. This increase is usually related to the charge on the substrate and/or carrier, diffusion effects, and, in some cases, tertiary changes in enzyme configuration. However, in some cases, no change in K_m is observed (Table 4).

3. *STABILITY*

Like all other proteins, enzymes are susceptible to thermal denaturation, whether they are immobilized or in the "free" state. In many cases, however, the rate of inactivation and denaturation of an immobilized enzyme is less than that of the free enzyme. The thermal stability of the enzyme papain is shown in Fig. 3. Inacti-vation occurred at lower temperatures with the free enzyme. Similar results have been observed with many other enzymes. Table 3 gives the activation energies for several soluble and immobilized enzymes. Enzymes which show excellent thermal stability do not necessarily show excellent operational stability because the operational stabil-ity of immobilized enzymes is not only a function of thermal stabil-ity, but of such factors as carrier durability and organic inhibitors and inhibitor concentrations, including that of heavy metals. The clogging of the carrier also affects half-life. Enzyme half-lives under controlled operating conditions should be temperature-dependent, but this does not mean that the activity decrease is denaturation-dependent. Table 5 gives an example of the dependency of half-life on operating temperature for an immobilized enzyme.

TABLE 5 Half-life and Temperature Versus Productivity for Immobilized
Glucoamylase Covalently Coupled to ZrO_2-Coated Porous Glass[*]

Temperature (°C)	Half-life (days)	Relative reaction rate (% of that at 60°C)
60	13	100
50	100	70
45	645	30
40	900	25

[*]When plotted as half-life versus 1/T (°K), a positive slope is observed. From the slope the deactivation can be calculated.

IV.   EXAMPLES OF EXPERIMENTAL PREPARATION OF
      IMMOBILIZED ENZYMES

A.   Preparation of Polyacrylamide-Entrapped Enzymes

Dissolve N,N'-methylenebisacrylamide (Eastman Chemical Co.) (1.15 g)
in phosphate buffer (0.1 M, pH 6.8, 40 ml) by heating to 60°C. Cool
the solution to 35°C and add acrylamide monomer (Eastman Chemical
Co.) (6.06 g). After mixing, filter the solution into a 50-ml volumetric flask containing riboflavin (Eastman Chemical Co.) (5.5 mg)
and potassium persulfate (5.5 mg) and dilute to the mark. The resulting gel solution is stable for months if stored in the dark. Prepare
the immobilized enzyme by mixing 0.10 g of enzyme with 1.0 ml of gel
solution. Remove oxygen by purging with N_2. Polymerize by irradiating with a 150-W photoflood lamp for 1 hr. Freeze-dry the preparation to form a powder or prepared slices for use.

B.   Preparation of Covalently Bound Enzymes

     1. VIA ACRYL AZIDE DERIVATIVE OF POLYACRYLAMIDE

     This preparation is a modification of one given by Inman and
Dintzis [69].

     a. Preparation of Polyacrylamide Beads

     Acrylamide (21 g), N,N'-methylenebisacrylamide (0.55 g), tris-
hydroxymethylaminomethane (Sigma Chemical Co.) (13.6 g), N,N,N'N'-
tetramethylethylenediamine (Eastman Chemical Co.) (0.086 ml), and

HCl (1 M, 18 ml) are dissolved in 300 ml of distilled water. Ammonium persulfate (0.22 g) is then added as a catalyst. Free-radical polymerization is effected by stirring magnetically under a nitrogen atmosphere and irradiating with a 150-W projector spotlight (Westinghouse). After polymerization is complete, the stirring bar is removed and the polyacrylamide is broken into large pieces. The polymer is then made into small spherical beads by blending at high speed with 300 ml water for 5 min. The resulting bead suspension is stored under refrigeration.

b. *Preparation of Hydrazide Derivative*

Into a siliconized, three-neck, round-bottom flask is placed 300 ml of the polyacrylamide bead suspension. The flask and a beaker containing anhydrous hydrazine (Matheson, Coleman and Bell) (30 ml) are immersed in a 50°C constant temperature oil bath. After about 45 min, the hydrazine is added to the gel, the flask is stoppered, and the mixture is stirred magnetically for 12 hr at 50°C. The gel is then centrifuged and the supernate discarded. The gel is washed with 0.1 M NaCl by stirring magnetically for several minutes, centrifuging, and discarding the supernate. The washing procedure is repeated until the supernates are essentially free of hydrazine, as indicated by a pale violet color after 5 min when tested by mixing 5 ml with a few drops of a 3% solution of sodium 2,4,6-trinitrobenzenesulfonate (Eastman Chemical Co.) and 1 ml of saturated sodium tetraborate (Mallinckrodt). The hydrazide derivative is then stored under refrigeration.

c. *Coupling of Enzyme*

The hydrazide derivative (100 mg) is placed in a plastic centrifuge tube and is washed with 0.3 M HCl. It is then suspended in HCl (0.3 M, 15 ml), cooled to 0°C, and sodium nitrite (1 M, 1 ml), also at 0°C, is added. After magnetic stirring for 2 min in an ice bath, the azide derivative formed is rapidly washed with phosphate buffer (0.1 M, pH 6.8) at 0°C by centrifugation and decantation until the pH of the supernate is close to 6.8. The azide gel is then suspended in phosphate buffer (0.1 M, pH 6.8, 10 ml) containing 100 mg of enzyme

to be bound. The mixture is stirred magnetically for 60 min at 0°C,
after which time glycine (10 ml, 0.5 M) is added to couple with unreacted azide groups. After stirring for an additional 60 min at
0°C, the enzyme gel is washed several times with phosphate buffer
(0.1 M, pH 6.8) and stored under refrigeration.

The general reaction scheme for this preparation is given by
the equations below:

$$CH_2=CHC{\overset{O}{\underset{NH_2}{\diagup}}} + (CH_2=CHCONH)_2CH_2 \xrightarrow{(NH_4)_2S_2O_8}$$

$$\vdash C{\overset{O}{\underset{NH_2}{\diagup}}} \tag{7}$$

$$\vdash C{\overset{O}{\underset{NH_2}{\diagup}}} + H_2NNH_2 \xrightarrow[\text{12 HRS}]{50°C} \vdash C{\overset{O}{\underset{NHNH_2}{\diagup}}} \tag{8}$$

$$\vdash C{\overset{O}{\underset{NHNH_2}{\diagup}}} + HNO_2 \xrightarrow[\text{2 MIN}]{0°C} \vdash C{\overset{O}{\underset{N_3}{\diagup}}} \tag{9}$$

$$\vdash C{\overset{O}{\underset{N_3}{\diagup}}} + {\overset{NH_2}{\underset{\boxed{\text{ENZYME}}}{|}}} \xrightarrow{0°C} \vdash C{\overset{O}{\underset{NH}{\diagup}}} \atop \boxed{\text{ENZYME}} \tag{10}$$

2. VIA DIAZONIUM DERIVATIVE OF POLYACRYLIC ACID

a. Polymerization of Acrylic Acid

Approximately 50 ml of reagent grade acrylic acid (Aldrich
Chemical Co.) is dissolved in 20 ml hexane and placed in a round-
bottom flask. A few milligrams of ammonium persulfate is added as
a free-radical initiator and the system is kept in a dry nitrogen
atmosphere. The flask is heated with a heating mantle until rapid
polymerization is observed. The mantle is then quickly removed and
the flask allowed to cool to room temperature.

b. Preparation of Copolymer

The polymer is broken into small particles and neutralized with
sodium hydroxide. The sodium salt is evaporated to dryness in a
rotary evaporator and ground to a fine powder. The powder (approximately 3.6 g) is suspended in 6 ml of hexane and is cooled to approximately 4°C. The acid is converted to the acyl chloride by the addition

of $SOCl_2$ (2.8 ml) with stirring in an ice bath for 1 hr, removing generated gases by suction. The acyl chloride polymer is washed with ether and dried under vacuum. Then p-nitroaniline (0.5 g) and ether (6 ml) are added and the mixture is allowed to stir overnight. The product formed is filtered, washed with ether, and air-dried.

   c. *Coupling of Enzyme*

   The p-nitroaniline derivative (150 mg) is dissolved in 10 ml of distilled water and the solution is adjusted to pH 5 with dilute acetic acid. Ethyelenediamine is added slowly with stirring until a fine white precipitate is observed. The precipitate is then washed three times with distilled water and suspended in 5 ml of distilled water. The polymer is then reduced by the addition of $TiCl_3$, and the product is washed several times with distilled water by centrifugation and decantation. The reduced derivative is converted to a diazonium salt by addition of nitrous acid (0.5 M, 10 ml) at approximately 4°C with stirring for 2 min. The diazonium salt intermediate is flushed with cold distilled water and is rapidly washed several times with phosphate buffer (0.1 M, pH 6.8) by centrifugation and decantation. It is then mixed with a phosphate buffer solution (0.1 M, pH 6.8) containing 100 mg of pure enzyme to be bound at approximately 4°C for 1 hr. The resulting gel is washed several times with buffer and stored under refrigeration.

   The general reaction scheme for this preparation is given by the equations below:

$$CH_2=CHC{\overset{\displaystyle O}{\underset{\displaystyle OH}{<}}} \xrightarrow[\text{(NH}_4)_2S_2O_8]{80-100°C} {\dashv}C{\overset{\displaystyle O}{\underset{\displaystyle OH}{<}}} \tag{11}$$

$${\dashv}C{\overset{\displaystyle O}{\underset{\displaystyle OH}{<}}} \xrightarrow{SOCl_2} {\dashv}C{\overset{\displaystyle O}{\underset{\displaystyle Cl}{<}}} \tag{12}$$

$${\dashv}C{\overset{\displaystyle O}{\underset{\displaystyle Cl}{<}}} + H_2N{-}\!\!\bigcirc\!\!{-}NO_2 \longrightarrow {\dashv}C{\overset{\displaystyle O}{\underset{\displaystyle NH}{<}}}{-}\!\!\bigcirc\!\!{-}NO_2 \tag{13}$$

$${\dashv}C{\overset{\displaystyle O}{\underset{\displaystyle NH}{<}}}{-}\!\!\bigcirc\!\!{-}NO_2 \xrightarrow{TiCl_3} {\dashv}C{\overset{\displaystyle O}{\underset{\displaystyle NH}{<}}}{-}\!\!\bigcirc\!\!{-}NH_2 \tag{14}$$

$$\text{(15)}$$

$$\text{(16)}$$

3. VIA GLUTARALDEHYDE CROSSLINKING

To a solution containing phosphate buffer (0.1 M, pH 6.8, 2.7 ml) (note: not tris buffer) and bovine serum albumin (Nutritional Biochemicals Corp.) (17.5%, 1.5 ml) is added pure enzyme to be bound (50 mg). The resulting solution is rapidly mixed with glutaraldehyde (Sigma Chemical Co.) (2.5%, 1.8 ml) for several seconds and is then immediately applied to the electrode surface. The general reaction for this preparation is given by the equation below.

$$\text{(17)}$$

4. USING ENZACRYL AA

Enzacryl AA (Aldrich Chemical Co.) (200 mg) is placed into a 50-ml plastic centrifuge tube and allowed to stir magnetically overnight with HCl (2 M, 20 ml). It is then cooled to 0°C in an ice

bath, and an ice cold sodium nitrite solution (4%, 8 ml) is added.
The mixture is stireed for 15 min, then washed four times with phosphate buffer (0.1 M, pH 7.8) at 0°C by centrifugation and decantation. To the diazonium salt formed is added a phosphate buffer solution (0.1 M, pH 6.8, 2 ml) containing high-purity enzyme (80 mg).
The mixture is allowed to stir at 0°C for up to 48 hr. The bound
enzyme is then washed several times with phosphate buffer (0.1 M,
pH 6.8) and stored under refrigeration.

V.   REVIEWS ON IMMOBILIZED ENZYMES

A number of general reviews on immobilized enzymes have been published. Bernath et al., for example, wrote a review on the preparation, properties, and uses of immobilized enzymes [126]. A review
on the methods of preparation of insolubilized enzymes was prepared
by Gryszkiewicz [127], and White and Kennedy [128] wrote an excellent
coverage of popular matrices for enzymes and other immobilizations.
The best method for immobilization of a protein is described when
several possibilities exist, and some guidelines are given to facil-
itate selection of the method appropriate to the molecular type to
be immobilized. A review on magnetic supports for immobilized enzymes and bioaffinity adsorbents was prepared by Halling and Dunhill
[129].

A very comprehensive review was written by Zaborsky [130] which
has 482 references to scientific papers, 3 government reports, and
70 patents. Methods for covalent bonding of enzymes, preparation of
water-insoluble derivatives with multifunctional reagents, of water-
insoluble enzyme-adsorbent conjugates, of lattice-entrapped enzyme
conjugates, and of enzymes immobilized with permanent semipermeable
microcapsules and with ultrafiltration cells are described.

VI.   ANALYTICAL APPLICATIONS OF IMMOBILIZED ENZYMES

In the ensuing chapters, a comprehensive discussion of the use of
immobilized enzymes for various analytical applications will be
given: enzyme and microbial electrodes probes (Chapter 3), enzyme

reactors (Chapter 4), enzyme thermistor probes and luminescence methods (Chapter 5), immunological probes (Chapter 3), and instrumental systems using immobilized enzymes (Chapter 6).

Books on analytical applications have been written by Guilbault [4,18], Carr and Bowers [141], and Mosbach [2].

Some very good general reviews on the use of immobilized enzymes in analytical chemistry have been prepared by Blaedel and Hicks [6], Bowers and Carr [7], Berezin and Klesov [8], Gray et al. [9,10], Guilbault and Sadar [11,12], Ngo [131], Free [132], Kulys [133], Werner [134], Everse et al. [135], Weetall [120], Guilbault [19,138-140,163], Thomas and Caplan [155], Enfors and Mol [156], Scheller and Pfeiffer [157], Barker and Somers [158], Gaugh and Andrade [159], Suzuki et al. [160], Rechnitz [161], and Vallon and Pegon [162].

REFERENCES

1. J. M. Nelson and E. G. Griffin, *J. Am. Chem. Soc. 38,* 1109 (1916).

2. K. Mosbach, in *Immobilized Enzymes* (S. P. Kolowick and N. O. Kaplan, eds.). Methods in Enzymology, Vol. 44. Academic Press, New York, 1978.

3. G. G. Guilbault, *Anal. Chem. 38,* 537R (1966); *40,* 459R (1968); *42,* 334R (1970).

4. G. G. Guilbault, *Enzymatic Methods of Analysis.* Pergamon Press, Oxford, 1970.

5. H. V. Bergmeyer, *Methods of Enzymatic Analysis,* 2nd ed. Verlag Chemie, Weinhein, 1965.

6. W. J. Blaedel and G. P. Hicks, *Adv. in Analytical Chemistry and Instrumentation,* Vol. 3 (C. N. Reilley, ed.). Interscience, New York, 1964, pp. 105-140.

7. L. D. Bowers and P. W. Carr, *Anal. Chem. 48,* 544-559A (1976).

8. I. V. Berezin and A. A. Klesov, *Zh. Anal. Khim. 31,* 786 (1976).

9. D. N. Gray, M. H. Keyes, and B. Watson, *Anal. Chem. 49,* 1067A (1977).

10. D. N. Gray and M. H. Keyes, *Chemtech. 7,* 642 (1977).

11. G. G. Guilbault and M. H. Sadar, *Acc. Chem. Res. 12,* 344 (1979).

12. G. G. Guilbault and M. H. Sadar, *Techniques in the Life Sciences.* Elsevier, Amsterdam, 1978, Ch. 17.

13. T. Murachi, *Hakko Kogyo 35,* 386 (1977).

14. W. E. Hornby and G. A. Noy, in Ref. 2.

15. J. Campbell and W. E. Hornby, in *Biomedical Applications of Immobilized Enzymes and Proteins* (T. M. S. Chang, ed.). Plenum Press, New York, 1977.

16. R. C. Bogulaski and R. S. Smith, in Ref. 2.

17. R. S. Schifreen, D. A. Hann, L. D. Bowers, and P. W. Carr, *Anal. Chem. 49*, 1929 (1977).

18. G. G. Guilbault, *Handbook of Enzymatic Analysis*. Marcel Dekker, New York, 1977.

19. G. G. Guilbault, Use of enzyme electrodes in biomedical investigations. In *Medical and Biological Application of Electrochemical Devices* (J. Koryta, ed.). John Wiley and Sons, New York, 1980.

20. B. Vasta and V. Usdin, *Immobilized Enzymes*. Melpar, Inc., Falls Church, Va., Final Report, Contract DA 18-108-405-CML.

21. E. K. Bauman, L. H. Goodson, G. G. Guilbault, and D. N. Kramer, *Anal. Chem. 37*, 1378 (1965).

22. G. G. Guilbault and D. N. Kramer, *Anal. Chem. 37*, 1675 (1965).

23. G. P. Hicks and S. J. Updike, *Anal. Chem. 38*, 726 (1966).

24. S. J. Updike and G. P. Hicks, *Science 158*, 270 (1967).

25. S. J. Updike and G. P. Hicks, *Nature (London) 214*, 986 (1967).

26. T. Wieland, H. Determan, and K. Buenning, *Z. Naturforsch. 21*, 1003 (1966).

27. A. D. McLaren and G. Peterson, *Soil. Sci. Soc. Am. Proc. 22*, 239 (1958).

28. L. B. Barnett and H. B. Bull, *Biochim. Biophys. Acta 36*, 244 (1959).

29. A. Nikolaev and S. R. Mardashev, *Biokhimiya 26*, 565 (1962).

30. T. M. Chang, *Science 146*, 524 (1964).

31. P. Bernfeld and J. Wan, *Science 142*, 678 (1963).

32. H. Brown, A. Patel, and S. Chattapadhyaz, *J. Chromatogr. 35*, 103 (1968).

33. P. Van Duijn, E. Pascoe, and M. Van der Ploeg, *J. Histochem. Cytochem. 15*, 631 (1967).

34. Z. Lojda, M. Van der Ploeg, and P. Van Duijn, *Histochemistry 11*, 13 (1967).

35. T. M. S. Chang, *Science 146*, 524 (1964).

36. T. M. S. Chang, Ph.D. thesis, McGill University, Montreal, 1965.

37. T. M. S. Chang, F. C. MacIntosh, and S. G. Mason, *Can. J. Physiol. Pharmacol. 44*, 115 (1966).

38. T. M. S. Chang, *Science J.*, 62 (*July 1967*).

39. T. M. S. Chang, *Sci. Tools 16*, 35 (1969).

40. T. M. S. Chang, *Biochem. Biophys. Res. Commun. 44*, 1531 (1961).

41. P. R. Rony, *Biotechnol. Bioeng. 13*, 431 (1971).

42. I. H. Silman and E. Katchalski, *Ann. Rev. Biochem. 35*, 873 (1966).

43. J. P. Hummel and B. S. Anderson, *Arch. Biochem. Biophys. 112*, 443 (1965).

44. G. Lindau and R. Rhodius, *Z. Physik. Chem. A172*, 321 (1935).

45. E. S. Vorobeva and O. M. Poltorak, *Vestn. Mosk. Univ. Khim. 21*, 17 (1966).

46. M. G. Goldfeld, E. S. Vorobeva, and O. M. Poltorak, *Zh. Fiz. Khim. 40*, 2594 (1966).

47. E. F. Gale and H. M. R. Epps, *Biochem. J. 38*, 232 (1944).

48. A. Y. Nikolayev, *Biokhimiya 27*, 843 (1962).

49. T. Tosa, T. Mori, N. Fuse, and I. Chibata, *Biotechnol. Bioeng. 9*, 603 (1967).

50. G. Hamoir, *Experientia 2*, 257 (1946).

51. L. Goldstein and E. Katchalski, *Fermentation Advances* (D. Perlman, ed.). Academic Press, New York, 1969, p. 391.

52. L. Goldstein and E. Katchalski, *Fresenius' Z. Anal. Chem. 243*, 375 (1968).

53. A. H. Sehon, *Symp. Ser. Immunobiol. Scand. 4*, 51 (1967).

54. E. F. Jansen and A. C. Olson, *Arch. Biochem. Biophys. 129*, 221 (1969).

55. R. Haynes and K. A. Walsh, *Biochem. Biophys. Res. Commun. 36*, 235 (1969).

56. A. F. S. A. Habeeb, *Arch. Biochem. Biophys. 119*, 264 (1967).

57. F. M. Richards, *Ann. Rev. Biochem. 32*, 268 (1962).

58. I. H. Silman, M. Albu-Weissenberg, and E. Katchalski, *Biopolymers 4*, 441 (1966).

59. R. Goldman, I. H. Silman, S. R. Caplan, O. Kedem, and E. Katchalski, *Science 150*. 758 (1965).

60. F. J. Wold, *Biol. Chem. 236*, 106 (1961).

61. G. Manecke and G. Gunzel, *Naturwissenschaften 54*, 647 (1967).

62. H. Zahn and H. Meinenhofer, *Makromol. Chem. 26*, 126 and 153 (1958).

63. D. J. Herzig, A. W. Rees, and R. A. Day, *Biopolymers 2*, 349 (1964).

64. H. F. Schick and S. J. Singer, *J. Biol. Chem.* *236,* 2447 (1961).

65. G. Kay and E. M. Crook, *Nature* *216,* 514 (1967).

66. B. P. Surinov and S. E. Manoylov, *Biokhimiya* *31,* 514 (1967).

67. H. H. Weetall, *Nature* *223,* 959 (1969).

68. H. H. Weetall, *Science* *166,* 615 (1969).

69. J. K. Inman and H. M. Dintzis, *Biochemistry* *8,* 4074 (1969).

70. S. A. Barker, P. J. Somers, R. Epton and J. V. McLaren, *Carbohyd. Res.* *14,* 287 (1970).

71. R. P. Patel, D. V. Lopiekes, S. R. Brown, and S. Price, *Biopolymers* *5,* 577 (1967).

72. B. F. Erlanger, M. F. Isambert, and A. M. Michelson, *Biochem. Biophys. Res. Commun.* *40,* 70 (1970).

73. A. B. Patel, S. N. Pennington, and H. D. Brown, *Biochim. Biophys. Acta* *178,* 26 (1969).

74. W. E. Hornby, H. Fillipuson, and A. McDonald, *FEBS Lett.* *9,* 8 (1970).

75. W. E. Hornby and H. Fillipuson, *Biochim. Biophys. Acta* *220,* 343 (1970).

76. D. J. Inman and W. E. Hornby, *Biochem. J.* *129,* 255 (1972).

77. K. P. Wheller, B. A. Edwards, and R. Whittam, *Biochim. Biophys. Acta* *191,* 187 (1969).

78. G. R. Craven and V. Gupta, *Proc. Natl. Acad. Sci. USA* *67,* 1329 (1970).

79. D. Gabel, P. Vretbald, R. Axen, and J. Porath, *Biochim. Biochim. Biophys. Acta* *214,* 561 (1970).

80. R. Axen and J. Porath, *Nature* *210,* 367 (1966).

81. L. Goldstein, *Meth. Enzymol.* *19,* 935 (1970).

82. M. L. Green and G. Crutchfield, *Biochem. J.* *115,* 183 (1969).

83. D. Gabel and B. Hofsten, *Eur. J. Biochem.* *15,* 410 (1970).

84. K. Mosbach and B. Mattiasson, *Acta Chem. Scand.* *24,* 2093 (1970).

85. G. Kay and M. D. Lilly, *Biochim. Biophys. Acta* *198,* 276 (1970).

86. B. L. Vallee and J. F. Riordam, *Ann. Rev. Biochem.* *38,* 733 (1969).

87. J. Sir Ram, M. Bier, and P. H. Mauerer, *Adv. Enzymol.* *24,* 105 (1962).

88. H. Fraenkel-Conrat, in *The Enzymes* (P. D. Boyer, H. Lardy, and K. Myrback, eds.). Academic Press, New York, 1959, Vol. 1, p. 589.

89. G. J. H. Melrose, *Rev. Pure Appl. Chem.* *21,* 83 (1971).

90. D. Fraser and H. G. Higgins, *Nature* *172,* 459 (1953).

91. M. Tabachnick and H. Sabotka, *J. Biol. Chem.* *235,* 1061 (1960).

92. L. Goldstein, M. Pecht, S. Blumberg, D. Atlas, and D. Levin, *Biochemistry 9,* 2322 (1970).

93. J. Salak and Z. Vodrazka, *Coll. Czeck. Chem. Commun. 32,* 3451 (1967).

94. H. G. Higgins and D. Fraser, *Aust. J. Sci. Res. (Series A) 5,* 737 (1952).

95. H. G. Higgins and K. J. Harrington, *Arch. Biochem. Biophys. 85,* 409 (1959).

96. H. Ozawa, *J. Biochem. (Japan) 62,* 419 (1967).

97. R. Axen and J. Porath, *Acta Chem. Scand. 18,* 2193 (1964).

98. R. P. Patel and S. Price, *Biopolymers 5,* 583 (1967).

99. K. L. Carraway and R. B. Triplett, *Biochim. Biophys. Acta 160,* 272 (1970).

100. K. L. Carraway and R. B. Triplett, *Biochim. Biophys. Acta 200,* 564 (1970).

101. H. D. Brown, A. B. Patel, S. Chattopadhyay, and S. N. Pennington, *Enzymology 35,* 215 (1968).

102. Y. Levin, M. Pecht, L. Goldstein, and E. Katchalski, *Biochemistry 3,* 1905 (1964).

103. E. F. Jansen and A. C. Olson, *Arch. Biochem. Biophys. 129,* 221 (1969).

104. A. Schejter and A. Bar-Eli, *Arch. Biochem. Biophys. 136,* 325 (1970).

105. F. A. Quiocho and F. M. Richards, *Biochemistry 5,* 4062 (1966).

106. K. Ogata, M. Ottesen, and I. Svendsen, *Biochem. Biophys. Acta 159,* 403 (1968).

107. F. M. Richards and J. R. Knowles, *J. Molec. Biol. 37,* 231 (1968).

108. A. F. S. A. Habeeb and R. Hiramoto, *Arch. Biochem. Biophys. 126,* 16 (1968).

109. R. Axen, J. Porath, and S. Ernbach, *Nature 214,* 1302 (1967).

110. I. Cebra, D. Givol, H. Silman, and E. Katchalski, *J. Biol. Chem. 236.* 1720 (1961).

111. L. Goldstein, Y. Levin, and E. Katchalski, *Biochemistry 3,* 1913 (1964).

112. Y. Levin, M. Pecht, L. Goldstein, and E. Katchalski, *Biochemistry 3,* 1905 (1964).

113. I. H. Silman, M. Albu-Weissenberg, and E. Katchalski, *Biopolymers 4,* 441 (1966).

114. E. Katchalski, Tech. Report AFOSR Grant 67-2025, June 30, 1967.

115. H. H. Weetall, *Nature (London) 223,* Aug. 30 (1969).

116. H. H. Weetall, *Science 166,* 615 (1969).

117. H. H. Weetall and L. S. Hersh, *Biochim. Biophys. Acta, 185,* 464 (1969).

118. H. H. Weetall and L. S. Hersh, *Biochim. Biophys. Acta 206,* 54 (1970).

119. L. S. Hersh, H. H. Weetall, and I. W. Brown, *J. Biomed. Mater. Res. 3,* 471 (1969).

120. H. Weetall, *Anal. Chem. 46,* 602A (1974).

121. A. Bar-Eli and E. Katchalski, *J. Biol. Chem. 238,* 1690 (1963).

122. E. Katchalski, in *Solid Phase Proteins: Their Preparation, Properties and Applications* (A. S. Hoffman, ed.). Battelle Seattle Research Center, 1971, p. 1.

123. I. H. Silman and E. Katchalski, *Ann. Rev. Biochem. 35,* 873 (1966).

124. G. J. H. Melrose, *Rev. Pure Appl. Chem. 21,* 83 (1971).

125. L. Goldstein and E. Katchalski, *Fresenius' Z. Anal. Chem. 243,* 375 (1968).

126. F. Bernath, K. Venkatasubramanian, and W. R. Vieth, in *Annual Reports on Fermentation Processes* (D. Perlman, ed.), Vol. 1. Academic Press, New York, 1977, pp. 235-266.

127. J. Gryszkiewicz, *Folia Biologica 19,* 119 (1971).

128. C. White and J. F. Kennedy, *Enzyme Microb. Technol. 2,* 82 (1980).

129. P. J. Halling and P. Dunhill, *Enzyme Microb. Technol. 2,* 2 (1980).

130. O. Zaborsky, *Immobilized Enzymes.* CRC Press, Cleveland, 1973.

131. T. T. Ngo, *Int. J. Biochem. 11,* 459 (1980).

132. A. H. Free, *Ann. Clin. Lab. Sci. 7,* 479 (1977).

133. J. J. Kulys, *Enzyme Microb. Technol. 3,* 344 (1981).

134. M. Werner, *Clin. Biochem. 12,* 200 (1979).

135. J. Everse, C. Ginsburg, and N. Kaplan, *Meth. Biochem. Anal. 25,* 135 (1977).

136. See Ref. 120.

137. H. Weetall, Immobilized enzymes: Some applications of foods and beverages. In *Food Prod. Dev.* (May 1973).

138. G. G. Guilbault, *Appl. Biochem. Biotechnol. 7,* 85 (1982).

139. G. G. Guilbault, in *Enzyme Engineering,* Vol. 6 (L. Wingard, ed.). Plenum Press, New York, 1982.

140. G. G. Guilbault, in *Ion-Selec. Electrode Rev. 4*, 187-231 (1982).

141. P. Carr and L. Bowers, *Immobilized Enzymes in Analytical and Clinical Chemistry: Fundamentals and Application.* John Wiley and Sons, New York, 1980.

142. K. Venkatasubramanian, R. Saini, and W. R. Vieth, *J. Food Sci. 40*, 109 (1975).

143. B. Adu-Amankwa, A. Constantinides, and W. R. Vieth, *Biotechnol. Bioeng. 23*, 2609 (1981).

144. P. R. Coulet, J. H. Julliard, and D. C. Gautheron, *Biotechnol. Bioeng. 16*, 1055 (1974).

145. P. R. Coulet and D. C. Gautheron, *L'Actualite Chimique (May 1981)*, 21.

146. P. R. Coulet and D. C. Gautheron, *J. Chromatography 215*, 65 (1981).

147. P. R. Coulet and D. C. Gautheron, in *Analysis and Control of Immobilized Enzyme Systems* (D. Thomas and J. P. Kerneveg, eds.). North-Holland, Amsterdam, 1976.

148. D. C. Gautheron and P. R. Coulet, *Enzyme Engineering,* Vol. 4 (G. R. Brown, G. Manecke, and L. Wingard, eds.). Plenum Press, New York, 1978.

149. P. R. Coulet and D. C. Gautheron, *Biochimie 62*, 543 (1980).

150. P. R. Coulet, *Ann. N.Y. Acad. Sci. 326*, 271 (1979).

151. J.-M. Engasser and P. R. Coulet, *Biochim. Biophys. Acta 485*, 29 (1977).

152. P. R. Coulet, C. Godinot, and D. C. Gautheron, *Biochim. Biophys. Acta 391*, 272 (1975).

153. J. Julliard, C. Godinot, and D. C. Gautheron, *FEBS Lett. 14*, 185 (1971).

154. P. R. Coulet, J. Julliard, and D. C. Gautheron, French Patent 73-23283 (1980).

155. D. Thomas and N. O. Caplan, Enzyme Membranes, in *Immobilized Enzymes* (S. P. Kolowick and N. O. Kaplan, eds.). Methods in Enzymology, Vol. 44. Academic Press, New York, 1978.

156. S.-O. Enfors and N. Molin, *Process Biochem. (Feb. 1978)*, 9-12.

157. F. Scheller and D. Pfeiffer, *Z. Chim. 18*, 50 (1978).

158. A. S. Barker and P. J. Somers, *Enzyme Electrodes and Enzyme Based Sensors.* Topics in Enzyme and Fermentation Technology (A. Wiseman, ed.), Vol. 2. E. Horwood, 1978, Ch. 3.

159. D. Gough and J. Andrade, *Science 180*, 380 (1973).

160. S. Suzuki, I. Satoh, and I. Karube, *Appl. Biochem. Biotechnol. 7*, 147 (1982).

161. G. Rechnitz, *Science 214*, 287 (1981).

162. J. J. Vallon and Y. Pegon, *Le Pharmacien Biologiste 14,* 531 (1980).

163. G. G. Guilbault, *Enzyme Microb. Technol. 2,* 258 (1980).

164. H. Weetall, N. Havelvala, W. Pitcher, C. Detar, W. Van, and S. Yaverbaum, *Biotech. Bioeng. 16,* 295 (1974).

165. H. Weetall, *Food Prod. Develop.,* Part I, April 1973.

3

Enzyme Electrode Probes

I. GENERAL DISCUSSION

The advent of another 50 ion-selective electrodes onto the analytical market has brought about a revolution in the area of electroanalytical chemistry. These electrodes generally work well for all types of cations and anions, but show good selectivity for inorganic species only. The ion-selective electrodes for organic species are in most cases nonselective, irregular in response, and not very useful.

In clinical analysis today, all electrolytes (Na^+, K^+, Ca^{2+}, Cl^-, H^+, HCO_3^-) are measured with ion-selective electrodes in almost all instruments. What could then be more natural than to place into these instruments electrode probes for metabolites, such as glucose, urea, uric acid, cholesterol, and creatinine? Such "enzyme electrodes" could be fashioned by taking an ion-selective electrode for O_2, CO_2 or NH_3, and coating it with a layer of stabilized, immobilized enzyme. The substrate to be measured diffuses into the enzyme layer where it is converted to a product with the uptake or release of O_2, CO_2, or NH_3. The latter is then measured with an appropriate ion-selective electrode with the potential or current produced linearly related to the product concentration.

In this chapter all analytical methods in which a soluble or immobilized enzyme is used together with an electrochemical probe as the measurement technique will be described.

-

II. USE OF ELECTRODES TO MEASURE ENZYME REACTIONS

A. General Discussion

The determination of enzyme activity is of great importance in clinical studies and in the area of biochemistry. Electrochemical methods
have been applied for these assays for many years. Probably the most
common electrochemical method for the assay of an enzyme which produces or consumes an acid is to follow the pH change of the reaction
mixture as a measure of the activity of the enzyme. This method is
not generally employed directly since the activity of a given enzyme
is affected by changes in pH. Instead a "pH stat" method is used in
which the pH of the assay mixture is maintained by the addition of an
acid or base. The rate of the addition of reagent gives the reaction
velocity.

B. O_2 Electrode

The activity of an enzyme in a system in which oxygen is consumed can
be determined using an oxygen electrode. The electrode is a gold
cathode separated by an epoxy casting from a silver anode. The inner
sensor body is housed in a plastic casing and comes in contact with
the assay solution only through the membrane. When oxygen diffuses
through the membrane it is electrically reduced at the cathode by an
applied potential of 0.8 V. This reaction causes a current to flow
between the anode and cathode which is proportional to the partial
pressure of oxygen in the sample. The rate of uptake of oxygen can
be related to the activity of the enzyme or the concentration of substrate in the assay mixture. Good correlation between glucose values
determined in blood by a measurement of oxygen uptake and those obtained by standard chemical tests was found by Kadish and Hall [1]
and by Makino and Koono [2].

     Cheng and Christian [3] described an amperometric method for
measuring enzyme reactions involving air oxidation of reduced NAD
and an O_2 electrode. Dehydrogenase reactions are monitored with a
membrane O_2 electrode, using aerobic oxidation of the reduced form
of NAD, NADH, in the presence of horseradish peroxidase and cofactors.

$$2NADH + O_2 + 2H^+ \xrightarrow{\text{peroxidase}} 2NAD + 2H_2O$$

A linear ratio between the concentration of NADH and the maximum rate of O_2 consumption is observed. The feasibility of using this method for enzyme reactions was demonstrated with the alcohol dehydrogenase system.

Szabo et al. [4] used an O_2 electrode for the assay of glucose using soluble glucose oxidase. A reproducibility of ±1.1% was obtained.

Nikolelis and Mottola [5] have developed conditions for the repetitive determination of amylase and some disaccharides by coupling enzyme-catalyzed reactions yielding glucose as a product with the glucose oxidase-catalyzed oxidation of this sugar. Determinations were performed by sample injection into a continuous circulating reagent mixture and monitoring of O_2 depletion with a three-electrode system. Maltose, sucrose, and lactose in the range 50-500, 10-250, and 25-250 µg/100 ml, respectively, and amylase in the range of 50-500 U/100 ml have been analyzed with a standard deviation of 2%. The maximum determination rate is 240 injections/hr for maltose, 700/hr for sucrose and lactose, and 120/hr for amylase.

Polarographic methods for free and esterified cholesterol in plasma and other biological preparations using the oxygen electrode were described by Dietschy et al. [6] and by Kameno et al. [7]. The dissolved oxygen consumed in the reaction media during the enzymic reactions:

$$\text{Cholesterol ester} \xrightarrow{\text{hydrolase}} \text{cholesterol} \tag{1}$$

$$\text{Cholesterol} + O_2 \xrightarrow{\text{oxidase}} H_2O_2 \tag{2}$$

is measured and related to the concentration of substrate. Noma and Nakayama [8] found a linear relationship between O_2 consumption and cholesterol concentration up to 8 g/liter; only 10 µl of serum is required and the C.V. was only 1.3%. Bilirubin and ascorbic acid were without effect in the present method, unlike the calorimetric methods.

Cheng and Christian [9] developed a rapid assay of galactose in blood serum and urine by an amperometric measurement of O_2 consumption on a Beckman glucose analyzer. From 30 to 200 mg/dl can be assayed with good reproducibility.

Borrebaeck and Mattiasson [10] evoked the principle of competitive binding assay in combination with an immobilized lectin (concanavalin A), in close proximity to an oxygen sensor, to quantify carbohydrates and determine the association constants for lectin-carbohydrate interactions. Methyl α-D-mannopyranoside was determined down to 0.5 μg/ml.

C.   Gas Membrane Electrodes

Alexander and Seegopaul [166] utilized a sulfur dioxide probe for the determination of glucose oxidase. The analytical method is based on the coupled indicator reaction scheme in which the product of the enzymatic reaction (hydrogen peroxide) oxidizes hydrogen sulfite

$$\beta\text{-D-glucose} + O_2 \longrightarrow H_2O_2 \tag{3}$$

$$H_2O_2 + 2HSO_3^- \longrightarrow S_2O_6^{2-} + 2H_2O \tag{4}$$

$$HSO_3^- + H^+ \longrightarrow SO_2 + H_2O \tag{5}$$

ions; the change in concentration of hydrogen sulfite is then measured with the SO_2 probe. As little as 0.01 U/ml of enzyme is detectable with few interference problems at sampling rates up to 90 samples/hr.

A rapid and sensitive potentiometric assay for monoamine oxidase using an ammonia electrode was described by Meyerson et al. [12]. Treatment of the substrate (tyramine or serotonin) with enzyme results in liberation of NH_3 gas, which is measured. The amount of NH_3 formed agreed with that equivalent to the depletion of the substrate (measured by an HPLC method). The specific activity to various substrates was found to be the following:

Substrate	Sp. activity
Serotonin	6.31
Tryptamine	6.07
Tyramine	5.63
Dopamine	4.03
Kynuramine	3.46
Benzylamine	2.68
Octamine	2.07
Norepinephrine	0.63

Tran-Mingh and Brown [15] and Szabo et al. [4] described methods for the assay of urea in solution, using soluble urease and a CO_2 electrode. The response is 1-2 min and the reproducibility approximately 1.5%.

Katz [16] and Katz and Rechnitz [17] have described a potentiometric method for urease. A Beckman cationic-sensitive glass electrode that responds to $[NH_4^+]$ is used to follow the course of the reaction. Guilbault et al. [18] used an ammonium ion-selective electrode for the automatic assay of urea and urease. The rate of change in the potential of the electrode $\Delta E/min$ is proportional to the concentration of urea over the range 0.1-50 $\mu g/ml$.

Guilbault and Shu [19] used a CO_2 electrode to measure the gas liberated after urease hydrolysis. A linear response of potential versus urea concentration was observed between 10^{-1} and 10^{-4} M. Only acetic acid shows any response at the CO_2 electrode, and this is very slight. Llenado and Rechnitz [20] developed an automated system with a flow-through NH_3 gas electrode as sensor for the assay of urea in blood.

Thompson and Rechnitz described a method for creatinine using creatininase and an NH_3 electrode to sense the product of the reaction [21].

L-Amino acids have been measured in solution using NH_3 or CO_2 electrodes. In one paper Guilbault et al. [18] used an NH_4^+ ion-selective electrode to measure the NH_3 liberated in the reaction:

$$\text{L-Amino acid} + O_2 \xrightarrow{\text{oxidase}} \alpha\text{-keto acid} + NH_3 + H_2O_2 \qquad (6)$$

Shu and Guilbault [19] used a CO_2 electrode to measure the gas liberated when tyrosine decarboxylase is used for the quantitative assay of L-tyrosine.

The enzyme L-phenylalanine ammonia lyase (EC 4.3.1.5) has been proposed for the assay of L-phenylalanine; the NH_3 liberated was measured with an NH_3 gas electrode by Guilbault and Hsiung [22]. The enzyme is highly selective with no interference from other amino acids.

D. Solid-Membrane Electrodes

Solid-membrane ion-selective electrodes have been used to determine the activity of rhodanase and cholinesterase. In the rhodanase system, a cyanide-selective electrode followed the decrease of cyanide ion during the reaction:

$$CN^- + S_2O_3^{2-} \xrightarrow{\text{rhodanase}} SCN^- + SO_3^{2-} \qquad (7)$$

which is catalyzed by rhodanase [23]. Results obtained by this method are comparable to those obtained by spectrophotometric procedures. The method is easily adapted to automated systems. The cholinesterase assay was performed using a sulfide-selective electrode to monitor the amount of thiocholine released under the influence of cholinesterase:

$$\text{Acetylthiocholine} + H_2O \xrightarrow{\text{cholinesterase}} \text{thiocholine} + CH_3COOH$$
$$(8)$$

The amount of thiocholine released is proportional to the activity of cholinesterase [24]. Llenado and Rechnitz [25] used systems very similar to those described above for the assay of β-glucosidase, rhodanase, and glucose oxidase. An ion-selective electrode for cyanide was used to follow the production of cyanide ion in the assay of β-glucosidase:

$$\text{Amygdalin} + H_2O \xrightarrow{\text{β-glucosidase}} \text{benzaldehyde} + \text{glucose} + HCN$$
$$(9)$$

and the consumption of cyanide in the rhodanase system as was described previously [23]. An iodide-selective electrode was used with the glucose oxidase assay:

$$\beta\text{-D-glucose} + H_2O + O_2 \xrightarrow{\text{glucose oxidase}} \text{D-gluconic acid} + H_2O_2 \qquad (10)$$

$$H_2O_2 + 2H^+ + 2I^- \xrightarrow{Mo(VI)} 2H_2O + I_2 \qquad (11)$$

to measure the decrease in iodide concentration resulting from oxidation of iodide to iodine by hydrogen peroxide.

Mascini and Palleschi [26] used an iodide-selective electrode to determine L-amino acids and alcohols with oxidase enzymes and a tubular iodide-selective electrode. From 10^{-2} to 10^{-4} M concentrations can be determined. The curves are very sharp over one decade; to change the sensitivity of the system, the concentration of iodide is changed.

A similar system was used by Papastathopoulos and Rechnitz [27] for cholesterol with a double-enzyme procedure in an automated analysis system.

$$\text{Cholesterol esters} + H_2O \xrightarrow{\text{hydrolase}} \text{free cholesterol} \qquad (12)$$

$$\text{Free cholesterol} + O_2 \xrightarrow{\text{oxidase}} H_2O_2 \qquad (13)$$

$$H_2O_2 + 2I^- + 2H^+ \xrightarrow{Mo(VI)} H_2O + I_2 \qquad (14)$$

A specially constructed flow-through membrane electrode was used to monitor the change in I^- concentration.

Diamandis and Hadjiioannou [28] described a new kinetic method for potentiometric determination of creatinine in serum based on the reaction with picrate in alkaline media (Jaffe reaction). The reaction is monitored with a picrate selective electrode, and the increase in potential of the probe is measured and related directly to the creatinine concentration. Small cation-exchange columns are used to separate creatinine from the free NH_4^+ present in solutions. The recovery was 100.7% with good agreement to the spectrophotometric method ($r = .994$).

Hara et al. [29] described a potentiometric method using a
solid silver phosphate electrode to continuously measure the activity
of alkaline phosphatase in solution. The phosphate liberated in the
enzymatic hydrolysis of phenyl phosphate is measured and related to
the activity of phosphatase. As expected, the measurement of phosphate ion is seriously interfered with by protein and halide ion, but
the interferences could be removed by covering the electrode surface
with a cellulose acetate membrane and by separating the chloride ion
from solution by gel chromatography. Satisfactory correlation with
the conventional method was obtained.

Certain antibiotics, such as amphotericin B and mystatin, are
known to interact selectively with cholesterol in bilayer lipid membranes, resulting in changes in the transmembrane electrical properties. The possibilities for use of this effect in selective trace
organic analysis are demonstrated by experiments performed with simple
electrical circuitry including a pH meter. The limit of detection is
10^{-9} M of stimulant; transmembrane resistance-based responses correlate with aqueous antibiotic concentration are rapid and reversible.

E. Oxidation-Reduction Methods

The electrochemical measurement of peroxide liberated in an enzymatic
reaction has been described by many authors. For example, Clark et
al. [31] described a 1-min electrochemical enzymatic assay for cholesterol in biological materials.

In this rapid and specific microscale electrochemical enzymatic
assay for cholesterol and cholesterol esters, 10 μl of standard or
sample is injected directly into a heated (50°C) thermostated, oxy-
stated cuvet containing pH 7.25 buffer, cholesterol oxidase (EC
1.1.3.6), and cholesterol esterase (EC 3.1.1.13). The cholesterol
esters are hydrolyzed by the esterase, and the cholesterol is simultaneously oxidized by the oxidase. The hydrogen peroxide produced
from oxidation of the unesterified cholesterol is measured by a
polarographic anode covered with an acetate/polycarbonate membrane.
The membrane allows hydrogen peroxide to diffuse to the platinum
anode, where it is oxidized, but prevents the diffusion of ascorbic

acid, uric acid, and bilirubin to the electroactive surface. Turbidity does not interfere. The correlation (r) between results by this method and the Abell-Kendall method for 105 samples of serum was .9994 and for 105 samples of plasma was .9997. The method is convenient for the analysis of high-density lipoprotein cholesterol in plasma and serum supernates and in many kinds of tissue homogenates.

Hahn and Olsson [32] described an amperometric assay for total cholesterol in human serum using cholesterol esterase, cholesterol oxidase, peroxidase, and ferricyanide. The ferricyanide produced is measured with a tubular carbon electrode by reduction:

$$Fe(CN)_6^{4-} + H_2O_2 \xrightarrow{\text{peroxide}} H_2O + Fe(CN)_6^{3-} \tag{15}$$

at a voltage of -75 mV. The method shows good reproducibility, accuracy, and sensitivity. Bilirubin and hemoglobin show no interference up to 4 mg/dl.

Razumas et al. [33] determined alkaline phosphatase by using o-hydroxyphenyl or p-aminophenyl phosphate as substrate for alkaline phosphatase, or p-aminophenol-β-D-glucoside for β-glucosidase. The products, p-aminophenol and catechol, are measured amperometrically; the limit of detection of the hydrolase is about 2.5×10^{-3} U/ml.

Wallace and Coughlin [34] described a highly sensitive electrochemical assay of lactate dehydrogenase, alcohol dehydrogenase, and malate dehydrogenase using an improved amperometric detection of NADH recently developed [35]. An anodic current sensitivity of 750 μA/mmole of NADH was obtained with a Pt electrode in an H cell. As little as 2×10^{-3} of enzyme can be determined.

Matsumaga et al. [36] proposed an electrode system for determination of microbial populations. Two Pt anode/silver peroxide cathode electrodes are used; the response time is 15 min, and the current difference between the two electrodes is proportional to the populations of microbial cells in cultures of *Saccharomyces cerevisine* and *Lactobacillus fermentum*. A microbioassay of vitamin B was described by these same researchers [37] using the same experimental

system. A linear relationship was observed between the steady-state
current and the concentration of vitamin B in the culture broth. The
assay takes 6 hr with 8% error. The mechanism of reaction was believed to involve oxidation of the FADH or NADH at the anode.

F. Conductimetric Methods

The first mention of a conductimetric method for urea was by Hanss
and Rey [38,39]. The NH_3 liberated produces a high conductivity in
citrate buffer. A few nanomoles of urea could be assayed. The results are identical to the classical colorimetric procedure; advantages are simplicity, speed, and sensitivity. This methodology has
been incorporated into the Beckman analyzers for urea and creatinine
using the soluble enzymes urease and creatininase. Excellent results
have been obtained [40].

A rapid determination of cholinesterase in serum by a conductimetric procedure was described earlier by Rey and Hanss [41]. The
increase in conductivity which appears during the enzymatic hydrolysis of acetylcholine is used to measure the serum cholinesterase
activity. The measurement is simple and takes only a few minutes;
only 20-40 μl of serum is needed, and the results are similar to
those obtained with the Ellman method. The precision and sensitivity
are unaffected by the color or turbidity of the samples.

III. ENZYME ELECTRODE PROBES

A. General Discussion

In addition to determinations of enzyme activity, electrochemical
methods have been combined with enzymatic systems to provide highly
selective and sensitive probes for the determination of the concentration of a given substrate. This is possible because under controlled conditions the rate of an enzyme catalyzed reaction is proportional to the concentration of substrate. The concept of using
an enzyme as a reagent in conjunction with an electrode was introduced by Clark and Lyon [42] and the first working enzyme electrode
was reported by Updike and Hicks [43] using glucose oxidase immobi-

lized in a gel over a polarographic oxygen electrode to measure the
concentration of glucose in biological solutions and in tissues. By
immobilizing the enzyme, the amount of the material required to perform routine analysis is greatly reduced and the need for frequent
assay of the enzyme preparation is not necessary. Furthermore, the
stability of the enzyme is often improved when it is incorporated in
a suitable gel matrix. An electrode for the determination of glucose
prepared by covering a platinum electrode with glucose oxidase chemically bound has been used for over 300 days [44].

Of the two methods used to immobilize an enzyme (the chemical
modification of the molecules by the introduction of insolubilizing
groups and the physical entrapment of the enzyme in an inert matrix
such as starch or polyacrylamide gels), the technique of chemical
immobilization is the best to make electrode probes. The method of
chemical bonding with the bifunctional reagent, glutaraldehyde, is
very simple and quite useful. The enzyme is treated with the aldehyde and an inert support (albumin, glass beads, etc.); a rigid layer
of bound enzyme results, which is quite stable for several months and
thousands of assays [45]. The immobilized enzyme is then placed over
the sensor of an electrode which is sensitive to the product of the
enzyme-substrate reaction. When the enzyme electrode is placed in a
solution which contains the substrate for which the electrode is designed, the substrate diffuses into the enzyme layer, where the enzyme-
catalyzed reaction takes place producing an ion which is detected by
the electrode. Excellent chemical analysis can be performed with
enzymes, which are biological catalysts; the real advantages of immobilized enzymes are many in analyses using electrochemical probes
or other methods of analysis. One advantage of the immobilized enzyme is a pH shift (i.e., the pH optimum can be shifted to that region at which one wants to make a measurement, by choosing the right
support for immobilization). Take an enzyme with a narrow pH range
of, say, 6-8; this can be shifted on immobilization down to the acidic
side or, conversely, up to the basic side. The enzymes are furthermore much more stable. In some work at Edgewood Arsenal, Maryland,
we actually heated our enzyme probes to 150°F and brought them back

down to room temperature with very little loss of activity. No
soluble enzyme could be treated in this fashion [45].

One advantage often overlooked is that better selectivity can
be realized with the enzyme when immobilized; this insolubilized
reagent becomes much more selective for an inhibitor, and only the
most powerful inhibitor can actually attack the enzyme. We demonstrated this several years ago in an immobilized cholinesterase alarm
for the assay of organophosphorus compounds in air and water. No
other common interferants disturbed the alarm; it responded only to
organophosphorus compounds. This factor is quite important for an
electrode probe.

In 1961 at Edgewood Arsenal, I first experimented with some
soluble enzymes, such as glucose oxidase, and developed an electrochemical assay for glucose. This led to the use of immobilized
enzymes with a commercially available ion-selective electrode sensor
to form one self-contained sensor that could be used to measure
either organic or inorganic compounds which are primary or secondary
substrates for the immobilized enzyme. The base sensor can be glass,
that is, the cation response can be measured (the ammonium ion, for
example), or the pH change in a penicillin electrode can be measured,
as done by Mosbach [46] and Papariello [47]. Or a gas membrane can
be used as a base sensor, such as the ammonia or the CO_2 membrane.
Next are the polarographic sensors, which measure peroxide or oxygen,
or any of the solid membrane electrodes (e.g., the cyanide or iodide
electrodes). For example, the enzyme can be placed on top of a flat
glass electrode sensor. A membrane is then put over the outside of
this sensor to hold the enzyme in and keep things like catalase and
bacteria out. This protects the enzyme from bacterial spoilage,
which is one of the primary causes of loss of enzyme activity.

With potentiometric devices, the response can be measured either
by a steady-state (i.e., equilibrium) method measuring millivolts or
microamperes, or by a rate method which senses the change in millivolts or microamperes per minute. Measurements of substrate can be
performed by either a steady-state or a rate method. But measurements of enzyme activity must be done by a rate method. This point

is often hazy in the literature; one can find many claims of measuring enzymes by steady-state methods. This is impossible by basic definition of enzyme activity. Enzymes are catalysts and have to be measured by a rate method, but this may be either an interrupted or a continuous measurement of rate. In Table 1 is presented a rather complete compilation of enzyme electrodes. It is an expansion of a table that was published in a recent book of mine, *Handbook of Enzymatic Analysis* [45]. In this table are listed the enzymes that act on these various materials and some of the base sensors that might be useful. Take as a typical example glucose, which can be assayed with glucose oxidase:

$$O_2 + glucose \xrightarrow{\text{oxidase}} H_2O_2 + gluconic\ acid \qquad (16)$$

One can measure the uptake of oxygen with a gas membrane electrode, a technique pioneered by Clark [42] and perfected by Hicks and Updike [43], or record the peroxide or oxygen polarographically. There are other ways: one can measure the gluconic acid by a pH change, as Mosbach [46] showed very nicely at low buffer capacity, or use an iodide membrane (the latter is much less recommended). The point I would like to make is that there are many ways to measure a particular substrate, and one should choose the one best for the application. For example, one would not choose to measure urea in biological fluid with an ammonium cation electrode because of the interference of potassium and sodium. One would choose, preferentially, an ammonia gas membrane electrode in which there is no interference from sodium and potassium.

Among the basic characteristics of enzyme electrodes is the five-step process of their operation. First, the substrate must be transported to the surface of the electrode. Second, the substrate must diffuse through the membrane to the active site. Third, reaction occurs at the active site. Fourth, product formed in the enzymatic reaction is transported through the membrane to the surface of the electrode, where, fifth, it is measured. The first step--transport of the substrate--is most critically dependent on the stirring rate

of the solution, so that rapid stirring will bring the substrate very
rapidly to the electrode surface. If the membrane is kept very thin,
using highly active enzyme, then steps 2 and 4 are eliminated or minimized; since step 3 is very fast, the theoretical response of an enzyme electrode should approach the response time of the base sensor.

B. Construction of Enzyme Electrodes

There are four steps to follow in the construction of an enzyme electrode. Let us consider each of these factors in detail.

Step 1. Select an enzyme that reacts with the substance to be
determined. From information in standard reference books on enzymology, such as Biochemists' Handbook, find an enzyme system which is
suitable for your determination. In the ideal case, this will involve
the use of the primary function of the enzyme (i.e., the main substrate-
enzyme reaction). For example, for a glucose electrode, glucose oxidase would be used; for a urea electrode, urease; for a L-glutamic
acid electrode, L-glutamate dehydrogenase. In other cases, this may
necessitate using an enzyme that acts on the compound of interest as
a secondary substrate (i.e., alcohol oxidase or malic dehydrogenase
for acetic acid [48]). Of course, this latter case will introduce
more interferences and less selectivity into the assay.

Note that in some cases there are several enzymes which act on
the substrate of interest via different reactions. For example, L-
tyrosine could be determined using L-tyrosine decarboxylase and measuring the CO_2 liberated [49], or using L-amino acid oxidase using a
Pt electrode [50] or an NH_4^+ electrode [51,52]. The latter enzyme,
although less selective, can be obtained commercially in high purity;
the former is available in low purity and has to be purified before
use. Hence, the scientific capabilities of one's laboratory might
dictate the choice of enzyme.

Step 2. Obtain the enzyme. Having found the enzyme to be used
for your application, check the catalogs of commercial suppliers (see
complete list, Appendix 3) to see if the enzyme can be purchased, and
its purity. The latter may or may not pose a problem. Many enzymes

TABLE 1 Enzyme Electrode Probes for Substrates

Substrate	Enzyme	Sensor	Ref.
Acetic, formic acids	Alcohol oxidase	Pt (O_2)	93
Acetylcholine	Acetylcholinesterase	Choline	83,84
		pH	227,228
Acetyl-β-methylcholine	Acetylcholinesterase	Acetylcholine	85-87
Adenosine	Adenosine deaminase	NH_3	224
AMP	5-Adenylate deaminase	NH_4^+	88,89,226
ATP	Glucose oxidase/hexokinase	O_2	90,223
Alcohols[a]	Alcohol dehydrogenase	Pt	91,335
		C	188,209
		O_2	186
	Alcohol dehydrogenase/diaphorase	Pt	92
	Alcohol oxidase	Pt (O_2)	93
		Pt (H_2O_2)	94
Aldehydes[b]	Alcohol oxidase	Pt (O_2)	93
Amines (mono)[c]	Monoamine oxidase	Pt (O_2)	244
Amines (di)[d]	Diamine oxidase	Pt (O_2)	277
D-Amino acids[e]	D-Amino acid oxidase	NH_4^+	52,95
L-Amino acids[f]	L-Amino acid oxidase	Gas (NH_3)	96
		Gas (O_2)	56,199
		Pt (O_2)	55,56
		Pt (H_2O_2)	50,134,206

Substrate	Enzyme	Detected	References
	Decarboxylases	NH_4^+	51,52,97-99,101,199
a. L-Arginine	Arginase	I^-	100,101
	Arginine decarboxylase/diamine oxidase	CO_2	49,102
b. L-Asparagine	Asparaginase	NH_4^+	66
		O_2	184
c. L-Glutamic acid	Glutamate dehydrogenase	NH_4^+	52,103
		NH_4^+	105
		Pt (NADH)	192,214
	Glutamate decarboxylase	CO_2	104,106
d. L-Histidine	Histidinase	NH_4^+	107
		NH_3	185
		CO_2	332
e. L-Lysine	Lysine decarboxylase	CO_2	108,109,182,183
	Lysine decarboxylase/diamine oxidase	O_2	184
f. L-Methionine	Methionine ammonia lyase	NH_3	110
g. L-Phenylalanine	Phenylalanine ammonia lyase	NH_3	98
		CO_2	182,183
		NH_4^+	101
		I^-	101
h. L-Tyrosine	Tyrosine decarboxylase	CO_2	49,102,111,182,183
	Tyrosinase	Gas (O_2)	112
Amygdalin	β-Glucosidase	CN^-	60,67,68
Ascorbic acid (vitamin C)	Ascorbate oxidase	O_2	113

TABLE 1 (continued)

Substrate	Enzyme	Sensor	Ref.
Butyrylthiocholine	Cholinesterase	Pt (SCh)	114
Catechol	Catechol-1,2-oxygenase	O_2	115,116,187
Cholesterol (free)	Cholesterol oxidase	Pt (O_2)	122
		Pt (H_2O_2)	194,195
Cholesterol (total)	Cholesterol esterase/cholesterol oxidase	Pt (O_2)	118-121
		Pt (H_2O_2)	117,219,336
Copper(II)	Tyrosinase (apoenzyme) (activation)	O_2	234
Creatine	Creatine hydrolase	Pt (H_2O_2)	341
Creatinine	Creatininase	NH_4^+	123
	Creatininase (purified)	NH_3	124
	Creatininase/sarcosine oxidase	Pt (H_2O_2)	341
Ethanol	Alcohol dehydrogenase	Pt (NADH)	125,214
Fluoride	Urease (inhibition)	Gas (CO_2)	174,175
Galactose	Galactose oxidase	Gas (O_2)	126
		Pt (H_2O_2)	194,205
D-Gluconate	Gluconate kinase/dehydrogenase	Gas (CO_2)	181
Glucose	Glucose oxidase	pH	46
		Pt (H_2O_2)	54,69,75,127,128,134-141,178,193-196,218
		Pt (quinone)	129,130
		Pt (DCIP)	131,132,243

Substance	Enzyme	Electrode	References
		Pt (O_2)	56,69,93,133,177,180
		Pt (zero current)	202,203
		Pt C (disks)	217,278
		Gas (O_2)	42,43,197,199,204,276
		I^-	76
	Glucose oxidase/isomerase	Gas (O_2)	200
	Glucose oxidase/peroxidase	Pt (O_2)	142
		Pt [Fe(CN)$_6$]$^{4-}$	198,201,275
Glutamine	Glutaminase	NH_4^+	104
		NH_3	229,230
Guanine	Guanase	NH_3	143
	Guanase/xanthine oxidase	I^-	144
Lactic acid	Lactate dehydrogenase	Pt [Fe(CN)$_6$]$^{4-}$	129,145,146,189,211
		Pt (NADH)	125,192,214
		C (NADH)	147,209
		Gas (O_2)	334
	Lactate oxidase	Gas (O_2)	340
	Cyt-b_2 LDH	Pt	212,213,243
Lactose	β-Galactosidase/glucose oxidase	Gas (O_2)	153,154
		Pt (H_2O_2)	194
Lectin	Glucose oxidase	Gas (O_2)	155
Malate	Malate dehydrogenase	Pt (NADH)	125

Table 1 (continued)

Substrate	Enzyme	Sensor	Ref.
Maltose	Maltase/glucose oxidase	Gas (O_2)	153,154
		Pt (H_2O_2)	194
Mercury(II)	Urease (inhibition)	NH_3	233
NAD	Glucose dehydrogenase/glucose oxidase	O_2	223
NADH	Alcohol dehydrogenase	Pt	225
Nitrate	Nitrate reductase/nitrite reductase	NH_4^+	156
	Nitrate reductase (NAD-dependent)	Pt (NADH)	79
Nitrite	Nitrite reductase	Gas (NH_3)	78
	Glucose oxidase (inhibitor)	O_2	231
Oxalic acid	Oxalate decarboxylase	Gas (CO_2)	157,215
Penicillin	Penicillinase	pH	46,47,62,158,159, 220,221
Peroxide	Catalase	Pt (O_2)	160
Phenol	Phenol hydroxylase	Gas (O_2)	161-163,187
	Tyrosinase	Pt $[Fe(CN)_6]^{4-}$	190
	Polyphenol oxidase	O_2	210
Phosphate	Phosphatase/glucose oxidase	Pt (O_2)	77
	Phosphorylase/phosphoglucomutase, etc.	Pt (NADH)	173

Pyruvate	Lactate dehydrogenase	Pt	192
Semicarbazide	Glucose oxidase (inhibition)	O_2	231
Succinic acid	Succinate dehydrogenase	Pt (O_2)	165
Sucrose (saccharose)	Invertase/mutarotase/glucose oxidase	Pt (H_2O_2)	167,194
		Pt (O_2)	153,164
Sulfate	Aryl sulfatase	Pt	191
Thiosulfate	Rhodanase	CN^-	168
Tyramine	Monoamine oxidase	O_2	244
Urea	Urease	NH_4^+	53,58,63-65,74,80, 199,216
		pH	46,82
		Gas (NH_3)	66,73,169,170
		Gas (CO_2)	49,171
Uric acid	Uricase	Pt (O_2)	55
		CO_2	172

[a]Methanol and ethanol are primary substrates; low activity to butanol and isopropanol.
[b]Formaldehyde, acetaldehyde respond.
[c]Tyramine only substrate tested.
[d]Electrode responds to putrescine, cadaverine, hexamethylene diamine, histamine, and spermidine.
[e]Electrode responds to D-phenylalanine, D-alanine, D-valine, D-methionine, D-leucine, D-norleucine, and D-isoleucine.
[f]Electrode responds to L-cysteine, L-leucine, L-tyrosine, L-tryptophan, L-phenylalanine, and L-methionine.

are stable in an impure state (i.e., jack bean urease and glucose
oxidase from the food industry, General Mills) and can be used satis-
factorily in a pseudo-"immobilized" form (e.g., as a liquid covered
with a dialysis membrane) for up to a week. In other cases, the impure enzyme has too low an activity to be useful without further
purification (many of the decarboxylases available from Sigma),
involving further work, possibly assistance from others.

    In still other cases, one may find that the enzyme that one
wants to use is not available commercially. In this case there are
two possibilities:

    1. Contact a large biochemical supply house and inquire whether
        it will isolate and purify the enzyme you want. Many will,
        for a suitable fee. New England Enzyme, for example, spe-
        cializes in such a service.
    2. Look up the enzyme in the literature or standard biochemistry-
        enzymology reference books, ascertain the isolation and puri-
        fication methods used, and perform the purification yourself.
        This we have done ourselves, in many cases with excellent
        results, and in most cases the techniques are simple enough
        to be carried out by a person with reasonable scientific
        training. The results are frequently well worth the effort.

    Step 3. Immobilize the enzyme. A simple rule-of-thumb to follow
is that the better the enzyme is immobilized, the more stable it is,
and hence, the longer it can be useful and the more assays that are
possible from one batch. Let us now consider the various possibilities and the characteristics of the product.

    (1) Commercially available immobilized enzyme. This is the ideal
case and is the first choice, if possible. There are a number of enzymes available in the immobilized form. Most of these are fine products, as good or better immobilization than most scientists can do
themselves. A listing of these can be found in Appendix 2. I have
probably used the products of Boehringer and Aldrich with good success. Of the enzymes available that are likely to be of most use to
the reader one can mention urease (Boehringer), glucose oxidase

(Boehringer or Aldrich), ribonuclease (Boehringer), and uricase (Aldrich).

(2) *Soluble "immobilized" enzyme*. The second choice available, which is the easiest for the novice if choice 1 is not possible, is to use the soluble enzyme in construction of the electrode. A thick paste of the enzyme powder is made with a little water (1-2 µl); this paste is spread over the surface of the electrode, and the layer is covered with a 20- to 25-µm-thick cellophane membrane (Will Scientific or Arthur H. Thomas, USA, about 100-µm pore size). Such soluble enzyme electrodes are stable for up to about a week if refrigerated between uses. Electrodes with the more crude enzymes, urease or glucose oxidase mentioned above, might be stable for longer periods of time.

(3) *Physically entrapped enzymes*. In ease of preparation, this is the next choice. Many enzymes have been physically bound in polyacrylamide gels, which are crosslinked polymers with the enzyme trapped inside. A typical preparation is described above, and similar preparations can be effected by anyone with a minimum of effort. The stability of the final product depends on the degree of care taken and the control of experimental conditions, as has been carefully pointed out by Guilbault and Montalvo [53], but can be as long as 3-4 weeks or about 50-100 determinations.

(4) *Chemically bound enzymes*. These are the most difficult to prepare, although synthesis can be effected by anyone who has had a year's course in organic chemistry. The products are most stable and can be used for 200-10,000 assays and stored at room temperature for over a year between assays [54]. The best actual method involved will depend on the individual enzyme. In my experience, the polyacryl acid diazo coupling [50,54] and the glutaraldehyde methods [55, 56] have yielded extremely satisfactory results. The covalent binding to polyacrylamide crosslinked polymer, used by Boehringer in its commercial preparations, is also quite satisfactory.

Reactive intermediates for direct coupling of enzymes via only one or two steps are available from Corning (Corning, N.Y.), Aldrich

(Milwaukee, WI), and Koch-Light (England). These are recommended to anyone interested in making chemically bound enzymes. The glutaraldehyde method is also quite simple to effect (glutaraldehyde is available from Sigma, St. Louis, MO).

   Step 4. Place the enzyme onto the appropriate electrode. In order to develop an electrode for the substrate of interest, one must have as the base sensor an electrode that responds to either one of the reactants A or B in Eq. (17), or to one of the products C or D:

$$A + B \xrightarrow{\text{enzyme}} C + D \tag{17}$$

The sensor can be a gas electrode (to measure all O_2-consuming reactions, NH_3- or CO_2-liberating enzymes), a glass electrode (to follow H^+ changes in reactions that liberate acid, or NH_4^+-producing enzymes), a Pt electrode (to follow all enzyme reactions involving electroactive species or O_2), or some other ion-selective electrode (e.g., the CN^- electrode for amygdalin, a NH_4^+ antibiotic electrode for deaminase enzymes, an I^- electrode for oxidative enzymes coupled with the $I^- \rightarrow I_2$ indicator reaction, the S^2-electrode for cholinesterase substrates). In most cases, the limiting factor in design of an enzyme electrode will be the availability of a sensor to monitor the reaction. Of course, there are other possibilities for monitoring enzyme reactions; for example, a thermistor can be covered with enzyme (see Chapter 4) and the temperature change resulting from the enzyme reaction monitored [57]. Considerable research is being performed in this area.

   Assuming that a sensor is available and the enzyme has been obtained and immobilized, let us now describe the preparation of typical enzyme electrodes (Fig. 1):

   (1) *Type A--Dialysis membrane electrode.* Turn the ion-selective electrode upside down and place 10-15 U of the soluble enzyme, physically entrapped enzyme, or chemically bound enzyme (after immobilization of the enzyme the preparation should be freeze-dried to form a powder) onto the electrode. Take a piece of cellophane dialysis membrane (20-25 μm thick, obtained from either Will Scientific or Arthur H. Thomas, USA), which has been cut into a circular piece with the

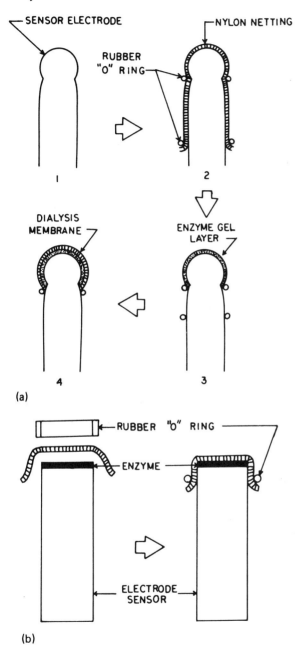

FIG. 1   Preparation of enzyme electrodes.   (a) Physically entrapped enzyme electrodes.   (b) "Soluble" or chemically bound enzyme electrodes.

diameter about twice the size of the electrode sensor, and wrap the
cellophane around the electrode, taking care that the powder is evenly
spread over the surface of the electrode in a thin layer [this may be
as conveniently done by placing a thick paste of the enzyme in water
onto the tip of a flat electrode while it is held upside down (Fig.
1b) and coating the enzyme onto the surface with a spatula]. Place
a rubber O ring, with a diameter that fits the electrode body snugly,
around the cellophane (Fig. 1b) and gently push it onto the electrode
body so that the cellophane-enzyme stays on the bottom of the electrode and is held tight and flat (Fig. 1b). Place the electrode in
a buffer solution for a few hours or overnight to allow penetration
of buffer into the enzyme layer and permit loss of entrapped air.
Store the electrode in buffer between use [50,54,58,60].

(2) *Type B electrode--Physical entrapment onto surface.* Hold
the electrode sensor upside down (Fig. 1a) and cover it with a thin
nylon net (about 90 μm thick; a sheer nylon stocking obtained from
any ladies shop is satisfactory) which is secured with a rubber O
ring in the same manner as above. This serves as a support for the
enzyme gel solution. Prepare the enzyme gel solution by mixing 0.1 g
of enzyme (purity about 10-50 U/mg) with 1.0 ml of gel solution:
1.15 g of N,N'-methylenebisacrylamide (Eastman Organics, Rochester,
NY), 6.06 g of acrylamide monomer (Eastman), 5.5 mg of potassium persulfate, and 5.5 mg of riboflavin in 50 ml of water. Gently pour the
enzyme gel solution onto the nylon net in a thin film, making sure
all the pores of the net are saturated; 1 ml of this solution should
be enough for several electrodes. Place the electrode in a water-
jacketed cell at 0-5°C and remove oxygen, which inhibits the polymerization, by purging with N_2 before and during polymerization. Complete
the polymerization by irradiating with a 150-W Westinghouse projector
spotlight for 1 hr. At the end of this time the enzyme layer should
be dry and hard. Place a piece of dialysis membrane over the outside
of the nylon net for further protection and secure with a second
rubber O ring. Soak the electrode in buffer solution overnight and
store in buffer between use [53,54,61-65].

(3) *Type C electrode--Direct polymerization onto the membrane*.
This can be effected by a direct attachment of the enzyme to the
surface of the electrode, if glass, by the Corning technique, or by
direct chemical attachment on the electrode surface, as was done by
Anfalt et al. [66] in the case of the Orion NH_3 electrode. In the
latter study, membranes were prepared by dropping 0.1 ml of soluble
urease solution (0.5 U) onto the surface of the gas diffusion membrane of the electrode. The membrane was set aside for 12 hr at 4°C
to allow evaporation of the solvent, and glutaraldehyde solution was
then added dropwise (2.5% in phosphate buffer, pH 6.2). The membrane
was set aside for a further 1.5 hr at 5°C and then rinsed carefully
with water in order to remove free enzyme and buffer. Note: Some
activity was lost over a 20-day period indicating that insufficient
enzyme was used [66]. At least 5-10 U of urease would have been
better in this case, again in 0.1 ml of solution.

Of the three types of electrodes described above, type A is good
for ease of preparation (once the bound enzyme is obtained) and for
long-term stability (if the enzyme is chemically bound). The type B
electrodes are not difficult to prepare but are time consuming, re-
quiring 1 hr of polymerization time per electrode, and have a maximum
stability of only 3-4 weeks or 50-100 assays. The type C electrode
with direct attachment to the electrode is recommended also because
it yields a very stable electrode sensor. In the type A and B electrodes, one can easily replace the enzyme layer, when it is no longer
useful, with a new layer; the sensor can be reused until its lifetime
is exhausted.

C. Performance Characteristics of Electrodes

Now that the electrode has been made, let us consider some of the
factors that affect its response and stability.

1. STABILITY

Some of the factors that affect the stability of enzyme electrodes are listed in Table 2. The stability of an enzyme electrode
is difficult to define because an enzyme can lose some activity,

TABLE 2   Factors that Affect the Stability of Enzyme Electrodes

---

1. Type of entrapment

    a.   Soluble + dialysis membrane: 1 week or 25-50 assays

    b.   Physical: 3-4 weeks, 50-100 assays

    c.   Chemical: 4-14 months, 200-1000 assays

2. Content of enzyme in gel and purity

3. Optimum conditions of enzyme

4. Stability of base sensor

---

resulting in a shift of the calibration curve downward. Yet if the
slope remains constant,as is frequently the case, the electrode is
still useful, needing calibration only daily. This is seldom a
problem since all who use electrodes of any type, such as glass,
reset the pH or potential of the electrode at least once a day.
This should be done with the enzyme electrode also using serum (i.e.,
Monitrol, Dade, Miami). Another problem in the definition of stability is that many workers measure the potential of their electrode
occasionally over a long period of time and report these data as the
stability. This may mean that the electrode was used one time a day
or week, 10 times a day, or 100 times a day. Naturally, the more
the use, the shorter will be the overall lifetime. The first factor
that affects the stability is the type of entrapment used. As a
general rule, a "soluble" electrode is useful for about 1 week or
25-50 assays, provided the electrode is kept refrigerated between
uses. The physically entrapped polyacrylamide electrodes are good
for about 3 weeks or 50-100 assays, depending crucially on the degree
of care exercised in the preparation of the polymer. The chemically
attached enzyme can be kept definitely if not used very much even at
room temperature (see Table 2 and Fig. 2), as long as 14 months for
glucose oxidase, greater than 4-6 months for L-amino acid oxidase or
uricase. One can expect to get about 200-1000 assays per electrode,
again depending on how a synthesis is effected. Although the electrode can be stored at room temperature, it is recommended that all

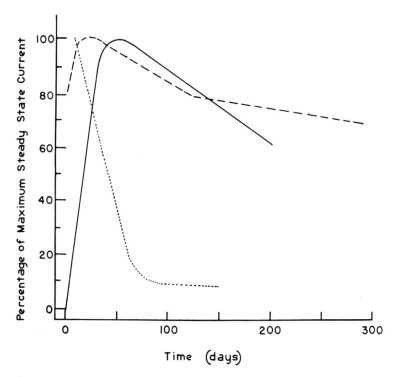

FIG. 2 Long-term stability of glucose electrodes by the steady-state
method. (---) Type 1 electrode, (——) Type 2 electrodes, (···) Type 3
electrode. Type 1: chemically bound, glutaraldehyde; type 2: chemically bound, polyamide; type 3: physically bound. (From Ref. 44.)

electrodes be kept in a refrigerator and covered with a dialysis membrane to prevent the action of bacteria, which tend to feed on the
enzyme, destroying its activity. The dialysis membrane (mol wt exclu-
sion about 1500, pore size about 35 μm) prevents the enzyme from getting out and bacteria from getting in.

The stability of the physically entrapped enzyme varies greatly
with experimental conditions and a thorough study of these factors
has been made by Guilbault and Montalvo [53].

To determine the effect of physical immobilization parameters
on the stability of the urea electrode, a series of enzyme electrodes
was prepared while varying one immobilization parameter and maintaining

all of the parameters constant. To determine the stability of the
immobilized urease coating on the surface of the cation electrode,
the steady-state potential was obtained for a given urea substrate
concentration at periodic time intervals. If the steady-state potential is constant within a certain period of time, no loss of activity
of the immobilized enzyme has occurred. All stability data reported
were obtained with the electrode stored at 25°C in tris buffer between
measurements.

The maximum stability that could be achieved with the physically
entrapped enzyme electrode was obtained with the following immobilization parameters: photopolymerizing for 1 hr at 28°C with a No. 1
150-W photoflood lamp and a gel layer thickness of 350 μm; and an
enzyme concentration in the gel of 175 mg/cm^3 gel (10 U). The slope
of the stability curve, $\Delta mV/\Delta t$, shows that the measured stability
depends on the substrate concentration used in the stability measurements. When the urea concentration used in the stability measurements
was high enough that the steady-state response was independent of the
substrate concentration, $\Delta mV/\Delta t$ was 0.2 mV/day over a 14-day period.
At 8.33 x 10^{-2} M urea, the upper limit of substrate concentration
which can be measured with the enzyme electrode, the steady-state
response falls by only 0.7 mV during the 14-day operation at 25°C.
After 14 days, the loss in activity was much greater for both substrate concentrations.

To study the effect of the activity of immobilized urease on
enzyme gel stability, physically entrapped enzyme electrodes were
prepared with activity of enzyme from 375 to 3500 Sumner U/g of
enzyme. No appreciable change in stability occurred with this relatively large change in enzyme activity. In contrast, highly purified
urease is known to be very unstable in solution. A similar trend in
stability would be expected with immobilized urease.

Greater stability with the enzyme electrode was always obtained
when the gel solution was less than 2 days old. Gel solutions were
stored without added polymerization catalysts when the storage period
was greater than 2 days. The solutions were always stored in the

dark at room temperature. The stability of the urease electrode was
studied as a function of enzyme gel layer thickness in the range 30-
350 μm. The stability increased with thickness of the enzyme gel
layer, but response time increased.

Several experiments were run to determine quantitatively the
effect of photopolymerization light intensity and time on enzyme
electrode stability. When a high-intensity photoflood lamp is substituted with a 60-W domestic lamp, the loss in activity rises from
0.2 to 4.2 mV/day for 8.33 x 10^{-2} M urea. A similar loss in activity
for the electrode was obtained when only the photopolymerization time
was reduced from 1 hr to 15 min.

To study the effect of photopolymerization temperature and water
content of the gel layer during photopolymerization on the electrode
stability, a series of enzyme electrodes were prepared with temperatures ranging from 4 to 43°C. The water content of the gel layer
over the electrode surface was also varied when the photopolymerization temperature was changed; this is because the rate of evaporation
of water from the thin enzyme gel layer varies directly with temperature. When the photopolymerization temperature and water content of
the gel were varied to study electrode stability, the other immobilization parameters were adjusted to give maximum stability. The stability, measured with 8.33 x 10^{-2} M urea, showed a loss of only
0.2 mV/day at 28°C photopolymerization temperature; upon lowering
the immobilization or photopolymerization temperature to 6°C, the
loss in electrode activity is much higher, 3.7 mV/day. At 6°C the
rate of evaporation of water from the enzyme gel layer during photopolymerization is so slow that the gel layer is still damp to the
touch after immobilization is complete. This large loss in activity
is due to leaching of enzyme from the gel layer. Enzyme which had
leached out of the gel layer could easily be detected in the buffer
solution used to store the electrode. At 28°C the rate of evaporation of water from the gel layer is sufficiently rapid so that when
the polymerization is complete, the electrode is dry to the touch.
The enzyme electrode is now more stable because a less porous polymer

is formed. At higher polymerization temperature, such as 43°C, the resulting electrode is again less stable than when the polymerization temperature is 28°C. Therefore, maximum stability is obtained with this enzyme electrode when the photopolymerization temperature is 25-28°C.

To determine the effect of a film of cellophane (dialysis membrane) on enzyme electrode stability, an electrode was made by placing a thin film of cellophane over the enzyme gel layer. The cellophane was permeable to the urea substrate but not the high molecular weight enzyme. Polymerization parameters were the same as those used to obtain the maximum stability for the cellophane electrode. Enzyme electrode stability, measured with either 8.33×10^{-2} or 1×10^{-3} M urea, showed no measurable loss in activity for 21 days (electrode stored between measurements in tris buffer at 25°C). After 21 days, the electrode began to lose activity. The increased stability of this electrode over the membraneless electrode is apparently due to the cellophane, which prevents any enzyme from leaching out of the enzyme gel layer. The membrane also keeps bacteria from getting into the enzyme gel.

Another factor that affects the apparent stability of all electrodes, especially the "soluble" and physically entrapped electrodes, is the content of enzyme in the reaction layer. As will be shown later, a certain amount of enzyme is required to yield a Nernstian calibration curve. Many times it is advantageous to add more enzyme (e.g., twice as much); in this case more enzymatic activity can be lost, yet a linear Nernstian plot is still obtained.

Still another factor that affects the stability of an electrode is the choice of operating conditions. An example of this is the comparison of the results obtained by Rechnitz and Llenado [67,68] and those of Mascini and Liberti [60] for the amygdalin electrode. Amygdalin is cleaved by β-glucosidase to give CN^- ions, which are sensed by a CN^- ion-selective electrode. Since this electrode responds best to free CN^- ions, obtainable only at pH values >10, Rechnitz and Llenado used this high pH for operation of their electrode. Even though the enzyme was physically bound, it lost activity

continually and showed a lifetime of only a few days. It is known
that almost all enzymes will lose activity at pH values <3 and >9,
and undoubtedly this was one contributant to the poor stability.
Mascini and Liberti used only a soluble enzyme at a pH of 7 and
found not only better stability (1 week, which is still all that can
be expected from a soluble enzyme) but also faster response times.
Another reason for the poor stability of Rechnitz and Llenado [68]
is the "sausage" polymerization these authors tried, in which large
pieces of physically entrapped enzyme are prepared and slices cut
for each assay. From our own experience and that of others with
such a technique, the sausage obtained is like a roast beef placed
in an oven for 30 min: it is well done on the outside and raw on the
inside. The reader is advised not to attempt such a large-scale
entrapment but to prepare individual small batches of polyacrylamide
enzyme gels.

A thorough comparison of the stability of the three types of
immobilized electrodes--soluble (type 1), physically entrapped (type
2), and chemically bound (type 3)--was shown by Guilbault and Lubrano
[54] for glucose oxidase (Fig. 2). The long-term stabilities of the
types 1, 2, and 3 electrodes were studied by testing the response of
each type of electrode to 5 x 10^{-3} M glucose in phosphate buffer,
pH 6.6, at least once a week for several months. When not in use
the electrodes were stored in phosphate buffer at 25°C. The results,
shown in Fig. 2, show that the long-term stability decreased in the
order chemically bound > physically bound > solubilized. Not only
did the type 1 electrode response decrease drastically with time, it
also decreased with each determination of glucose. This is a serious
problem and as a result the type 1 electrode is of little use analytically, except with frequent calibration and use of a large excess of
enzyme. This problem is not encountered with the types 2 and 3 electrode consisting of immobilized enzyme. The activity of these two
electrodes actually increases for the first 20-40 days before beginning to decrease; this is probably due to the establishment of diffusion channels in the matrix over a period of time, with concomitant
increase in apparent activity until the channel formation ceases and

only denaturation is observed. Likewise, it could be due to changes
in the conformation of the fraction of enzyme immobilized in a nonactive conformation to the more stable and preferred conformation.
Immobilization in an unfavorable conformation can be due to pH, temperature, or stirring effects during the immobilization process.
The decrease in response is due to a decrease in activity of the
enzyme layer because of slow denaturation and possibly also slow
irreversible inhibition. The type 2 and 3 electrodes eventually
reach a stability change of -0.25 and -0.08% of maximum response
per day, respectively. The physically bound enzyme lost 50% of its
activity in 7 months, but the chemically bound one had lost only 30%
of its activity in 400 days (13 months). Of course, this stability
would have been much less if the electrodes had been subjected to
considerable use each day; actually, about 200 assays for the type
2 electrode and almost 1000 for the type 3 enzyme electrode are
possible.

Still another factor affecting the stability of some enzyme
electrodes is the leaching out of a loosely bound cofactor from the
active site, a cofactor which is needed for the enzymatic activity.
Such was found by Guilbault and Hrabankova [52] in the case of D-
amino acid oxidase in a polyacrylamide membrane. The bond between
protein and coenzyme (flavine adenine dinucleotide, FAD) is very
weak in D-amino acid oxidase, and FAD is easily removed by dialysis
against buffer without FAD. Without FAD in the solution used to
store the electrode all activity is lost in 1 day; using a 4×10^{-4} M
solution of FAD in tris buffer, pH 8.0, to store the electrode between
use resulted in a 3-week stability with little loss in activity.

Finally, the stability of the enzyme electrode will depend on
the stability of the base sensor. This, in most cases, is not the
limiting factor in the stability, the sensor having a longer stability than the immobilized enzyme. This factor should be considered,
however, in use of some of the shorter lifetime electrodes, such as
liquid membrane electrodes.

2. *RESPONSE TIME*

There are many factors that affect the speed of response of an enzyme electrode, and these are listed in Table 3. To obtain a response, the substrate must (1) diffuse through solution to the membrane surface, (2) diffuse through the membrane and react with enzyme at the active site, and (3) the products formed must then diffuse to the electrode surface where they are measured. Let us consider each of these factors in detail and see how the response time can be optimized. Mathematical models describing this effect can be derived as was done by Blaedel et al. [70] and by Mell and Maloy [71].

a. *Rate of Diffusion of the Substrate*

In simplest, practical terms the rate of substrate diffusion depends on the stirring rate of the solution, as has been shown experimentally by Mascini and Liberti [60] for the amygdalin electrode and as is described in Fig. 3. In an unstirred solution the substrate gets to the membrane surface, albeit slowly, so that long response times are observed. At high stirring rates the substrate quickly diffuses to the membrane surface where it can react. The difference can be as much as a decrease of response time from 10 min to 1-2 min, or less. With rapid stirring for the urea electrode [53,63], a response time less than 30 sec was achieved. Of importance also is the relationship of stirring rate to the equilibrium potential observed.

TABLE 3   Factors Affecting the Response Time of an Enzyme Electrode[a]

1.	Stirring rate of solution
2.	Concentration of substrate $10^{-1} > 10^{-3} > 10^{-5}$
3.	Concentration of enzyme
4.	pH optimum
5.	Temperature (most effect on rate)
6.	Dialysis membrane

[a]A fast response is defined as a low response time.

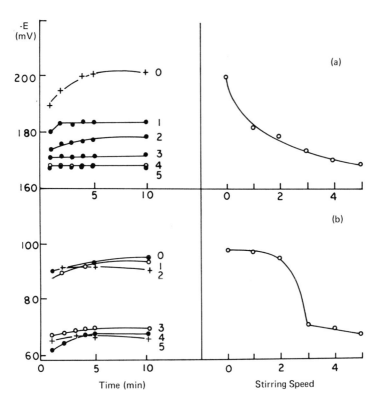

FIG. 3 Stirring effect. Response time and equilibrium value with different stirring speed. The number 0 corresponds to unstirred solutions; numbers 1-5 represent increasing stirring speed and are arbitrary numbers. [Amygdalin] = (a) 10^{-2} M, (b) 10^{-4} M; enzyme amount = 1 mg. (From Ref. 60.)

As shown in Fig. 3, the potential shifts as a function of stirring rate due to the changes in the amount of substrate brought to the electrode surface and the degree of its reactivity. Hence, for fast response time and steady reproducible values it is recommended that a fast stirring of the solution be effected with a constant stirring rate (i.e., set the speed on your stirrer and use this same setting for all readings).

b. *Reaction with Enzyme in Membrane*

The rate of reaction will depend, according to the Michaelis-Menten equation:

$$V = \frac{k_3 [E] [S]}{K_m + [S]}$$

on the activity of enzyme and factors that affect it (i.e., pH, temperature, inhibitors, and the concentration of substrate). The equilibrium potential obtained, however, should be dependent only on the substrate concentration and the temperature (since this term appears in the Nernst equation). The response rate also depends on the thickness of the membrane layer in which reaction occurs [71] and on the size of the dialysis membrane used to cover the enzyme layer, if one is used. Let us consider each of these factors separately.

3. FACTORS THAT AFFECT RESPONSE TIME

a. Effect of Substrate

Two typical examples of the effect of substrate concentration on response rate are shown in Figs. 4 and 5. Figure 4 shows the response of a β-glucosidase membrane electrode to amygdalin at

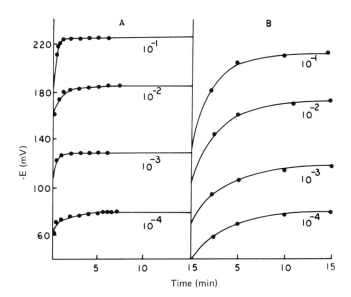

FIG. 4 Amygdalin response-time curves for an electrode containing 1 mg of β-glucosidase immobilized by a dialysis paper. (A) At pH 7. (B) At pH 10. (From Ref. 60.)

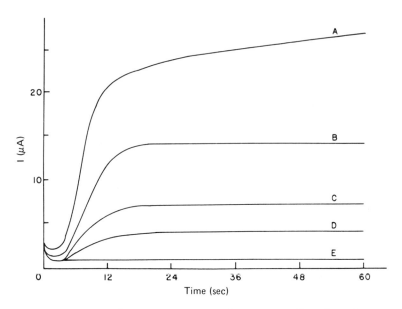

FIG. 5 Family of current versus time curves for the glucose electrode poised at 0.6 V. Glucose solutions are in phosphate buffer,
pH 6.0; ionic strength 0.1. A = 2.0 x 10^{-2} M; B = 1.0 x 10^{-2} M;
C = 5.0 x 10^{-3} M; D = 2.5 x 10^{-3} M; E = 5.0 x 10^{-4} M. (From Ref.
54.)

various concentrations [60] and Fig. 5, the response of a glucose
oxidase membrane electrode to glucose [54]. In both cases, the rate
of reaction increases (as indicated by the increased inflection of
the E versus time or i versus curve) as the substrate concentration
increases and a faster response time is observed (i.e., 1 min for
10^{-1} M amygdalin, 5 min for 10^{-4} M amygdalin). As an alternative
to waiting until an equilibrium potential or current is reached, the
rate of change in the current or potential (Δi or $\Delta E/\Delta t$) can be measured and equated with the concentration of substrate. This was done
by Guilbault and Lubrano in the case of the glucose electrode [54]
(Fig. 5).

   b. *Effect of Enzyme Concentration*
   The activity of enzyme in the gel will have two effects on an
enzyme electrode: (1) it will ensure that a Nernstian calibration

plot is obtained and (2) it will affect the speed of response of the
electrode. However, this effect is a tricky one, inasmuch as an increase in the amount of enzyme also affects the thickness of the membrane. This is demonstrated in Fig. 6, taken from the results of
Mascini and Liberti [60], for the amygdalin electrode. As the amount
of enzyme is increased from 0.1 to 2.5 mg of β-glucosidase, a shorter
response time is observed; yet when 5 mg of enzyme was used, the response time becomes quite longer. This latter effect is due to a
further thickening of the membrane layer by the use of more weight
of enzyme, resulting in an increase in the time required for the substrate to diffuse through the membrane (see Ref. 71 for a mathematical
discussion of this effect). If one weight of enzyme had been chosen
and the activity of enzyme increased at constant mass, a steady increase in the rate of response would have been observed followed by
a gradual leveling off in response time. Hence, for best results it
is recommended that as active an enzyme as possible be used, to ensure rapid kinetics, in as thin a membrane as obtainable.

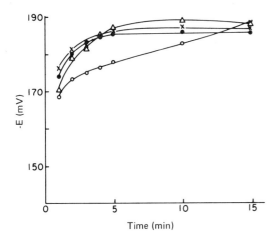

FIG. 6 Response versus time curves with different amounts of enzyme.
pH 7; [Amygdalin] = 10^{-3} M. (●) 0.1 mg enzyme. (x) 1 mg enzyme.
(Δ) 2.5 mg enzyme. (○) 5 mg enzyme. (From Ref. 60.)

c. Effect of pH

Every enzyme has a maximum pH at which it is most active and a certain range of pH in which it demonstrates any reactivity. The immobilized enzyme has a different pH range from the range of the soluble enzyme because of its environment, as has been discussed. The pH range for immobilized glucose oxidase is about 5.8-8.0 [54] (solution enzyme 5-7), for β-glucosidase about 5-8 [60]. Hence, for fastest responses one should work at the pH optimum. This is not always possible, however, because the sensor electrode may not respond optimally at the pH of the enzyme reaction. Thus, a compromise generally is necessary between these two factors. However, one should be careful not to be trapped into forcing the enzyme system to conform with the requirements of the sensor, as was done by Llenado and Rechnitz in the case of the amygdalin electrode [67,68]. These authors tried a pH of 10, which has been shown to be optimum for the electrode sensor, the CN⁻ electrode. Longer response times were obtained at pH 10 (Fig. 4) compared with pH 7 because the enzyme has very little activity at this pH. Furthermore, the enzyme rapidly loses activity at high pH values (>9-10), and Rechnitz and Llenado found their immobilized enzyme electrode to be very unstable. Yet by working at pH 7, Mascini and Liberti [60] found their soluble enzyme electrode to be still useful after a week. Similar effects are noted in other studies; Guilbault and Shu [72], for example, found the response time of a glutamine electrode decreased from pH 6 to pH 5, the latter being optimum for the enzyme reaction.

Further examples of making the electrode conform to the enzyme system, instead of vice versa, are the results of Anfalt et al. [66] and Guilbault and Tarp [73] on the NH_3 sensor used to monitor the urea-urease reaction. Although at the optimum pH for this enzyme reaction (7-8.5) there is very little free NH_3 present to be sensed by a gas-type electrode, which would predict poor results for urea assay, both groups found that the sensitivity of the sensor was more than sufficient for each measurement at these low pH values. This is partly ascribed to the fact that there is a buildup of larger

amounts of product in the reaction layer than in solution, and hence
an increase in sensitivity is obtained for the sensor, which sits
close to the enzyme layer.

d. *Effect of Temperature*

One would predict a dual effect of temperature: an increase in
the rate of reaction resulting in a faster response time as well as
a shift in the equilibrium potential by virtue of the temperature
coefficient in the Nernst and Van't Hoff equations.

This was demonstrated by Guilbault and Lubrano [54] for the
glucose electrode, in which the effect of temperature on the electrode response was studied from 10 to 50°C. Linear plots of the log
rate and log total current versus 1/T were observed as predicted by
the Van't Hoff [$\ln k = \ln C - (\Delta H/RT)$, K being the equilibrium con-
stant], and the Arrhenius [$(\ln k = \ln A - (E_a/RT)$, k being the rate
constant] equations. Practically, this means that the temperature
of the enzyme electrode should be carefully controlled for best results, although the effect of temperature is most pronounced on reaction rate measurement. Similar effects were noted by Guilbault
and Lubrano for amino acid oxidase [50].

Guilbault and Hrabankova [52] found that the response of the
D-amino acid oxidase electrode to D-methionine showed only very
small effects at increasing temperature (25-40°C), although the
theoretical Nernstian slope is 61.74 mV/decade at 37°C. Similarly,
Papariello et al. [47] found that although the response time of his
penicillin electrode was somewhat more rapid at 37°C than at 25°C,
no great improvement was observed.

Hence, the user of enzyme electrodes is advised to control the
temperature if he is making kinetic measurements, but not to bother
in making equilibrium measurements; simply use room temperature or
about 25°C for convenience.

e. *Thickness of the Membrane*

The time required to reach a steady-state potential or current
reading is strongly dependent on the gel layer thickness. This is
due to an effect on the rate of diffusion of the substrate through

the membrane to the active sites of the enzyme, and on the rate of
diffusion of the products through the membrane to the electrode
sensor where they are measured. A mathematical model relating the
thickness of the membrane d, the diffusion coefficient D, the Michaelis constant K_m, and the maximum velocity of the enzyme reaction V_{max}
(= k_3E_0) was prepared by Mell and Maloy [71]:

$$V = \frac{k_3E_0d^2}{DK_m} \tag{18}$$

where V compares the rate of chemical reaction in the membrane to the
rate of diffusion through the membrane. The larger V, the faster the
enzyme catalysis is compared with diffusion. Guilbault and Montalvo
[53] observed that the time interval for 98% of the steady-state response was about 26 sec with a 60-μm-thick net of urease and about
59 sec with a 350-μm net for 8.33×10^{-2} M urea and an enzyme concentration of 175 mg/cm^3 of gel. Similarly, Anfalt et al. [66] in the
case of a urea electrode with glutaraldehyde-bound enzyme and Mascini
and Liberti [60] using a β-glucosidase amygdalin electrode observed
an increase in response time in going from thin to thick membranes.

Thus it is recommended that as thin a membrane as possible be
used for best results; this can be achieved using a highly active
enzyme.

   f. *Effect of Dialysis Membrane*

In most cases, it is advantageous to use a dialysis membrane to
cover the electrode, as has been previously pointed out. This membrane serves to protect the enzyme and prolong the stability of the
electrode. Guilbault and Montalvo [53] noted that the cellophane
coatings had little effect on the response time of the urea electrode.

A thorough study of the effect of the thickness of the dialysis
membrane on the response rate was made by Mascini and Liberti [60]
for the amygdalin electrode. Thier results are shown in Fig. 7,
indicating that the response time and the equilibrium values are
altered by varying the thickness of the membrane. A thin 20-μm

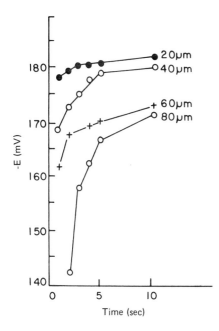

FIG. 7 Effect of thickness of dialysis paper on response time of
an equilibrium value. [Amygdalin] = 10^{-2} M enzyme amount = 1 mg.
(From Ref. 60.)

membrane (Arthur H. Thomas) or a 25-μm membrane (Will Scientific)
has essentially no effect on the response time and both are recommended for use. Additional thicknesses of membrane will cause an
increase in the time required for response, however.

   g. *Electrode Base Sensor*

   The final factor that affects the speed of response is the
electrode sensor itself and how fast it reaches a potential or current proportional to the amount of product or reactant it has sensed.

   In most cases of enzyme electrodes in which rapid stirring is
used to minimize factor a and a thin membrane of highly active enzyme
is kept under optimum conditions to minimize factor b, the determining
factor becomes the response time of the sensor.

   Guilbault and Montalvo [53] observed a response time of 26 sec
with 60-μm-thick enzyme layer of urease in a urea electrode, compared

with a response time of 23 sec for an uncoated cation electrode.
Anfalt et al. [66] observed response times of their urea probe of
30 sec to 1 min, almost the same as those for the uncoated NH_3 gas
electrode. Mascini and Liberti [60] noted a 1-min response time
for their CN^--based amygdalin electrode, at high substrate concentration (10^{-1} M), results similar to those for the uncoated CN^-
electrode.

At low substrate concentrations (10^{-2} to 10^{-5}) factor b becomes
the rate-limiting factor on the response time.

4. OTHER CHARACTERISTICS

a. Shape of Response Curve

The concentration of enzyme in the membrane will have two
effects on the electrode response. The first, the effect on the response time, was discussed above. The second is on the shape of the
response curve. This is best indicated in Fig. 8, which is taken
from the work of Guilbault and Montalvo [53] on the urea electrode.

To study the effect of enzyme concentration on the enzyme gel
layer activity, gels were prepared with enzyme concentrations ranging from 3 to 110 U of urease/cm^3 of gel. The steady-state response
of each enzyme-coated electrode when dipped in urea solutions from

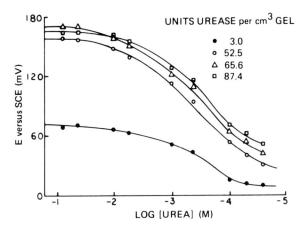

FIG. 8 Effect of enzyme concentration on electrode response; 350-μm
netting. (From Ref. 53.)

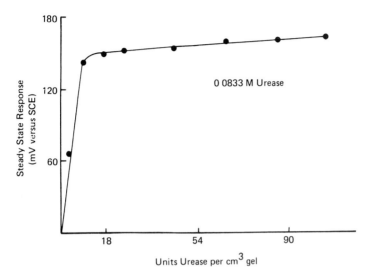

FIG. 9 Dependence of gel layer activity on enzyme concentration;
350-μm netting. (From Ref. 53.)

5×10^{-5} to 1.6×10^{-1} M was measured. The results are shown in
Fig. 8. The slope of each curve increases with the amount of enzyme
in the gel layer on the electrode until, with larger enzyme concentrations, only a small increase in activity of the gel membrane is
obtained. Figure 9 is a plot of the steady-state response for
8.33×10^{-2} M urea against the amount of urease in the gel membrane.
There is a rapid increase in response or activity up to 7.5 U urease/
cm^3 of gel. Above 7.5 U urease/cm^3 of gel a large increase in enzyme
concentration gives only a small increase in activity of the enzyme
gel membrane on the cation electrode. Optimum enzyme concentration
in the gel, considering only the economy of enzyme, is obtained at
about 7.5 U urease/cm^3 of gel.

A series of urease electrodes was prepared with the same urease
concentration (65.6 U of urease/cm^3 of gel) but with different gel
compositions to determine if the activity of the gel layer depended
on the gel composition. With a 350-μm gel layer over a cation electrode, variation gave less than a 2% difference in response with
8.33×10^{-2} M urea. Variation of the crosslinking material from

5 to 19% at constant gel concentration likewise gave a very small difference in response.

With a urease concentration of 65.6 U/cm^3 of gel, the steady-state response to 8.33×10^{-2} M urea decreased by only 2% upon decreasing the gel layer thickness from 350 to 60 μm. However, the response time decreased significantly.

Similar results are obtained in the case of every enzyme electrode. The amount of enzyme to be used depends on the individual electrode and the curve, similar to Fig. 9, that is observed. However, as a rule-of-thumb, 10-20 U per membrane is generally sufficient to give an excellent response curve. In the case of the more unstable soluble and physically entrapped enzyme electrodes, an excess of enzyme should be used (i.e., 50 U/cm^2 of gel as shown in Fig. 9) so that a loss of enzyme does not affect the potentials observed.

Likewise, purified enzymes should be used (at least 1 U/mg) so as to keep the thickness of the membrane to a minimum for fast response rates.

Table 4 lists the approximate amount of enzyme to be used in making enzyme electrodes.

b. Wash Time of the Electrode

Because there is a buildup of product in the enzyme membrane, every type of enzyme electrode described requires a washing after use in order to rereach the baseline potential. This wash time varies from only 20 sec in the case of urease and the air-gap electrode [73] to as long as 10 min for urease with a pH electrode sensor [47] (Table 4). The wash time will increase with the thickness of the enzyme membrane, as expected, and also with the nature of enzyme and the base sensor. This latter effect is due to the charge on the enzyme, which in the case of urease, for example, is negative, with a resulting attraction for the positive NH_4^+ ions formed. The glass likewise has an attraction for these ions, resulting in longer wash times required.

The electrode can be washed in an automatic electrode washer as
described by Montalvo and Guilbault [74], by simply placing the electrode in water or by rinsing it under a distilled water tap. The
type of washing will depend on the type of electrode. Soluble and
physically entrapped enzymes must be washed gently because they can
be easily washed out of their matrix. The more sturdy chemically
bound enzymes can be washed under a water tap [73]. The latter, of
course, yields a much quicker return to baseline. Oxygen-based electrodes [55,56] will show a quick return to baseline if simply placed
in a fresh buffer solution. Other electrodes, such as the Pt-based
amperometric enzyme electrodes [50,54,59,69], have to be pretreated
before first use by applying a potential (+0.6 V versus SCE) until
the anodic current decays to a low value; after each run, the electrode is washed in a stirring phosphate buffer solution until the
current decays to a low value (0.5-1 min) indicating the removal of
unreacted substrate and reaction products.

   c. *Range of Substrate Determinable*
   As indicated in Table 4, all enzyme electrodes sense substrate
in the general range of 10^{-2} to 10^{-4} M, with some electrodes useful
to as high as 10^{-1} M or as low as 10^{-5} M. In all cases, curves similar to those in Fig. 8 are obtained, approximately Nernstian in the
linear range with a slope of close to 59.1 mV/decade. All curves
level off at high substrate concentrations, as predicted by the
Michaelis-Menten equation, which states that the reaction becomes
independent of substrate at high substrate concentration. A leveling
off of the curve at low substrate concentration is also observed;
this is due to the limit of detection of the electrode sensor used.

D. Effect of Interferences

Any enzyme electrode will be only as good as its selectivity. Hence,
a consideration of possible interferences is now in order. Interferences fall into two categories: interferences in the electrode
sensor and interferences with the enzyme reaction itself.

TABLE 4   Various Electrodes and Their Characteristics

Type	Sensor	Immobilization[a]	Stability	Response time	Amount of enzyme (U)	Range[b]
1. Urea	Cation	Physical	3 weeks	30 sec-1 min	25	10^{-2}-5 x 10^{-5}
	Cation	Physical	2 weeks	1-2 min	75	10^{-2}-10^{-4}
	Cation	Chemical	>4 months	1-2 min	10	10^{-2}-10^{-4}
	pH	Physical	3 weeks	5-10 min	100	5 x 10^{-3}-5 x 10^{-5}
	Gas (NH_3)	Chemical	>4 months	2-4 min	10	5 x 10^{-2}-5 x 10^{-5}
	Gas (NH_3)	Chemical	20 days	1-4 min	0.5	10^{-2}-10^{-4}
	Gas (CO_2)	Physical	3 weeks	1-2 min	25	10^{-2}-10^{-4}
2. Glucose	pH	Soluble	1 week	5-10 min	100	10^{-1}-10^{-3}
	Pt (H_2O_2)	Physical	6 months	12 sec kinetic[c]	10	2 x 10^{-2}-10^{-4}
	Pt (H_2O_2)	Chemical	>14 months	1 min, steady state	10	2 x 10^{-2}-10^{-4}
	Pt (H_2O_2)	Soluble		1-2 min	10	10^{-2}-10^{-4}
	Pt (quinone)	Soluble	<1 week[d]	3-10 min	10	2 x 10^{-2}-10^{-3}
	Pt (O_2)	Chemical	>4 months	1 min	10	10^{-1}-10^{-5}
	I^-	Chemical	>1 month	2-8 min	10	10^{-3}-10^{-4}
	Gas (O_2)	Physical	3 weeks	2-5 min	20	10^{-2}-10^{-4}
	Gas (O_2)	Chemical	>3 weeks	2-5 min	10	2 x 10^{-2}-10^{-4}
3. L-Amino acids[f] General	Pt (H_2O_2)	Chemical	4-6 months	12 sec kinetic[c]	10	10^{-3}-10^{-5}
	Gas (O_2)	Chemical		2 min	10	10^{-2}-10^{-4}
	Pt (O_2)	Chemical	>4 months	1 min	10	10^{-2}-10^{-4}
	Cation	Physical	2 weeks	1-2 min	10	10^{-2}-10^{-4}
	NH_4^+	Chemical	>1 month	1-3 min	10	10^{-2}-10^{-4}
	I^-	Chemical	>1 month	1-3 min	10	10^{-3}-10^{-4}
L-Tyrosine	Gas (CO_2)	Physical	3 weeks	1-2 min	25	10^{-1}-10^{-4}

No.	Substrate	Sensor	Immobilization	Stability	Response	Enzyme units	Range[b]
	L-Glutamine	Cation	Soluble	2 days[d]	1 min	50	10^{-1}–10^{-4}
	L-Glutamic acid	Cation	Soluble	2 days[d]	1 min	50	10^{-1}–10^{-4}
	L-Asparagine	Cation	Physical	1 month	1 min	50	10^{-2}–5×10^{-5}
4.	D-Amino acids[g] (general)	Cation	Physical	1 month	1 min	50	10^{-2}–5×10^{-5}
5.	Lactic acid	Pt [Fe(CN)$_6$]$^{4-}$	Soluble	<1 week	3–10 min	2	2×10^{-3}–10^{-4}
6.	Succinic acid	Pt (O$_2$)	Physical	<1 week	1 min	10	10^{-2}–10^{-4}
7.	Acetic, formic acids	Pt (O$_2$)	Chemical	>4 months	30 sec	10	10^{-1}–10^{-4}
8.	Alcohols	Pt (H$_2$O$_2$)	Soluble	1 week	12 sec kinetic[c]	10	0.5–100 mg%
		Pt (H$_2$O$_2$)	Soluble	1 day[d]	1 min	~1	0.5–50 mg%
		Pt (O$_2$)	Chemical	>4 months	30 sec	10	0.5–100 mg%
9.	Penicillin	pH	Physical	1–2 weeks	0.5–2 min	400	10^{-2}–10^{-4}
			Soluble	3 weeks	2 min	~1000	10^{-2}–10^{-4}
10.	Uric acid	Pt (O$_2$)	Chemical	4 months	30 sec	~10	10^{-2}–10^{-4}
11.	Amygdalin	CN$^-$	Physical	3 days[e]	10–20 min	100	10^{-2}–10^{-5}
12.	Cholesterol	Pt (H$_2$O$_2$)	Soluble		2 min		10^{-2}–10^{-4}
13.	Phosphate	Pt (O$_2$)	Chemical	4 months	1 min	10 each	10^{-2}–10^{-4}
14.	Nitrate	NH$_4^+$	Soluble		2–3 min	10	10^{-2}–10^{-4}
15.	Nitrite	NH$_3$ (gas)	Chemical	3–4 months	2–3 min	10	5×10^{-2}–10^{-4}
16.	Sulfate	Pt	Chemical	1 month	1 min	10	10^{-1}–10^{-4}

[a]"physical" refers to polyacrylamide gel entrapment in all cases; "chemical" is attachment to glutaraldehyde with albumin, to polyacrylic acid, or to acrylamide chemically followed by physical entrapment.

[b]Analytical useful range either linear or with constant change if curvature is observed.

[c]"kinetic," rate of change in current measured after 12 sec; "steady state," current reaches a maximum in 1 min.

[d]Preparation lacks stability as evidenced by constant decrease in signal each day.

[e]Time required for signal to return to baseline before reuse.

[f]Electrode responds to L-cysteine, L-leucine, L-tyrosine, L-tryptophan, L-phenylalanine, and L-methionine.

[g]Electrode responds to D-phenylalanine, D-alanine, D-valine, D-methionine, D-leucine, D-norleucine, and D-isoleucine.

1. INTERFERENCES IN THE ELECTRODE SENSOR

If possible, the electrode used to sense the products of or reactants in the enzyme reaction should be one that will not react with other substances present in the sample being assayed.

The urea electrode probes originally described by Guilbault and Montalvo [53,63] could not be used for assaying urea in blood or urine because the cation glass sensor used to measure the NH_4^+ produced also responds to Na^+ and K^+ present in blood or urine.

In an attempt to eliminate this effect, a cell using a glass electrode (Beckman Electrode 39137 or 39047) as the reference unit was tried by Guilbault and Hrabankova [64]. Calibration curves for urea were found to be the same when the cell with the SCE reference electrode was used. Also, the interferences of monovalent cations in solution are smaller in this case because both electrodes are sensitive to these ions. However, concentrations of Na^+ and K^+ higher than 10^{-4} mole/liter considerably decreased the electrode response. This effect could be explained as a decrease of activity coefficients in the presence of other ions or as a decrease of the enzymatic reaction rate caused by a higher ionic strength.

Combining the uncoated glass reference electrode cell with a cation exchanger, the determination of urea in blood and urine is possible with a deviation of less than 3%.

A further improvement was described by Guilbault and Nagy [65], in which a solid antibiotic nonactin electrode was used as the sensor. This electrode has a selectivity of NH_4^+/K^+ of 6.5 and NH_4^+/Na^+ of 7.5×10^2, thus partially eliminating the response to these ions by the sensor. By using a three-electrode system (a chemically bound urease over a nonactin solid electrode versus an uncoated nonactin electrode and a calomel electrode as reference) and diluting to a constant interference level of K^+ (Na^+ does not interfere because of the high selectivity coefficient of the sensor), urea in blood was assayed with an accuracy of 2-3% or better. In this procedure, a standard calibration plot of E versus urea was prepared in the presence of KCl, at its highest level present in blood. Before each run, the blood sample to be assayed was brought to the same potential as that observed in the KCl-buffer solution by adding more KCl to the sample, using the uncoated electrode versus SCE. The potential of the urease sensor versus SCE was then read and the urea determined.

Anfalt et al. [66] used an Orion NH_3 electrode, which is a glass electrode with a layer of NH_4Cl and a gas-permeable membrane over the outside, to sense the NH_3 produced from the urea-urease reaction. In theory, the use of the gas NH_3 electrode, which senses only NH_3 and not K^+, Na^+, or other ions or substrates, should provide the desired sensor selectivity. However, at the optimum pH for the urea-urease reaction (7-8.5) the free NH_3 present is quite small compared with NH_4^+ ions. Yet, because the enzyme layer is placed directly onto the gas electrode, there was sufficient NH_3 produced at the electrode surface to give nice linear calibration plots at pH values of 7, 7.4, 8, and 9 with slopes close to 59.1 mV/decade (actually the slope at pH 7 was 69.5 mV/decade, higher than Nernstian).

In the Pt electrode devices described for measuring oxidative enzyme systems, two approaches have been taken. One is to measure the peroxide produced by monitoring the total current change [54,69, 75] or the rate of change in the current [54,69] at +0.6 V versus SCE. The other is to measure the uptake of oxygen by the enzymatic reaction [55,76] by measuring the reduction of O_2 at -0.6 V versus SCE. In either system, compounds that are either oxidized or reduced at a Pt electrode at ±0.6 V versus SCE would interfere. Clark [75] eliminates this problem by using a second uncoated Pt electrode, held at the same potential as the enzyme electrode in order to measure any compounds present in blood and also by covering the Pt enzyme electrode with a membrane that prevents diffusion of electroactive substrates (such as ascorbic acid). His system, as modified, forms the basis of an instrument for measuring glucose sold commercially by Yellow Springs Instrument Company (Yellow Springs, Ohio). The normal value of such interferences (ascorbic acid or uric acid are examples of oxidizable compounds) is generally <5% of the glucose signal. Another way to eliminate the effect is to measure very quickly (about 21 sec) the rate of change in the current. This was done by Guilbault and Lubrano [54,69], who found that assays of more than 200 blood samples could be performed with an accuracy and precision of better than 2%. Since all other compounds present in blood that consume O_2 can be subtracted before measurement with the enzyme sensor [55,76], still better selectivity is obtained using the measurement of O_2 at -0.6 V.

In other electrodes, such as the CN^- solid precipitate electrode used by Llenado and Rechnitz [67,68] and Mascini and Liberti [60] for amygdalin, any ions capable of forming insoluble silver salts will interfere because they form a precipitate on the electrode surface. Also, substances capable of reducing silver ion will interfere, as will heavy-metal ions and transition metal ions capable of forming cyanide complexes. In using an iodide sensor to measure glucose, Nagy et al. [76] found several types of interferences which make this type of measurement approach very limited: interferences at the iodide electrode (thiocyanate, sulfide, CN^-, and Ag^+), and interferences from oxidizable compounds present in blood, such as uric acid, tyrosine, ascorbic acid, and Fe(II), which compete in the oxidation of iodide to iodine in the peroxide-peroxidase system. These compounds had to be removed by sample pretreatment.

Finally, in the glass electrode systems for pH measurement described by Papariello et al. [47] and Mosbach and co-workers [46], any acidic or basic components present would interfere in the measurement. However, by adjusting to a definite pH before initiating the enzyme reaction and assuming only the enzyme reaction will give rise to a pH change, these effects can be minimized or eliminated.

2.   INTERFERENCES WITH THE ENZYME REACTION

Such interferences fall into two classes: substrates that can catalyze the reaction in addition to the compound to be measured and substances that either activate or inhibit the enzyme.

With some enzymes, such as urease, the only substrate that reacts at a reasonable rate is urea; hence, the urease-coated electrode is specific for urea [53,63]. Uricase, likewise, acts almost specifically

on uric acid [55]. Others, such as penicillinase [46,47], react with
a number of substrates; ampicillin, naficillin, penicillin G, penicillin V, cyclibillin, and dicloxacillin can all be determined with
a penicillinase electrode.

   Similarly, D-amino acid oxidase [52] and L-amino acid oxidase
[50,51] are less selective in their responses. The former in an
electrode gives good response to D-phenylalanine, D-alanine, D-valine,
D-methionine, D-leucine, D-norleucine, and D-isoleucine; the latter
to L-leucine, L-tyrosine, L-phenylalanine, L-tryptophan, and L-
methionine. Alcohol oxidase [59] responds to methanol, ethanol, and
allyl alcohol. Hence, in using electrodes of these enzymes, a separation must be used if two or more substrates are present, or the
total must be determined. In the case of L-amino acid assay, the
use of decarboxylative enzyme [49] which acts specifically on different amino acids is an attractive alternative. Enzyme electrodes of
this type are known for L-tyrosine, L-phenylalanine, L-tryptophan,
and others.

   Glucose oxidase acts on a number of sugars [76]: glucose and
2-deoxyglucose are the main substrates, but cellobiose and maltose
also react, probably due to the presence of other hydrolytic enzymes
in the glucose oxidase preparation.

   The activity of the enzyme can be adversely affected by the
presence of certain compounds, called inhibitors. Generally, these
are heavy-metal ions, such as Ag^+, Hg^{2+}, and Cu^{2+}, and sulfydryl-
reacting organic compounds, such as p-chloromercuribenzoate and
phenylmercury(II) acetate (due to their reaction with the free S—H
groups present at the active site of many enzymes, especially the
oxidase) [54]. One important point to realize, however, is that the
immobilized enzyme is much less susceptible to inhibitors, especially
weak or reversible inhibitors, due to the protection of the immobilization matrix. Thus by using the enzyme in a immobilized form,
most of the worries about inhibitors are eliminated. From personal
experiences in designing and using almost 20 different enzyme electrodes, not a case of enzyme inhibition interfering with an assay

has been noted. However, one should always be aware of this problem,
especially in assaying solutions containing heavy-metal ions, and
especially pesticides.

E. Assay of Inorganic Ions

To gain an overall view of what has been done and how broad the field
really is, we shall first discuss inorganic ion determination. Nonenzymatic electrodes have been designed for phosphate, sulfate, and
nitrate, but these have some undesirable characteristics. The phosphate and sulfate electrodes are totally useless except for titra-
tions of sulfate or phosphate, and other commercial electrodes, such
as the nitrate electrode, suffer serious interferences. An enzyme-
based phosphate electrode has been formulated [77] using alkaline
phosphatase and glucose oxidase immobilized in a membrane, with
measurement of the oxygen uptake by the glucose reaction:

$$\text{Glucose phosphate} \rightleftharpoons \text{phosphate + glucose} \tag{19}$$

$$\text{Glucose} + O_2 \xrightarrow{\text{glucose}\atop\text{oxidase}} \text{gluconic acid} + H_2O_2 \tag{20}$$

The amount of glucose is controlled by the phosphate, which is reacting with glucose in this reversible reaction, forming glucose
phosphate. So, by a measurement of the rate of oxygen uptake, one
can measure the phosphate concentration because the rate is proportional to the concentration of this anion. The curve is similar to
that of an inhibitor or an enzyme reaction and is extremely reproducible, but what is remarkable is the specificity. We ran about
50 anions and cations, and very few interfered. Tungstate and
arsenate are the only materials that gave an appreciable interference, and fortunately these are almost never found in blood or
estuaries. Molybdate responds somewhat; the selectivity is about
10:1. Borate and EDTA give a slight interference. The selectivity
of the electrode for phosphate over sulfate ion is about 500:1.
Other ions, chloride, acetate, bromide, etc., did not respond at
all. So indeed, the phosphate electrode is selective.

Two different enzyme electrodes were described for phosphate.
One uses phosphorylase a, phosphoglucomutase, and glucose-6-phosphate
dehydrogenase. The second incorporates glyceraldehyde phosphate dehydrogenase, phosphoglycerate kinease, and hexokinase. In both cases
the assay is conducted by monitoring the NADH at +0.65 V [173].

Another electrode we built [191] was a sulfate electrode based
on the hydrolysis of 4-nitrocatechol sulfate, catalyzed by aryl sulfatase, to produce 4-nitrocatechol and sulfate. We looked at the
polarographic wave for oxidation of the product of the enzymatic
reaction, 4-nitrocatechol, to the corresponding quinone. The rate
of oxidation to the quinone is proportional to the sulfate concentration. Such an enzymatic sulfate electrode is extremely difficult
to fabricate, and for one primary reason: the measuring technique is
critically dependent on both the type of substrate (some substrates
do not work at all) and the concentration of substrate. So all these
factors have to be considered in formulating the enzyme electrode.
However, the system does work very nicely. Figure 10 shows some
curves demonstrating the specificity of the electrode. The response
to sulfate is quite good. Note the line for phosphate, which gives
a slight response at 10^{-4} M to a very moderate response at 10^{-2} M.
Normally, the response to sulfate would be at least 10 times as great.
The response to cyanide depends on the concentration of nitrocatechol
sulfate. By using a low concentration of nitrocatechol sulfate, we
could eliminate much of the interference of such ions as cyanide.

Both nitrite and nitrate ions are important in food technology.
Recent studies have shown that besides being the cause of methemoglo-
binemia, nitrate and nitrite can react with secondary and tertiary
amines to form a carcinogenic reagent, N-nitrosamine. Thus, the
analysis of nitrate and nitrite in the environment has become in-
creasingly important. A specific and sensitive method has been
developed by using an enzymatic approach and newly developed electrochemical and fluorometric approaches. MVH- (methyl viologen, reduced form) nitrate reductase (EC 1.6.6.1), induced from *Escherichia
coli*, and MVH-nitrite reductase (EC 1.6.6.4), isolated from spinach

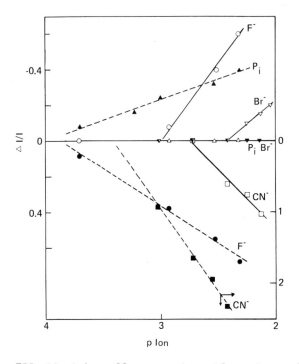

FIG. 10 Anion effects on the sulfate electrode. $\Delta I/I$ is plotted against the p ion. Electrode type IV; 10% (wt/wt enzyme in the active layer; pH 4.10; E = +0.8 V versus SCE. Open marks with the solid lines are at a 2×10^{-4} M NCS concentration; closed (solid) marks with dashed lines are at 3.8×10^{-5} M NCS concentration. ($\triangle$) Phosphate. ($\square$) Cyanide. ($\circ$) Fluoride. ($\triangledown$) Bromide. [G. G. Guilbault and T. Cserfalvi, *Anal. Chim. Acta 84*, 259 (1976); reprinted with permission of authors.]

leaves, were purified and immobilized. A newly developed air gap electrode was adapted to monitor the ammonia formed by reduction of nitrate and nitrite with MVH-nitrate and nitrite reductase. The analytical characteristics of both methods have been thoroughly studied. The detection limit using the air gap electrode is 5×10^{-5} M for nitrite [78] or nitrate [156]; using the fluorometric monitoring method the limit is 1×10^{-6} M nitrate. Meal samples and water samples have been analyzed. The results obtained are very satisfactory, especially those obtained by the fluorometric method [78,156].

Most recently an NADH-nitrate reductase (EC 1.6.6.1) induced
from *Chlorella vulgaris* [79] was highly purified by affinity chromatography and was used for a highly specific assay of NO_3^-:

$$NO_3^- + NADH \longrightarrow NO_2^- + NAD + H_2O \qquad\qquad (21)$$

The change of NADH was monitored spectrophotometrically, fluorometrically, or electrochemically.

A fluoride-sensitive enzyme electrode was constructed and optimized for use by Tranh-Minh and Beaux [174,175]. A urease electrode
was constructed using a silicone rubber membrane (part of the CO_2
electrode) with glutaraldehyde. Reversible inhibition by fluoride
was observed, with a response time of 5-6 min and linearity from
10^{-4} to 10^{-2} M concentrations of fluoride.

F. Assay of Organic Species

1. *UREA*

An enzyme electrode was prepared for the substrate urea by
immobilizing urease in a polyacrylamide matrix on 100-μm Dacron and
nylon nets. These nets were placed over the Beckman 39137 cation-
selective electrode (which responds to NH_4^+ ion) [63]. The resulting
"enzyme" electrode responded only to urea. The urea diffuses to the
urease membrane where it is hydrolyzed to NH_4^+ ion. This NH_4^+ ion is
monitored by the ammonium ion-selective electrode, the potential
observed being proportional to the urea content of the sample in the
range 1.0-30 mg of urea/100 ml of solution. This enzyme electrode
appears to possess stability (the same electrode has been used for
weeks with little change in potential readings or drift), sensitivity (as little as 10^{-4} mole/liter urea is determinable), and specificity. Results are available to the analyst in less than 100 sec
after initiation of the test, and the electrode can be used for
individual samples or in continuous operation.

In a later publication, Guilbault and Montalvo [80] described
an improved urea-specific enzyme electrode that was prepared by
placing a thin film of cellophane around the enzyme gel layer to
prevent leaching of urease into the surrounding solution. The

electrode could be used continuously for 21 days with no loss of
activity. A full discussion of the parameters that affect the polymerization of urease as well as of the stability of four types of
urease electrode was published by Guilbault and Montalvo [53].

Guilbault and Montalvo [74] described the preparation of a sensitized cation-selective electrode. By placing a film of urease
over the outside of an ordinary cation-selective glass electrode,
these workers obtained an electrode with increased sensitivity.

Because sodium and potassium ions interfered in the measurement,
Guilbault and Hrabankova [64] used an uncoated NH_4^+ ion electrode as
reference electrode to the urease-coated NH_4^+ electrode and added ion-
exchange resin in attempts to develop an urea electrode useful for
assay of blood and urine. Good precision and accuracy were obtained.

In attempts to improve the selectivity of the urea determination, Guilbault and Nagy [65] used a silicone rubber-based nonactin
ammonium ion-selective electrode as the sensor for the NH_4^+ ions liberated in the urease reaction. The selectivity coefficients of this
electrode were 6.5 for NH_4^+/K^+, 7.50 x 10^2 for NH_4^+/Na^+, and much higher
for other cations. The reaction layer of the electrode was made of
urease enzyme chemically immobilized on polyacrylic gel. A further
improvement was described by Guilbault et al. [58] using a three-
electrode system, which allowed dilution to a constant interference
level. Analysis of blood serum showed good agreement with spectrophotometric methods, and the enzyme electrode was stable for 4 months
at 4°C. Schindler and Gülich [216] described an urea electrode based
on use of an NH_4^+-selective electrode. Using catheter electrodes,
continuous measurements could be made at precisely localized positions of the living organism. Problems of toxicology, sterility,
and safety were discussed.

Still further improvement in the selectivity of this type of
electrode was obtained by Anfalt et al. [66], who polymerized urease
directly onto the surface of an Orion ammonia gas electrode probe by
means of glutaraldehyde. Sufficient NH_3 was produced in the enzyme
reaction layer even at pH values as low as 7-8 to allow direct assay

of urea in the presence of large amounts of Na^+ and K^+. A response
time of 2-4 min was observed.

Guilbault and Tarp [73] described a still better, totally inter-
ference-free, direct-reading electrode for urea using the air gap
electrode of Ruzicka and Hansen [81]. A thin layer of urease chemically bound to polyacrylic acid was used at a solution pH of 8.5,
where good enzyme activity was still obtained, yet where sufficient
NH_3 is liberated to yield a sensitive measurement with the air gap
NH_3 electrode. The urea diffuses into the gel; the NH_3 produced
diffuses out of solution to the surface of the air gap electrode,
where it is measured. A linear range of 3×10^{-2} to 5×10^{-5} M was
obtained with a slope of 0.75 pH unit/decade. The electrode could
be used continuously for blood serum analysis for up to 1 month (at
least 500 samples) with a precision and accuracy of less than 2%.
The response time is 2-4 min at pH 8.5, and the electrode is washed
under a water tap for 5-10 sec after each measurement. Absolutely
no interference from any levels of substances commonly presented in
blood was observed (Na^+, K^+, NH_4^+, ascorbic acid, and so on).

A urea electrode using physically entrapped urease and a glass
electrode to measure the pH change in solution was described by Mosbach et al. [62]. The response time of the electrode to urea was
about 7-10 min and had a linear range from 5×10^{-5} to 10^{-2} M with
a change of about 0.8 pH unit/decade. The electrode could be kept
at room temperature for about 2-3 weeks. The ionic strength and pH
were controlled using a weak (10^{-3} M) tris buffer and 0.1 M NaCl.

Still another possibility for a urea electrode is the use of a
CO_2 sensor to measure the second product of the urea-urease reaction,
HCO_3^-. Guilbault and Shu [49] evaluated the use of the CO_2 sensor and
found that a urea electrode, prepared by coupling a layer of urease
covered with a dialysis net with a CO_2 electrode, had a linear range
of 10^{-4} to 10^{-1} M, a response time of about 1-3 min, and a slight
response to only acetic acid. Na^+ and K^+ ions had no interference.

A potentiometric enzyme electrode was reported by Alexander and
Joseph [82] in which an enzyme immobilized in polyvinylchloride is

used to coat an antimony electrode to detect changes in pH when the
electrode is immersed in a solution of the enzyme substrate. As an
example, urea is determined in solution by using immobilized urease
on an antimony electrode, giving a linear concentration range of
5.0×10^{-4} to 1.0×10^{-2} M urea with a slope of 44 mV/decade change
in urea concentration. The response slope is stable for about 1 week,
with response times in the range 1-2 min, but with absolute potential
changes occurring from day to day.

A highly specific, reproducible enzyme electrode for urea was
developed by Mascini and Guilbault [169] which is vastly superior to
other electrodes. The enzyme urease is chemically bound and attached
to a new, improved Teflon membrane which is an integral part of the
NH_3 gas membrane electrode. From 200 to 1000 assays can be performed
on one electrode with a C.V. of 2.5% over the range 5×10^{-5} to 10^{-2}
M. At least 20 assays/hr can be made with excellent correlation with
the results obtained by the spectrophotometric diacetyl procedure.

Enzyme electrodes for urea using glutaraldehyde-bound urease
and either CO_2 or glass electrodes have been described by Tran-Minh
and Broun [199]. Using an NH_4^+ glass electrode a response of 10^{-1} to
10^{-5} M was obtained, similar to the results of Guilbault et al. [53,
63-65]. Using a CO_2 electrode linearity was observed from 5×10^{-4}
to 10^{-2} M.

2. GLUCOSE AND SUGAR ELECTRODES

The first report of an enzyme electrode was given by Clark and
Lyons [42], who proposed that glucose could be determined amperometrically using soluble glucose oxidase held between Cuprophane
membranes. The oxygen uptake was measured with an O_2 electrode:

$$\text{Glucose} + O_2 + H_2O \xrightarrow{\text{glucose oxidase}} H_2O + \text{gluconic acid} \quad (22)$$

The term *enzyme electrode* was introduced by Updike and Hicks
[43], who coated an oxygen electrode with layer of physically entrapped glucose oxidase in a polyacrylamide gel. The decrease in
oxygen pressure was equivalent to the glucose content in blood and
plasma. A response time of less than 1 min was observed. Clark

[75] proposed measuring the hydrogen peroxide produced in the enzymatic reaction with a Pt electrode. An instrument based on this concept using glucose oxidase held on a filter pad is now sold by Yellow Springs Instrument Co. (Yellow Springs, Ohio). Two platinum electrodes are used, one to compensate for any electrooxidizable compounds in the sample, such as ascorbic acid, the second to monitor the enzyme reaction.

Guilbault and Lubrano [54,69] described a simple, stable, rapid-reading electrode for glucose. The electrode consists of a metallic sensing layer (Pt-glass) covered by a thin film of immobilized glucose oxidase covered in place by means of cellophane. When poised at the correct potential, the current produced is proportional to the glucose concentration. The time of measurement required with this amperometric approach is less than 12 sec using a kinetic method. The electrode is stable for over 1 year when stored at room temperature with only a 0.1% change from maximum response per day. The enzyme electrode determination of blood glucose compares favorably with commonly used methods with respect to accuracy, precision, and stability, and the only reagent needed for assay is a buffer solution.

Lubrano and Guilbault [134] constructed enzyme electrodes using membranes containing immobilized glucose oxidase constructed in a simple, easily reproducible procedure. The enzyme is cocrosslinked with bovine serum albumin using the bifunctional agent glutaraldehyde. The best response was obtained using membranes of high activity and thin membrane construction. The electrodes could be used continuously for several months.

Nagy et al. [76] described a self-contained electrode for glucose based on an iodide membrane sensor:

$$\text{Glucose} + O_2 \xrightarrow{\text{glucose oxidase}} \text{gluconic acid} + H_2O_2 \qquad (23)$$

$$H_2O_2 + 2I^- + 2H^+ \xrightarrow{\text{peroxidase}} 2H_2O + I_2 \qquad (24)$$

The highly selective iodide sensor monitors the local decrease in the iodide activity at the electrode surface. The assay of glucose

was performed in a stream and at a stationary electrode. Pretreatment of the blood sample was required to remove interfering reducing
agents, such as ascorbic acid, tyrosine, and uric acid.

Nilsson et al. [46] described the use of conventional hydrogen
ion glass electrodes for the preparation of enzyme-pH electrodes by
either entrapping the enzymes within polyacrylamide gels around the
glass electrode or as a liquid layer trapped within a cellophane
membrane. In an assay of glucose, based on a measurement of the
gluconic acid produced, the pH response was almost linear from 10^{-4}
to 10^{-3} M with a ΔpH of about 0.85/decade. Electrodes of this type
were also constructed for urea and penicillin (see below). The ionic
strength and pH were controlled using a weak (10^{-3} M) phosphate buffer, pH 6.9, and 0.1 M Na_2SO_4.

Kamin and Wilson [217] utilized a rotating ring-disk enzyme
electrode (RRDEE) for glucose. An enzyme layer of glucose oxidase
was immobilized onto graphite oxide, Pt, and C paste disk electrodes.
The detection of H_2O_2 serves to measure the extent of reaction. At
speeds >1600 rpm the enzyme operates under catalysis control.

Shu and Wilson [278] prepared a glucose electrode by immobilizing glucose oxidase via an albumin glutaraldehyde copolymer onto a
rotating disk carbon paste electrode held at -0.2 V. The change in
the composition of I^-/I_2 was measured as the enzymatic reaction pro-
ceeded. At 400 rpm rotation speeds the response time was <50 sec.
The linear range was 10-75 mg/dl, and <15% decrease of activity was
observed after 1 month at 4°C; at higher rotation speeds greater
sensitivity was observed, but a lower linear range.

A glucose electrode was described by Yoshimo et al. [218] utilizing immobilized glucose oxidase and a Pt electrode to sense the
H_2O_2 produced. This electrode was used to sense α-amylase (see
below).

Williams et al. [129] described an amperometric method in which
1,4-benzoquinone (Q) oxidized the coenzyme $FADH_2$. The hydroquinone
produced (QH_2) was monitored amperometrically at a platinum electrode. A background current was produced by plasma components,

which reduced Q to QH_2; thus a differential measurement was required. Gorton and Bhatti [130] described a potentiometric determination for glucose based on oxidation by 1,4-benzoquinone with immobilized glucose oxidase as catalyst in an enzyme reactor. The electrode is preceded by an analytical dialysis unit to remove proteins. The ratio of quinone to hydroquinone was measured with a flow-through gold electrode. Another gold electrode preceded the enzyme reactor to correct for serum components such as ascorbic acid, which can also reduce quinone. The operating range is $0.04-10 \times 10^{-3}$ M β-D-glucose. A similar system was described by Gorton and Bhatti [130], and Mindt et al. [131,132] replaced the quinone by dichloroindophenol and developed an enzyme electrode system for glucose. Alternative acceptors include 2,6-dichlorindophenol, methylene blue, and pyocyanine perchlorate [243]; a patent on an enzyme electrode using these substrates has been issued.

Scheller and his group in Berlin have been very active in immobilizing enzymes to inert supports, such as gelatin. In one paper [196] they devised an amperometric glucose sensor that determines the H_2O_2 formed during the enzyme reaction. The linear measuring range is 10-180 mg/dl, the response time 15-30 sec. The C.V. is 5% within-day and 9% day-to-day. An instrument based on this concept is being marketed in Eastern Europe. In another paper [197] they described the manufacture of an enzyme electrode from a commercial pO_2 electrode and immobilized glucose oxidase. The response time is 15 sec to 3 min, with a regression coefficient of the calibration curve of .999.

Another group that has been very active in preparation of enzyme electrodes is that of Kulys et al. in Lithuania. In one paper [198] they described amperometric electrodes using glucose oxidase and peroxidase [Eqs. (25) and (26) below]. The first two types of electrodes employ a Pt or glass C electrode and a bienzyme membrane. In the third type of electrode peroxidase was adsorbed on the organic metal electrode, which was then covered with a glucose oxidase membrane. In the first two electrodes ferrocyanide was used. The

electrodes possess a linear dependence of the stationary current on
the glucose concentration in the range 0.01-1 mM; a stationary current is attained in 2-4 min. Types I and II retain their activity
for >100 days, type III 4-6 days only. No response to ascorbic acid
or other interferences was observed. A patent based on the use of
immobilized glucose oxidase and peroxidase held on a Pt electrode,
with measurement of the O_2 uptake, is held by Blaedel and Olson
[142].

Tran-Ninh and Broun [199] described the preparation of electrodes for glucose, urea, and L-amino acids in which the enzyme is
immobilized to albumin via glutaraldehyde on a glass plate, then the
membrane is fitted with a silicone membrane onto a gas electrode (O_2
or CO_2). The theoretical and experimental behavior of these enzyme
electrodes were compared using the analysis of product concentrations
resulting from the enzyme reaction diffusing into the active layer.
Guilbault and Nanjo also described the preparation of stable electrodes for glucose and L-amino acid based on use of an O_2 probe.
Excellent specificity and good sensitivity resulted [56].

The construction and response of the immobilized enzyme, chemically modified electrode as an amperometric sensor has been described
by Ianniello and Yacynych [127]. Glucose oxidase and L-amino acid
oxidase have been covalently bonded to a graphite electrode via a
cyanuric chloride linkage. Various response characteristics as well
as kinetic parameters have been evaluated and compared with previously
reported amperometric enzyme electrodes.

A similar immobilization directly onto the surface of a glassy
carbon cylinder was originally proposed by Bourdillon et al. [128].
The enzyme activity was measured amperometrically by the H_2O_2 produced from glucose.

A modular, computerized, feedback control system (Biostator)
for dynamic control of blood glucose concentrations in diabetics was
described by Fogt et al. [176]. This on-line glucose analyzer for
use with whole blood utilizes a novel enzyme- (glucose oxidase)
membrane configuration and an electrochemical cell to measure the
H_2O_2 generated. The analyzer exhibits both short- and long-range

stability, and instrument response and analyte concentration are
linearly related over the full range of clinical interest. The
response is fast, accurate, and precise, and permits determination
of blood glucose within 2 min from the moment the blood leaves the
patient. Correlation studies were completed to show the agreement
between the Biostator glucose analyzer and the FDA's recommended
hexokinase/glucose-6-phosphate dehydrogenase procedure on whole
blood (e.g., average percent recovered for 11 concentrations between
250 and 9000 mg/liter was hexokinase, 95.6%, Biostator analyzer,
95.9%, bias and SD_d, respectively, at low, normal, and high glucose
values were 12 and 41 mg/liter at the 500 mg/liter level; 4 and 52
mg/liter at the 1000 mg/liter level, and 4 and 128 mg/liter at the
4000 mg/liter level). No appreciable interference is observed with
above normal concentrations of bilirubin, uric acid, creatinine,
sodium salicylate, or dextran. Platelet adhesion, which tends to
decrease useful life of the membrane, has been significantly decreased.

A new principle for the construction of oxygen-dependent enzyme
electrodes was presented by Enfors [177]. The enzyme electrode is
based on a galvanic oxygen electrode which is furnished with an
electrolysis anode covered by immobilized enzyme and placed close
to the oxygen-sensing surface. An ordinary oxygen electrode is used
as a reference electrode. The enzymatic consumption of oxygen in
the enzyme electrode generates a potential difference between the
electrodes which is utilized to control electrolytic generation of
oxygen from water in such a way that a zero differential potential is
maintained. Thus, the enzyme electrode operates under ambient oxygen
tension and does not suffer from oxygen limitation. The electrolytic
current in this system gives a measure of the concentration of substrate surrounding the enzyme electrode. The electrode has been
applied for continuous D-glucose analysis in situ during batch culturing of Candida utilis.

A hydrogen peroxide permselective membrane with asymmetrical
structure was prepared and D-glucose oxidase (EC 1.1.3.4) was immobilized onto the porous layer by Tsuchida and Yoda [178]. An enzyme

electrode was constructed by combination of a hydrogen peroxide
electrode with the immobilized D-glucose oxidase membrane. The
enzyme electrode responded linearly to D-glucose over the concentration 0-1000 mg/dl within 10 sec. When the enzyme electrode was
applied to the determination of D-glucose in human serum, within-
day precision (C.V.) was 1.29% for D-glucose concentration with a
mean value of 106.8 mg/dl. The correlation coefficient between the
enzyme electrode method and the conventional colorimetric method
using a free enzyme was .984. The immobilized D-glucose oxidase
membrane was sufficiently stable to perform 1000 assays (2 to 4
weeks operation) for the determination of D-glucose in human whole
blood. The dried membrane retained 77% of its initial activity
after storage at 4°C for 16 months.

Gondo et al. [200] described an improved glucose sensor using
immobilized glucose oxidase and glucose isomerase. A better control
of the profile of the glucose sensor response was obtained together
with better stability. Solubilized collagen fibril was used as a
support. This observed improvement was not explained, but could be
due to the presence of glucose oxidase impurities in the isomerase.

Mor and Guaraccia [201] described a new glucose analyzer in
which the ferrocyanide produced in the indicator reaction is reoxi-
dized to ferricyanide.

$$\text{Glucose} + H_2O \xrightarrow[\text{oxidase}]{\text{glucose}} \text{gluconate} + H_2O_2 \qquad (25)$$

$$H_2O_2 + Fe(CN)_6^{3-} \xrightarrow{\text{peroxidase}} Fe(CN)_6^{4-} + H_2O \qquad (26)$$

In order to compensate for other reductive substances present in the
sample, a differential measurement is performed between the glucose
sensor and an auxiliary electrode. Linearity was observed for 20-
500 mg/dl of glucose.

Wingard and co-workers have been conducting research on biofuel
cells using immobilized enzymes, such as glucose oxidase. In their
system constant current voltammetry (potentiometry at zero current)
was used to evaluate electrodes formed containing immobilized oxidative enzymes as catalysts [202]. At zero current a potential is

produced which is proportional to the glucose concentration. The
glucose oxidase was physically entrapped in polyacrylamide gel. In
another paper [203] glucose oxidase, catalase, and bovine serum albumin were coimmobilized with glutaraldehyde around a platinum screen
or a Pt-Ir wire. The potential difference between this dual-enzyme
electrode and a Ag/Ag Cl reference electrode was proportional to the
log of the glucose concentration over the range 10-150 mg/100 ml of
solution. The results obtained suggest that these potentiometric
enzyme electrodes may have sufficient specificity for development
of a continuous in vivo glucose sensor. Glutaraldehyde crosslinking
was used as the method of enzyme immobilization.

Romette et al. [204] described a glucose oxidase electrode for
measurements of glucose in samples exhibiting high variability in
oxygen content. The enzyme is crosslinked with gelatin using glutaraldehyde and is used with an O_2 electrode. The enzyme electrode contains enough O_2 to compensate for the variability of the sample oxygen content. Measurements are made by collecting the derivative with
time of the signal given by the electrode.

Glucose oxidase shows greater specificity toward β-D-glucose
than toward α-D-glucose, but by adulteration of the enzyme with
aldose-1-epimerase the resulting mutarotation allows total glucose
to be determined [179].

Koyama et al. [180] described an improved enzyme sensor for
glucose with an ultrafiltration membrane and immobilized glucose
oxidase. The sensors consist of a lead anode and a platinum cathode
covered with a PTFE membrane over which are placed the immobilized
enzyme membrane and a protective ultrafiltration membrane (0.03 mm
thick). The decrease in current is proportional to the glucose
level.

Dr. Pierre Coulet and co-workers (Jacques Julliard, Danielle
Gautheron, and others) at the Laboratory of Biology and Membrane
Technology of University Claude Bernard in Lyon have been especially
active in the development of enzyme electrodes that are stable, self-
contained, and easily fabricated in large scale [135-140]. A generally useful mild coupling method for placement of the enzyme onto

the collagen membranes is used with acyl azide activation and
coupling as described in Chapter 2. For example, stable, very sensitive (10^{-8} M detectable) glucose sensors have been described that
are now sold commercially by Tacussel, Lyon, and are highly reliable.

The glucose sensor was prepared using glucose oxidase immobilized on collagen membranes by this technique [140] and consists of
a modified gas electrode in which the pH detector is replaced by a
platinum anode and the porous selective membrane by the enzyme membrane. The latter is tightly pressed against the anode by a screw
cap and is thus easily removable.

The electrode is immersed in a small vessel with the selected
buffer in which glucose-containing samples are injected. Enzymatically generated peroxide is detected by anodic oxidation at +650 mV
versus a Ag/Ag Cl reference. After an injection the current output
increases and reaches a steady state within a few minutes; differentiation gives a signal peak in 1 min. Linearity of signal and glucose concentration is obtained over the range 10^{-7} to 2×10^{-3} M.
A compensating electrode mounted with a noncollagen membrane permits
the detection of and correction for electrochemical interferences
when testing samples with high levels of electroactive species. The
glucose probe was used for rapid assays in blood [138] and food [193].

A two-enzyme electrode for maltose determination using the same
electrochemical detector has been designed with membranes prepared by
asymmetrical coupling [194]. The two enzymes involved were glucoamylase and glucose oxidase. Glucoamylase above was immobilized on
the membrane face exposed to the bulk phase into which the maltose-
containing samples were injected. The hydrolysis of maltose occurred
according to the reaction:

$$\text{Maltose} + \text{H}_2\text{O} \xrightarrow{\text{glucoamylase}} 2\text{-D-glucose} \qquad (27)$$

The glucose produced migrates through the membrane and was then oxidized on the inner face with immobilized glucose oxidase which is in
close contact with the platinum disk. As with the monoenzyme system
for glucose, the sensitivity and linearity were excellent.

The same base electrochemical sensor has also been used for the determination of various other species with collagen membranes bearing mono- or multienzyme systems. A single multipurpose electrode [195], where selected membranes bearing different oxidases can be easily replaced, has been described for assay of galactose (galactose oxidase), starch (glucoamylase), and sucrose (invertase). In all electrodes it was found that asymmetrical coupling improved the electrode performance and all sensors could be used for hundreds of assays. Linearity was approximately 10^{-7} to 10^{-2} M for maltose or galactose, 10^{-4} to 2×10^{-3} M for sucrose, and 6×10^{-5} to 10^{-3} for lactose.

Also, a new approach to heterogeneous enzymology is possible with such electrodes. As mentioned in Chapter 2, diffusional limitations with enzyme-matrix conjugates appear to be one of the main effects of immobilization. A very complex situation occurs with the motionless enzyme membrane fixed on electrodes. Considering glucose oxidase immobilized on both membrane faces, hydrogen peroxide is generated at the inner interface between the Pt anode and membrane and at the outer interface between the membrane and the reaction mixture. This product can be detected with the Pt anode in the immediate vicinity of the inner enzyme layer and with another sensor in the bulk phase. Two simultaneous monitorings of the immobilized enzyme activity and its dependence on glucose concentration and temperature, taking into account diffusional effects, were then possible [139].

Cheng and Christian [126] described a method for assay of galactose based on the aerobic oxidation of galactose by oxygen in the presence of galactose oxidase. The O_2 consumption is measured with an oxygen electrode on a Beckman glucose analyzer.

A rapid (40-sec), precise, and accurate micromethod for the determination of galactose in plasma and whole blood was described by Taylor et al. [205]. The method utilized an amperometric hydrogen peroxide electrode covered by a selective membrane of immobilized galactose oxidase. Linearity was obtained from 0-500 mg% in 0.07 M

phosphate buffer, pH 7.3, containing 10 mg% ferricyanide and 0.2 mg%
cupric chloride dihydrate. Of 39 compounds screened, only dihydroxy-
acetone interfered.

An enzyme electrode for sucrose was described by Satoh et al.
[167] using immobilized invertase, mutarotase, and glucose oxidase.

$$\text{Sucrose} + H_2O \xrightarrow{\text{invertase}} \alpha\text{-D-glucose} \qquad (28)$$

$$\downarrow \text{mutarotase}$$

$$H_2O_2 \xleftarrow[\text{glucose oxidase}]{O_2} \beta\text{-D-glucose}$$

The electrode was linear with sucrose concentrations of 1-10 mM.
Commercial instruments for galactose, glucose, sucrose, maltose, and
lactose have been marked by Leeds and Northrup (North Wales, PA) and
Yellow Springs Instrument Co. (Ohio).

An enzyme sequence electrode for D-gluconate has been described
by Jensen and Rechnitz [181] based on the following reactions:

$$\text{D-gluconate} + \text{ATP} \xrightarrow{\text{gluconate kinase}} \text{6-phospho-D-gluconate} + \text{ADP}$$

$$Mg^{2+} \quad NADP \left| \begin{array}{l} \text{6-phospho-D-gluconate} \\ \text{dehydrogenase} \end{array} \right.$$

$$NADPH + H^+ + CO_2 + \text{D-ribulose-5-phosphate}$$

$$(29)$$

The enzymes are entrapped in a dialysis membrane and also by
linking with glutaraldehyde to BSA. The assay exhibits linearity
from 5×10^{-5} to 5×10^{-3} M for the soluble enzyme and 6×10^{-5} to
8×10^{-4} M with a slope of 28 mV/decade and a response time of 8-15
min for the immobilized enzyme. A slight response to D-glucose-6-
phosphate is observed, probably due to a trace impurity of this
dehydrogenase in the enzyme preparation. No response is observed
to D-glucose, D-mannose, D-mannose-6-phosphate, or gluconic acid.

Cordonnier et al. [153] proposed magnetic enzyme membranes as
the active elements of electrochemical sensors for lactose, sucrose,
and maltose. In the device a magnetic film bearing the bienzyme
system (β-galactosidase/glucose oxidase for lactose, maltase/glucose

oxidase for maltose, invertase/glucose oxidase for sucrose) is fixed on a Clark pO_2 electrode bearing a cylinder magnet. The active bi-enzyme membrane is situated on the external surface of the gas-permeable membrane. The consumption of O_2 is measured and related to the concentration of lactose (1.4 x 10^{-3} to 6 x 10^{-3} M), sucrose (10^{-3} to 10^{-1} M), or maltose (0.5 to 2 x 10^{-3} M).

Cheng and Christian [154] described electrodes for lactose and maltose based on use of a gas O_2 electrode and the immobilized enzymes described above.

3. AMINO ACID ELECTRODES

Enzyme electrodes for the determination of L-amino acids were developed by Guilbault and Hrabankova [51], who placed an immobilized layer of L-amino acid oxidase over a monovalent cation electrode to detect the ammonium ion formed in the enzyme-catalyzed oxidation of the amino acid. Two different kinds of enzyme electrodes were pre-pared by Guilbault and Nagy for the determination of L-phenylalanine [101]. One of the electrodes used a dual-enzyme reaction layer-- L-amino acid oxidase with horseradish peroxidase--in a polyacrylamide gel over an iodide-selective electrode. The electrode responds to a decrease in the activity of iodide at the electrode surface due to the enzymatic reaction and subsequent oxidation of iodide.

$$\text{L-Phenylalanine} \xrightarrow{\text{L-amino acid oxidase}} H_2O_2 \tag{30}$$

$$H_2O_2 + 2H^+ + I^- \xrightarrow[\text{peroxidase}]{\text{horseradish}} I_2 + H_2O \tag{31}$$

The other electrode was prepared using a silicone rubber base nonactin-type ammonium ion-selective electrode covered with L-amino oxidase in a polyacrylic gel. The same principle of diffusion of substrate into the gel layer, enzymatic reaction, and detection of the released ammonium ion applies to the system. The CO_2' sensor was evaluated by Guilbault and Shu [49] for response to tyrosine when coupled with tyrosine de-carboxylase held in an immobilized form by a dialysis membrane. A linear range of 2.5 x 10^{-4} to 10^{-2} M was observed with a slightly faster response time than that observed with the urea electrode men-

tioned above. A slope of 55 mV/decade was obtained, compared with
59 mV/decade for the urea electrode.

Ianniello and Yacynych [206] constructed chemically modified
graphite electrodes with immobilized L-amino acid oxidase as a potentiometric sensor for L-amino acids. The electrodes displayed slopes
of 24-29 mV/decade over the range 10^{-2} to 10^{-5} M for L-phenylalanine,
L-methionine, and L-leucine. The electrode response was about 2 min,
and the slope degraded 33% after 78 days. A mechanism for the potentiometric response of such an electrode was presented and was shown
to be due to an interaction of H_2O_2 produced in the enzymatic reaction with the different surface-functional groups present on the
graphite. No potentiometric response was found for Pt or untreated
graphite C electrodes, but a pseudo-Nernstian response (79 mV/decade)
was obtained for a cyanuric chloride (CC) graphite electrode to H_2O_2
in the range 2.5×10^{-5} to 1.7×10^{-2} M, and 40 mV/decade with covalently attached L-amino acid oxidase to this CC graphite probe.

Electrodes specific for D-amino acids, which are oxidatively
catalyzed by D-amino acid oxidase, were reported by Guilbault and
Hrabankova [52]. The NH_4^+ ion produced is monitored with a cation
electrode:

$$\text{D-Amino acid} + O_2 \xrightarrow{\text{oxidase}} NH_4^+ + HCO_3^- \qquad (32)$$

The stability of these electrodes could be maintained for 21 days if
they are stored in a buffered FAD solution, since the FAD is weakly
bound to the active site of the enzyme and is needed for its activity.
Electrode probes suitable for the assay of D-phenylalanine, D-alanine,
D-valine, D-methionine, D-leucine, D-norleucine, and D-isoleucine were
developed. An electrode for asparagine was also developed using asparaginase as the catalyst [52]. No cofactor was necessary. Similar
results were obtained by Llenado and Rechnitz [95].

Guilbault and Shu [104] described an enzyme electrode for glutamine, prepared by entrapping glutaminase on a nylon net between a
layer of cellophane and a cation electrode. The electrode responds
to glutamine over the concentration range 10^{-1} to 10^{-4} with a response time of only 1-2 min.

Guilbault and Lubrano [50] prepared an electrode for L-amino acids by coupling chemically bound L-amino acid oxidase to a Pt electrode which senses the peroxide produced in the enzyme reaction:

$$\text{L-Amino acid} + O_2 + H_2O \longrightarrow COCOOH + NH_3 + H_2O_2 \qquad (33)$$

The time of measurement is less than 12 sec, using a kinetic measurement of the rate of increase in current per unit time, and the only reagent required is a phosphate buffer. The L-amino acids cysteine, leucine, tyrosine, phenylalanine, tryptophan, and methionine were assayed.

Mascini and Palleschi [100] determined L-amino acids, alcohols, and glucose with a tubular iodide-selective electrode and oxidase systems. The substrate, the appropriate enzyme, KI, and phosphate buffer solution are mixed in a Technicon AutoAnalyzer and the mixture is passed through a reaction coil (at 37°C for 12 min). To stop the reaction, $HClO_4$ is added as well as Mo(VI) to catalyze the oxidation of I^- by the H_2O_2 formed. After 6 min the solution reaches the tubular I^--selective electrode for measurement. From 0.1 to 10 mM amino acids and 1-10 mM alcohol samples can be assayed (20 samples in 1 hr) with a precision of 5-10%. Ianiello and Yancynych [206] described the construction and response of the immobilized enzyme chemically modified electrode as an amperometric sensor. L-amino acid oxidase was covalently bonded to a graphite electrode via a cyanuric chloride linkate. Various response characteristics as well as kinetic parameters were evaluated and compared with previously reported amperometric enzyme electrodes. The linear range was 10^{-2} to 10^{-5} for L-phenylalanine, L-methionine, and L-leucine. After 78 days 67% activity was retained.

Guilbault and Nanjo [56] proposed enzyme electrodes for L-amino acids based on the use of immobilized enzymes and a Pt-based O_2 electrode. The change in the dissolved O_2 level is monitored, and the electrode responds to L-methionine, L-leucine, L-phenylalanine, L-tyrosine, L-cysteine, L-lysine, and L-isoleucine. Excellent stability (>4 months), fast response times (<1 min), and greater selectivity over the peroxide based sensors were observed.

Tran-Minh and Broun [199] described two types of L-amino acid electrodes, one using an NH_4^+ ion electrode and another using an O_2 electrode. In the first type linear responses from 5×10^{-5} to 10^{-2} M were obtained for L-tryptophan, L-tyrosine, L-methionine, L-leucine, L-arginine, L-histidine, and L-phenylalanine. A one minute response time was obtained. For the O_2 electrode immobilized amino acid oxidase and catalase were used; 3 min was necessary for a steady-state response.

Enzyme electrodes for other L-amino acids have been developed using more specific enzymes.

Several workers have proposed enzyme electrodes for the assay of L-glutamic acid. Guilbault and Shu [104] and Ahn et al. [106] described the use of a CO_2 electrode with glutamate decarboxylase enzyme. The gas liberated is measured and related to the concentration of this amino acid. In other studies the enzyme glutamate dehydrogenase is used, linked to either an NH_4^+ ion electrode [105] or to a Pt probe [192] to measure NADH.

A totally specific enzyme electrode useful for the assay of L-lysine in grains and foodstuffs was described by White and Guilbault [108]. No response is noted with D-amino acid or any other L-amino acid. The electrode can be used for assay of this amino acid in mixtures, without the necessity for extensive separations and expensive instrumentation (i.e., the amino acid analyzer). The electrodes are quite stable, with a linear range of L-lysine concentration of 5×10^{-5} to 10^{-1} M. The only limitation is the long response time (5-10 min). After preparation, buffer is the only reagent needed.

Skogberg and Richardson [109] have later proposed a similar electrode based on immobilized L-lysine decarboxylase and a CO_2 probe to measure L-lysine in cereals and grains. Good results were obtained.

Berjonneau et al. [182] described electrodes prepared with three decarboxylative enzymes purified by affinity chromatography for the assay of L-lysine, L-tyrosine, and L-phenylalanine. The CO_2 is measured with a gas electrode; the reproducibility is ±0.5% and the precision 2% for the three electrodes. The range is 2×10^{-3} to 10^{-2} M phenylalanine. The enzymes were chemically fixed and stable at 4-37°C. At 37°C 90% of activity was maintained after 1-1/2 days, compared with only 50% in solution.

Calvot et al. [183] crosslinked enzyme molecules with albumin via glutaraldehyde to magnetic ferrite particles trapped in a membrane structure. Decarboxylases for tyrosine, phenylalanine, and lysine were used together with a CO_2 electrode bearing a cylindrical magnet. Linear calibration curves were obtained for L-lysine 2×10^{-4} to 1.3×10^{-3} M, tyrosine 2×10^{-4} to 1.5×10^{-3} M, and phenylalanine 3×10^{-3} to 1.5×10^{-2} M.

Guilbault and Nagy [101] described several enzyme electrodes for L-phenylalanine. One was based on the use of immobilized L-amino acid oxidase and peroxidase and an iodide membrane electrode. The second was based on an NH_4^+ nonactin electrode with L-amino acid oxidase. Linear curves were obtained from 10^{-2} to 2×10^{-5} M L-phenylalanine with the NH_4^+ electrode, 10^{-4} to 10^{-5} M with the iodide electrode. Responses to L-leucine and L-methionine was also observed.

Macholan [184] described a rapid determination of L-lysine and L-arginine using immobilized decarboxylases with diamine oxidase and an O_2 electrode. The action of the specific decarboxylase gives rise to the corresponding diamine, which is further oxidized by the diamine oxidase with uptake of O_2 as cosubstrate:

$$R\text{-CH(NH}_2)\text{COOH} \xrightarrow{\text{decarboxylase}} R\text{-CH}_2\text{-NH}_2 + CO_2 \qquad (34)$$

$$R\text{-CH(NH}_2) + O_2 + H_2O \xrightarrow[\text{oxidase}]{\text{amine}} RCHO + NH_3 + H_2O_2 \qquad (35)$$

L-lysine or L-arginine in the 10- to 100-mmole region can be determined; the assay of L-lysine is only interfered with by hydroxylysine; lysine interfered in the assay of arginine.

A potentiometric sensor for L-histidine has been made by immobilizing histidine ammonia lyase on an ammonia gas-sensing electrode by Walters et al. [185]. The enzyme was purified from *Pseudomonas* sp. The effect on the electrode response of thiol and metal ion activation of the enzyme was examined. The response of the electrode was linear between 3×10^{-5} and 10^{-2} M histidine with a slope

of 54 mV/decade. The sensor showed no decline in response over one
week of use. The response time was 3-8 min. The electrode responded
only to histidine and ammonia. The effect of inhibitors of the enzyme reaction on the electrode response was found to be diminished
by using high enzyme activities on the electrode. A method for pre-
venting leakage of the inner filling solution of the ammonia electrode was also presented. An electrode using histidinase and an NH_4^+
ion electrode was proposed by Ngo [107]. The NH_4^+ liberated in the
enzymatic reaction is monitored by the electrode. A highly sensitive
histidine electrode has been prepared by immobilizing the enzyme histidine decarboxylase at the surface of a CO_2 electrode [332]. The
enzyme was extracted from *Lactobacillus* 30a. The electrode showed
linearity from 3×10^{-4} to 1×10^{-2} M with a slope of 48-53 mV/decade
and a useful lifetime of over 30 days.

Fung et al. [110] proposed a totally selective electrode probe
for L-methionine. The electrode is prepared by immobilizing purified
L-methionine-γ-lyase (EC 4.4.1.11) onto an ammonia-specific electrode.
The α,γ elimination of L-methionine proceeds with formation of α-keto-
butyrate, methanethiol, and ammonia. Only L-methionine reacts with
the purified enzyme; a linear range of 10^{-2} to 10^{-5} M is observed.

A potentiometric L-tyrosine selective probe for the direct
determination of L-tyrosine in biological fluids and foods was described by Havas and Guilbault [111]. The sensor element of the
probe, based on a CO_2-sensitive gas membrane, is a layer containing
immobilized apo-L-tyrosine decarboxylase. A linear range 2.6×10^{-3}
to 4×10^{-5} M is observed, with no interference from any L-amino
acids or D-tyrosine. Kumar and Christian [112] have used a gas O_2
electrode and the enzyme tyrosinase to measure L-tyrosine. The decrease in the O_2 level serves as an indication of the concentration
of L-tyrosine.

4. ALCOHOL AND ACID ELECTRODES

Alcohol oxidase catalyzes the oxidation of lower primary ali-
phatic alcohols.

$$RCH_2OH + O_2 \xrightarrow{\text{alcohol}}_{\text{oxidase}} RCHO + H_2O_2 \qquad (36)$$

The hydrogen peroxide produced in these reactions may be determined amperometrically with a platinum electrode as in the determination of glucose above. Guilbault and Lubrano [94] used the alcohol oxidase obtained from a basidiomycete to determine the ethanol concentration of 1-ml samples over the range 0-10 mg/100 ml, with an average relative error of 3.2% in the 0.5-7.5 mg/100 ml range. This procedure should be adequate for clinical determinations of blood ethanol because normal blood from individuals who have not ingested ethanol ranges from 40 to 50 mg/100 ml. Methanol is a serious interference in the procedure, since the alcohol oxidase is more active for methanol than ethanol when H_2O_2 is measured (see below). However, the concentration of methanol in blood is negligible compared with that of ethanol.

A vastly improved alcohol electrode, selective for ethanol, was described by Guilbault and Nanjo [93]. The decrease in current as O_2 was depleted from solution in the enzymatic reaction [Eq. (36)] was measured at an applied potential of -0.6 V versus SCE.

From 0.4 to 50 mg% can be assayed with little interference. The electrode was also found to exhibit a totally different selectivity pattern from the H_2O_2 probe used above by Guilbault and Lubrano [94], ethanol being the preferred substrate. This is believed due to the total specificity of the O_2 probe, which provides a measurement independent of the chemical reaction that can occur between the aldehyde formed with peroxide to give the organic acid (which is also a substrate for the enzyme reaction). Thus, all previous selectivity values given in the literature for this enzyme are believed to be in error. The correct values are given below:

| | Relative reactivity | | |
| | | Enzyme electrode | |
Alcohol	Janssen[a]	O_2[b]	H_2O_2[c]
MeOH	100	3	0.98
EtOH	28	12,000	0.46
Allyl alcohol	17	190	1.14
n-Propanol	5.3	1,300	0.36
n-Butanol	2.1	2,500	0.21

[a]Colorimetric assay of H_2O_2 formed [207,208].
[b]Method of Ref. 93.
[c]Method of Ref. 94.

In addition, the enzyme reacts quite well with aldehydes and acids, as shown:

Substrate	Relative activity (O_2)
CH_3OH	1.0
HCOH	17,600
HCOOH	5,640
C_2H_5OH	1.00
CH_3COH	0.89
CH_3COOH	0.88
C_2H_5OH	1.00
Acetic acid	0.88
Formic acid	0.51
Lactic acid	0.37
Butyric acid	0.21
Pyruvic acid	0.03

Thus, acids like acetic or formic, and aldehydes like formaldehyde and acetaldehyde can be easily determined.

The assay of alcohol using alcohol dehydrogenase was proposed by Suzuki et al. [91], Johansson et al. [335], and by Cheng and Christian [186]. In the former procedure a Pt electrode was used to sense the electroactive NADH produced, whereas in the latter, the NADH is oxidized by O_2 and the uptake is measured [186]. Only 5 µl of sample is required, and assays are obtained in less than 1 min with linearity from 0.5 to 5 g/liter of EtOH.

$$\text{Alcohol} + 2\text{NAD} \xrightarrow{\text{ADH}} 2\text{NADH} \tag{37}$$

$$2\text{NADH} \xrightarrow[\text{peroxidase}]{O_2} 2\text{NAD} + 2H_2O \tag{38}$$

Smith and Olson [92] combined a Pt electrode with the bound alcohol dehydrogenase/diaphorase system to develop an enzyme electrode for alcohols.

The enzymatic electrochemical determination of ethanol and L-lactic acid with a C-paste electrode modified chemically with NAD using n-octaldehyde was described by Yao and Musha [209]. The NAD is converted to NADH by oxidation of ethanol and L-lactic acid catalyzed by their respective dehydrogenases, and the NADH formed is oxidized electrochemically to the original NAD, thus giving a well-defined linear-sweep voltammetric peak. The peak area was linearly related to the amount of ethanol or L-lactic acid in the range 0.05-2 x 10^{-9} moles. The reaction time is 6-8 min.

$$\vdash\!NAD^+ + C_2H_5OH \xrightarrow{\text{ADH}} \vdash\!NADH + CH_3CHO + H^+ \qquad (39)$$

$$\vdash\!NAD^+ + L\text{-lactate} \xrightarrow{\text{LDH}} \vdash\!NADH + \text{pyruvate} + H^+ \qquad (40)$$

$$\vdash\!NADH \xrightarrow{\hspace{1cm}} \vdash\!NAD^+ + 2e^- + H^+ \qquad (41)$$

The C.V. was about 6-7% at the 10^{-10} mole level and 2-3% at the 10^{-9} mole level. After the electrode had been used about five times, the response decreased about 6-10% compared with a fresh electrode.

Torstensson et al. [188] immobilized a liver alcohol dehydrogenase-NAD complex on the surface of a glass carbon electrode, and investigated the cyclic voltammetry in ethanol-containing buffers. One cycle was observed but repeated recycling could not be carried out, presumably due to catalytic decomposition of the coenzyme at the electrode surface.

Catechol finds important uses in the manufacture of dyes, drugs, rubbers, and antioxidants for lubricating oils and also in photography and as an agent for oxygen removal. Catechol is also found in materials of biological origin (e.g., urine), certain foods, as well as in cigarette smoke. Existing methods of catechol analysis are based on color reactions of phenol and polyphenols with a number of oxidants and complex-forming compounds. By their very nature, none of these methods is specific for catechol. Analysis thus requires a preceding separation from other phenolic compounds usually occurring together with catechol. Neujahr [187] described a simple, rapid, and sensitive method for determination of catechol by means of an oxygen probe coated with immobilized catechol-1,2-oxygenase (EC 1.13.1.1) isolated from yeast. This enzyme specifically cleaves the aromatic ring incorporating 1 mole of oxygen.

$$\text{(catechol)} + O_2 \longrightarrow \text{(cis,cis-muconate)} \tag{42}$$

Because the ring is cleaved and not just oxidized to an o-quinone, as is the case with, for example, tyrosinase (EC 1.14.18.1), side reactions such as polymerization of catechol oxidation products can be kept at a minimum.

Electrodes for assay of phenols have been proposed by several workers. Kyzlink and Macholan [161] used an oxygen electrode of the Clark type coated with a thin layer of phenol hydroxylase, which catalyzes the oxidation of phenols by O_2. A decrease in the electrode current is measured. Neujahr and Kjellen [162] used a Clark oxygen electrode coated with a thin paste of induced *Trichosporon cutaneum* to determine phenol in the range 0-15 mg/liter. The assay is completed 15 sec after adding the sample and the bioprobe was stable for 100 assays or 5 days.

In a later paper Kjellen and Neujahr [163] immobilized phenol hydroxylase to the surface of a Clark oxygen electrode. A linear range of 0.5-50 μM was obtained, and 150 assays could be performed with no loss of enzyme activity. It was necessary to incubate the enzyme electrode in a buffer containing NADPH for a few minutes before addition of sample.

Macholand and Schanel [210] used an active enzyme membrane prepared by immobilization of potato or mushroom polyphenol oxidase and albumin by glutaraldehyde on the surface of a polyamide netting. After the enzyme membrane had been stretched over the hydrophobic membrane of a Clark O_2 electrode, a sensor was obtained which can monitor phenolic substrates by the O_2 uptake in solution. The electrode was used for p-cresol, phenol, pyrocatechol, and pyrogallol in 20- to 200-mole concentrations in 3 ml of solution. The membranes with immobilized mushroom enzyme were used for a simple detection system for phenols in wastewater.

Schiller et al. [190] determined phenol concentrations with an electrode using immobilized tyrosinase. Tyrosinase catalyzes the oxidation of phenol in the presence of saturating levels of oxygen. The oxidation product, ortho-benzoquinone, is then chemically reduced in the presence of an excess of ferrocyanide ions. The coupled oxidation of ferrocyanide ions to ferricyanide results in a measurable potential difference in the electrochemical system. The resulting zero current potentials in these steady-state potentiometric measurements were shown to be directly proportional to the logarithm of phenol concentration in the range 3.8×10^{-7} to 1×10^{-4} M.

Electrochemical sensors for ethanol, lactate, and malate were designed by constraining a dehydrogenase enzyme and NAD^+ onto the surface of a platinum electrode by Blaedel and Engstrom [125]. Efficient electrochemical regeneration of NAD^+ from the enzymatically produced NADH then provided a current that was dependent on substrate concentration. An acetylated dialysis membrane was used to constrain both enzyme and NAD^+ and exhibited low permeability to $NAD^{\top}$ but fairly unrestricted permeability to the substrates. In a flow-through configuration, the sensors showed very high sensitivity, rapid response time, and good reproducibility. Response of the sensors was influenced by product inhibition of the enzymatic reaction, causing nonlinearity to appear at substrate concentrations below the Michaelis constant. Ethanol electrodes were applied to the analysis of plasma ethanol, resulting in excellent accuracy and precision in the physiologically significant range of ethanol concentrations.

An amperometric sensor for L-ascorbic acid has been made by immobilizing ascorbate oxidase in the reconstituted collagen membrane and mounting the enzyme-collagen membrane on a Clark oxygen electrode by Matsumoto et al. [113]. The enzyme was purified from the peel of cucumber, *Cucumis sativus*. The response of the electrode was linear between 5×10^{-5} and 5×10^{-4} M L-ascorbic acid and the precision was found to be better than 2.3% in 35 successive assays. The only limitation is that the electrode responds for D-isoascorbic acid (about 93% relative response). The sensor has a lifetime of 3 weeks. After

preparation, buffer is the only reagent needed. Its usefulness for
assay of L-ascorbic acid in food is also described.

A self-contained rapid reading electrode for uric acid was described by Nanjo and Guilbault [55]. The electrode was prepared by
placing a layer of glutaraldehyde-bound uricase over the tip of a
Beckman Pt electrode; the enzyme was then covered for support with
a thin layer of dialysis membrane. The decrease in the level of
dissolved oxygen in solution due to the enzymatic reaction was maintained at an applied potential of -0.6 V versus SCE.

$$\text{Uric acid} + O_2 \xrightarrow{\text{uricase}} \text{allantoin} \cdot H_2O_2 + H_2O \qquad (43)$$

The current observed is proportional to the level of uric acid at
concentrations of 10^{-5} to 10^{-1} M. By measuring the initial rate of
change in current an assay can be performed in less than 30 sec.

Further studies indicated the electrode could be used for the
assay of glucose and amino acids.

It was found that the peroxide produced in the reaction could
not be monitored at +0.6 V versus SCE as described in the method of
Guilbault and Lubrano for glucose [54,69], amino acids [50], and
alcohols [94] because (1) the polarographic curves for uric acid
and peroxide are too close to be separated at any pH useful for the
enzyme reaction, and (2) an allantoin-peroxide complex is the product
of the oxidation of uric acid, not free peroxide. Additionally, the
oxygen uptake method was found to be more sensitive, allowing the
assay of lower concentrations of substrates.

The use of a Pt electrode rather than the Clark-type oxygen
electrodes eliminated all the problems associated with gas membrane
electrodes, namely, slow response and blockage of the membrane by
substances present in blood [55].

A uric acid sensor based on the use of a carbon dioxide electrode and immobilized urate oxidase was described by Kawashima et al.
[172]. Glutaraldehyde and collagen immobilized electrodes on a gas
CO_2 sensor were prepared to monitor the production of CO_2 in the
enzyme-catalyzed reaction. Response curves with slopes of 58-65 mV/

decade were obtained, with a response time of 5-10 min. After 10
days of use the membranes could be restored with cupric sulfate
solution.

The use of regenerated cofactors in conjunction with enzyme
electrodes have been proposed by workers at the University of Lund,
Sweden. Jaegfeldt et al. [155] recycled NADH by oxidizing it coulometrically at a rotating platinum gauge electrode at 0.7 V versus
SCE. The NAD formed was reduced enzymatically in a reactor containing immobilized alcohol dehydrogenase. Several recyclings are possible with the same lot of NADH. An efficiency of 99.3% allowed
the use of this technology in fabrication of an alcohol electrode.

Williams et al. [129] used ferricyanide as a hydrogen acceptor
for lactic acid and described an enzyme electrode for lactate based
on the following reaction:

$$\text{Lactate} + 2\text{Fe(CN)}_6^{3-} \xrightarrow{\text{LDH}} \text{pyruvate} + 2\text{Fe(CN)}_6^{4-} \qquad (44)$$

$$2\text{Fe(CN)}_6^{4-} \xrightarrow{\text{Pt}} 2\text{Fe(CN)}_6^{3-} + 2e^- + H^+ \qquad (45)$$

By monitoring the electrooxidation of the ferrocyanide produced
at +0.4 V versus SCE at a Pt electrode covered with a porous or jelled
layer of lactate dehydrogenase and a dialysis membrane, a current was
produced proportional to the concentration of lactic acid. About 3-
10 min was required for measurement. Because of the low K_m value of
this enzyme ($K_m = 1.2 \times 10^{-3}$ M), it was necessary to dilute the sample
with buffered ferricyanide. A linear plot was obtained over the range
10^{-4} to 10^{-3} M.

An enzyme electrode for lactate proposed by Durliat et al. [145,
211] consisted of a platinum disk fitted inside Visking tubing and
containing a mixture of yeast lactate dehydrogenase (as used by
Williams) and potassium ferrocyanide. Response was observed up to
6 mM lactate. A similar system was used by Guillot et al. [146],
using the Lactate Analyzer 5400. The Fe(CN)_6^{4-} produced in reaction
(44) is oxidized at the detector electrode [Eq. (45)] and the current is proportional to the lactate concentration.

Chen and Liu proposed a potentiometric electrode for L-lactate, using LDH immobilized in polyacrylamide gel placed onto a Pt electrode. A coupled reaction in which ferrocyanide is produced is used to measure the L-lactate produced [189] [Eqs. (46) and (47)].

$$\text{Lactate + NAD} \xrightarrow{\text{LDH}} \text{NADH + pyruvate} \tag{46}$$

$$\text{NADH + Fe(CN)}_6^{3-} \longrightarrow \text{Fe(CN)}_6^{4-} \tag{47}$$

Cheng and Christian [334] used a coupled enzymatic method to measure blood lactate by amperometric monitoring of the rate of O_2 depletion with a Clark O_2 electrode.

$$\text{Lactate + 2NAD} \xrightarrow{\text{LDH}} \text{2NADH} \tag{48}$$

$$\text{2NADH + 2O}_2 \xrightarrow{\text{HRPO}} \text{2H}_2\text{O + 2NAD} \tag{49}$$

The maximum rate of O_2 depletion is monitored by an O_2 membrane electrode. Linearity is 1-12 mM; no sample pretreatment is necessary.

Davies and Mosbach [192] bound coenzymatically active dextran-bound NAD together with immobilized lactate dehydrogenase or glutamate dehydrogenase to develop electrodes for L-glutamate (linear range 10^{-3} to 10^{-4} M) and pyruvate (linear range 2×10^{-5} to 8×10^{-4} M). The dextran-bound NAD was also incorporated into model enzyme reactors using galactose dehydrogenase and alanine dehydrogenase, or lactate dehydrogenase and alanine dehydrogenase, both of which produced a constant level of alanine over a period of 6.5 hr and 90 cycles.

An L-lactate selective electrode consisting of an immobilized lactate oxidase layer and a Clark oxygen electrode was described for the sequential determination of L-lactate and lactate dehydrogenase (LDH) in the same sample [340]. L-lactate (5×10^{-6} to 5×10^{-4} M in the final concentration) is determined from the decrease in the electrode current after the addition of the sample to a buffer solution. LDH (1-300 IU/liter in the final activity) is then determined from the decreasing rate of the current which is induced by the enzymatic L-lactate production after the addition of pyruvate and NADH to the sample-containing solution. The sequential determination is completed within about 7 min. The precisions are 1.4% for L-lactate and 2.6% for LDH. The electrode can be used for more than 2 weeks and 140 sequential determinations. Its application for human sera is also described.

Shinbo et al. [212] developed a potentiometric enzyme electrode for lactate. The electrode was constructed by coating the sensor membrane of the redox electrode with a film of enzyme-gelatin gel layer. The enzyme used was LDH, which catalyzes the oxidation of lactate by ferricyanide. The change in the concentration of ferricyanide to ferrocyanide is monitored by the redox electrode, and a plot of E versus log [lactate] is an S-shaped curve.

Similarly, electrodes have been constructed by Mindt et al. [131,132] and by Durliat et al. [149-151], by Guillot et al. [152], and by Racine et al. [148] using cytochrome b_2 lactate dehydrogenase and a redox sensor to monitor the various electroactive species produced in the enzymatic reaction.

Kulys and Svirmickas [213] developed a reagentless lactate sensor based on cytochrome b_2 LDH adsorbed onto the semiconducting complex N-methyl-phenazinium-7,7,8,8-tetracyanoquinodimethane held on a Pt electrode. At an applied potential of -0.03-0.4 V versus Ag/AgCl steady-state currents are reached in 0.5-0.7 min. Lactate can be determined in the range 10^{-5} to 10^{-3} M, and the sensor is useful for 3-9 days, depending on the source.

Malinauskas and Kulys [214] have described alcohol, lactate, and glutamate sensors based on oxidoreductases with regeneration of NAD. Flow-through electrodes were developed in which the oxidoreductases are mixed with immobilized NAD cofactor held between a Pt electrode and a semipermeable membrane. The cofactor was easily regenerated by electrochemical oxidation or by phenazine methosulfate (best). Calibration curves were linear to 0.5, 1.5, and 100 mM of glutamate, lactate, and ethanol, respectively. The sensitivity of the alcohol and lactate sensors decreased 50% within 60 hr; glutamate 50% in 6 hr.

An electrode system using chromium (III) hexaantipyrine hexa-
cynoferrate (III) has been described [243], with lactate dehydro-
genase-Cyt b_2 and a semipermeable membrane. A response time of
1-3 min was claimed for 95% response and linearity was observed
from 0-180 mg/dl.

An enzyme electrode system for determination of oxalate was
described by Kobos and Ramsey [215] and by Yao et al. [157]. Oxalate
decarboxylase is immobilized onto a CO_2 electrode; the CO_2 liberated
in the enzymatic reaction is then proportional to the concentration
of oxalic acid present. A plot of E versus log [oxalate] is linear
from 2×10^{-4} to 1×10^{-2} M with a slope of 57-60 mV/decade [215].
The limit of detection is 4×10^{-5} M. The electrodes are not affected
by phosphate and sulfate, and show no decrease in activity after 1
month of operation [215]. The recovery of oxalate added to five ali-
quots of a human control urine sample averaged 97.7% with a C.V. of
4.5%.

5. CHOLESTEROL

An enzyme electrode for free cholesterol was described by Satoh
et al. [122]. The electrode comprised double membrane layers, of
which one was a cholesterol oxidase-collagen membrane and the other
an oxygen-permeable membrane, a platinum and lead anode. The assay
involved monitoring the decrease in dissolved oxygen [Eq. (51)].

$$\text{Cholesterol esters} \xrightarrow{\text{esterase}} \text{cholesterol} \tag{50}$$

$$\text{Cholesterol} + O_2 \xrightarrow{\text{oxidase}} \Delta^4\text{-cholestenone-3} + H_2O_2 \tag{51}$$

An assay of total cholesterol esters was proposed by Huang et
al. [117]. Chemical immobilization of the enzymes cholesterol esterase and cholesterol oxidase onto alkylamine glass beads provided a
stable enzyme stirrer to completely convert all the total cholesterol
esters into first cholesterol [Eq. (50)], then hydrogen peroxide
[Eq. (51)], which is measured by the current flow at a Pt electrode.

Clark et al. [219] described a cholesterol electrode by use of
a polarographic anode with multiple enzymes. Cholesterol ester

hydrolase and cholesterol oxidase are used to produce peroxide, which
is sensed by a Pt anode. Combining these two enzymes it is possible
to obtain the benefits of enzyme specificity and devise a system that
requires only small plasma samples. Since the enzyme electrode response is measured by the rate of reaction, results are achieved
within 4 min. The Yellow Springs Model 23 glucose analyzer was used.
The cholesterol oxidase was glutaraldehyde-bound to a collagen membrane but the esterase was soluble. Only a gradual small reduction
in sensitivity was observed with time of the electrode. The recovery
was 97% and the C.V. 5% with a range of 25-300 mg% cholesterol.

Coulet and co-workers [194,195] described an enzyme electrode
using collagen immobilized cholesterol oxidase for the microdetermi-
nation of free cholesterol. The electrode is poised at a potential
of +650 mV versus Ag/AgCl and detects the H_2O_2 produced in the enzymatic reaction. Very high sensitivity and a wide range of linearity
(10^{-7} to 0.8×10^{-4} M) results. The use of a nonenzymatic electrode
associated with the enzymatic one allowed the detection of, and correction for, electrochemical interferences when applied to human
sera for free-cholesterol determinations.

By contrast the upper limit of the electrode of Satoh et al.
[122] was 2×10^{-4} M free cholesterol, and the electrode of Huang
et al. [117] was linear from 0 to 5 g/liter (2.5×10^{-3} M).

The use of Pt-based electrodes with cholesterol esterase and
oxidase together with measurement of the uptake of O_2 [Eqs. (50) and
(51)] were described by Dietschy et al. [118], Kamerro et al. [119],
Kumar and Christian [120], and Nonna and Nakajama [121].

Clark et al. [336] described a 1-min electrochemical assay for
cholesterol in biological materials. Ten microliters of sample is
injected into a 50° thermostated cuvet containing soluble cholesterol
oxidase and esterase, and the peroxide produced is determined with a
polarographic anode covered with an acetate/polycarbonate membrane
(which prevents ascorbic acid, uric acid, or bilirubin from being
detected). The YSI 23C analyzer is used. The linearity is 100-
500 mg/dl (similar to the electrode of Huang et al. [117]).

6. AMYGDALIN ELECTRODE

An electrode specific for amygdalin based on a solid-state
cyanide electrode was reported by Rechnitz and Llenado [67,68]. The
enzyme β-glucosidase, immobilized in acrylamide gel, was used:

$$\text{Amygdalin} \xrightarrow{\text{β-glucosidase}} \text{HCN} + 2C_6H_{12}O_6 + \text{benzaldehyde} \qquad (52)$$

A linear range of 5×10^{-3} to 10^{-5} was reported, with a slope of
about 40 mV/decade and a response time of about 10 min at concentrations of 10^{-2} and 10^{-3} M amygdalin and 30 min at 10^{-4} to 10^{-5} M.
The electrode rapidly lost activity, an indication that an incomplete
physical entrapment had been effected.

One reason for this long response time and poor stability was
the high pH used (10.4), a pH at which the enzyme has low activity
and is denatured. This was recognized by Mascini and Liberti [60],
who improved the response time and other electrode characteristics
by working at a pH of 7. The electrode was prepared by spreading
the enzyme directly onto the membrane surface and covering it with
a thin dialysis membrane. Since the enzyme was not immobilized, a
stability of less than a week is obtained, but the response time
was only about 1-2 min at 10^{-1} to 10^{-3} M and 6 min at 10^{-4} M. Furthermore, a linear calibration was obtained from 10^{-1} to 10^{-4} M with
a slope of 53 mV/decade (compared with about 40 mV/decade at pH 10).

7. PENICILLIN ELECTRODE

The first attempt at design of a penicillin electrode was made
by Papariello et al. [47]. The electrode was prepared by immobilizing penicillin-β-lactamase (penicillinase) in a thin membrane of
polyacrylamide gel molded around, and in intimate contact with, a
glass (H^+) electrode. The increase in hydrogen ion from the peni-
cilloic acid liberated from penicillin is measured:

$$\text{Penicillin} \xrightarrow{\text{penicillinase}} \text{penicilloic acid} \qquad (53)$$

The response time of the electrode was very fast (<30 sec) and had
a slope of 52 mV/decade over the range 5×10^{-2} to 10^{-4} M for sodium
ampicillin. The reproducibility of the electrode was very poor,

probably because no attempt was made to control the ionic strength and pH.

Mosbach et al. [46] prepared a penicillin electrode by entrapping penicillinase as a liquid layer trapped within a cellophane membrane around a glass (H$^+$) electrode, yet controlled the ionic strength and pH by using a weak 0.005 M phosphate buffer, pH 6.8, and 0.1 M NaCl. Good results were obtained, in comparison with the results of Papariello et al.; the calibration plot was linear from 10^{-2} to 10^{-3} M with a ΔpH of 1.4 and as little as 5 x 10^{-4} M sodium penicillin could be determined. The electrode could be stored for 3 weeks and the average deviation was ±2% with a response time of about 2-4 min.

Papariello and co-workers [158] reported a revised model of their original penicillin electrode. The authors claimed it was critical that a membrane be placed between the enzyme layer and the glass electrode to achieve satisfactory results.

Nilsson and other [62] used a penicillin-sensitive enzyme electrode to analyze the concentration of benzylpenicillin in fermentation broth. The electrode response time was in the region of 2 min, and the response to penicillin concentration was linear within the range of 1-10 mM. At low buffer capacity the sensitivity of the enzyme electrode to penicillin was very high. Constant calibration curves were obtained with the electrode when used for 2 hr daily in a fermentation medium over a 6-day period.

Similarly, Enfors and Nilsson [220] and Hewetson et al. [221] described the use of an immobilized penicillinase electrode in the monitoring of penicillin in fermentation broths. Enfors and Nilsson [220] described the purpose of their research as the construction of an autoclavable enzyme electrode, and to describe some characteristics of a penicillin electrode built according to this principle; the response time of the electrode was 1 min. Hewetson et al. [221] described the development of an on-line electrode suitable for use in a fermentation environment. Both groups used chemically bound (glutaraldehyde) enzyme on a flat-surface, glass pH electrode.

Olliff et al. [159] described electrodes sensitive to penicillin
with a response time of <2 min based on covalent linkage of penicillinase to the glass of a pH electrode.

8. CREATININE

A new kinetic method for potentiometric determination of creatinine in serum based on the creatinine-picrate reaction in alkaline
solution (Jaffe reaction) was described by Diamandis and Hadjiioannou
[222]. The reaction is monitored with a picrate electrode, and the
increase in electrode potential during 270 sec is measured and related
directly to the creatinine concentration. Small cation-exchange columns were used to separate creatinine from interfering substances.

Thompson and Rechnitz [123] described the use of unpurified
creatininase and an NH_4^+ probe for an electrode for creatinine, and
a creatinine enzyme electrode was described by Chen and Guilbault
[124].

$$\text{Creatinine} \xrightarrow{\text{creatininase}} \text{N-methylhydantoin} + NH_3 \qquad (54)$$

The enzyme creatininase was immobilized onto alkylamine glass beads,
then placed in a stirrer. The free NH_3 produced was measured with a
gas membrane electrode. Low levels of creatinine could not be measured because of the NH_4^+ present in the sample--these could be removed by several techniques described. An improved direct-reading,
specific electrode for creatinine was developed by Guilbault and
Coulet [334] using a new enzyme from Carla Erba. Linearity from
1 to 100 mg% was obtained.

An enzyme electrode system for the determination of creatinine
and creatine was developed by utilizing three enzymes: creatinine
amidohydrolase (CA), creatine amidinohydrolase (CI), and sarcosine
oxidase (SO) [341]. These enzymes were coimmobilized onto the porous
side of a cellulose acetate membrane with asymmetrical structure,
which has selective permeability to hydrogen peroxide. Two kinds of
multienzyme electrodes were constructed by combining a polarographic
electrode for sensing hydrogen peroxide and an immobilized CA/CI/SO
membrane or CI/SO membrane for creatinine plus creatine or just creatine, respectively. The multienzyme electrodes responded linearly

up to 100 mg of creatinine and creatine per liter in human serum.
Response time was 20 sec in the rate method and the detection limit
was 1 mg/liter. Only 25 μl of serum sample is required. Analytical
recoveries, precisions, and correlations with the Jaffe method were
excellent, and the multienzyme electrodes were sufficiently stable
to perform more than 500 assays. No loss of activity of immobilized
enzymes was observed after 9 months of storage at 4°C in air.

9. COFACTOR ELECTRODES

Glucose oxidase bienzyme electrodes for ATP, NAD, starch, and
dissacharides were described by Pfeiffer et al. [223]. Enzyme-
catalyzed reactions that produce glucose by hydrolysis of saccha-
rides, as well as glucose-consuming systems, have been coupled with
the glucose oxidase reaction. For example, coimmobilization of glucose oxidase with hexokinase or glucose dehydrogenase gives a sensor
that indicates the concentration of ATP or NAD present. The whole
measuring time for one sample is 45-60 sec and up to 1000 assays
can be performed with a single electrode.

A 5'-AMP-sensing electrode was described by Rechnitz and co-
workers [88,89] using a highly selective deaminase enzyme in conjunction with an ammonia gas-sensing membrane electrode. The resulting nucleotide sensor is very highly specific for 5'-AMP with
good operating sensitivity and dynamic response.

Scheller and Pfeiffer [90] proposed a glucose oxidase-hexokinase
bienzyme electrode sensor for adenosine triphosphate. Membranes were
produced by photopolymerization of acrylamide containing glucose oxidase and hexokinase, either together or separately. The combined
membrane(s) are fixed to the surface of an oxygen electrode. Best
selectivity is obtained with the combined electrode, which can also
be applied for an assay of ATP pyrophosphatase.

An adenosine (adenosine riboside) selective electrode has been
devised by Deng and Enke [224] using the enzyme adenosine deaminase
in conjunction with an ammonia gas electrode. The electrode is capable of detecting adenosine at the micromolar level at pH 9.0 and
37°C. Operating variables have been critically examined to define
conditions for optimum linearity and sensitivity.

A flow-through NADH sensor was described by Malinauskas and Kulys [225]. Alcohol dehydrogenase was immobilized in a flow-through voltammetric electrode. The enzyme activity was monitored with an O_2 electrode. The ADH was bound on Sepharose by glutaraldehyde attached to albumin. Phenazine methosulfate (PMS) was used to recycle NADH to NAD.

$$NAD + ethanol \xrightarrow{\text{ADH}} pyruvate + NADH \tag{55}$$
$$\xleftarrow{\hspace{3cm}PMS\hspace{3cm}}$$

A newly developed [88,89] AMP electrode was used by Riechel and Rechnitz [226] to make direct binding measurements of the allosteric reaction between AMP and fructose-1,6-diphosphatase. The proposed technique was based on the ability of the enzyme electrode to distinguish between free and bound nucleotide.

10. ACETYLCHOLINE ELECTRODES

Baum proposed liquid membrane electrodes for acetylcholine and choline [83,85] that can form the basis of enzyme electrodes. By using acetylcholinesterase on a choline electrode, for example, an acetylcholine electrode results [83,84]. By placing the acetylcholinesterase on an acetylcholine electrode, one achieves an electrode for acetyl-β-methylcholine [86,87]. Tran-Minh and Beaux [227] used immobilized acetylcholinesterase on a pH electrode to develop a sensor for acetylcholine (from 10^{-4} to 10^{-2} M concentrations are determinable, with a 2-min response time). A similar technique was described by Durand et al. [227]. The acetylcholinesterase was immobilized in gelatin and then treated with glutaraldehyde to cross-link. Linearity was 10^{-5} to 10^{-4} M, with good stability over 35 days.

11. OTHER ELECTRODES

A guanine electrode based on guanase used with an ammonia gas membrane electrode was described by Nikolelis et al. [143]. Guanine in the range 10^{-4} to 10^{-2} M gives a linear Nernstian plot with a response time of 1.5-4 min. The results agree favorably with the xanthine oxidase method.

In a similar application, Gulberg and Christian [144] developed an immobilized xanthine oxidase system for the determination of guanase. The xanthine produced from guanine by the guanine deaminase reaction is passed through a tube containing xanthine oxidase immobilized on silica or titania. The resulting H_2O_2 is caused to react with KI in the presence of Mo(VI); the liberated I_3^- is measured amperometrically. A single determination took 15 min.

An enzyme electrode consisting of a monoamine oxidase-collagen membrane and an oxygen electrode was prepared for the determination of monoamines [244]. The monoamines are oxidized to aldehydes by the enzyme, and the consumption of O_2 is monitored amperometrically by the O_2 electrode. A linear relationship of current and tyramine concentration was observed in the range 50-200 μM.

A glutamine selective sensor consisting of glutaminase (porcine kidney tissue) immobilized at an ammonia gas electrode was proposed by Arnold and Rechnitz [229]. A linear range of 10^{-2} to 10^{-4} M was observed. Arnold and Rechnitz [230] compared bacterial, mitochondrial, tissue, and enzyme biocatalysts for glutamine-selective membrane electrodes. The electrode system using tissue slices had the most favorable combination of electrode properties and operating requirements of the electrodes tested.

The application of an immobilized enzyme electrode to inhibitor screening of β-D-glucose oxidase was described by Imai and Kawauchi [231]. Glucose oxidase was immobilized on the pyroxyline membrane of an oxygen electrode, which was used for inhibitor screening of the enzyme in vitro. Semicarbazide and nitrite showed competitive inhibition.

The principle of competitive binding assay in combination with an immobilized lectin in close proximity to an oxygen sensor has been used to quantify carbohydrates and to determine association constants for lectin-carbohydrate interactions as described by Borrebaeck and Mattiasson [232].

A novel application of the enzyme electrodes has been for the selective assay of anions and heavy-metal ions. Ogren and Johansson

[233] reported the determination of traces of mercury(II) by inhibition of an enzyme reactor electrode loaded with immobilized urease.
Similarly, F^- has been determined by inhibition of urease in an
electrode [174,175], and nitrite and semicarbazide by inhibition of
glucose oxidase [165].

The use of immobilized apoenzyme for quantitation of the cofactor needed for enzymatic activity was described by Mattiasson
et al. [234]. The sensitive part of an oxygen electrode was covered
by a nylon net onto which the apoenzyme was immobilized. The electrode response was proportional to the amount of apoenzyme activated
by the cofactor, resulting in a response proportional to the concentration of cofactor.

In the system discussed here Cu^{2+} was measured by the tyrosinase
apoenzyme; as little as 50 ppb was determinable.

G. Review

A number of excellent reviews have been written on immobilized
enzyme electrode probes and their applications. A list of these
references is given at the end of Chapter 2.

A flow-through electrochemical cell was constructed by Blaedel
and Wang [235] to evaluate an enzyme-porous electrode combination
composed of a Sepharose-bound enzyme which is packed into a reticu-
lated vitreous carbon (RVC) disk. Rapid response (15 sec) is obtained due to the intimate contact between the enzyme and the electrode surface. The characteristics of this cell with the use of
pulsed rotation voltammetry (PRV) to discriminate against major
background currents was described in a later publication [236].

A mathematical deviation for the exact solution of partial
differential equations describing the steady-state response of biocatalytic potentiometric membrane electrodes was described by Haneka
and Rechnitz [237]. The derivation is based on a previously proposed
model in which the steady-state response is seen to result from a
combination of diffusion and Michaelis-Menten kinetic steps. The
expressions derived were believed to provide further insight into

the behavior of biocatalytic membrane electrodes and permit convenient evaluation of key parameters as a function of major experimental variables.

A digital simulation to model the amperometric response of an
immobilized enzyme electrode following periods of electrode inac-
tivity was described by Mell and Maloy [238]. Agreement between
simulation theory and experimental behavior was observed using a
glucose oxidase electrode meeting diffusion-controlled criteria
established by the simulation model.

Carr [239] described the use of Fourier analysis of the partial
differential equations which govern the response of potentiometric
enzyme electrodes to determine the transient response behavior of
the electrode potential. When the enzyme reaction is first order,
the response was found to be governed by two dimensionless terms,
Dt/L^2 and $K_m D/VL^2$ (D = diffusion coefficient of the substrate,
L = thickness of the enzyme layer, K_m = Michaelis constant, t = time).
When the specific enzyme activities are reasonable, the electrode
will be within a millivolt of its final value when Dt/L^2 is greater
than 1.42. In the analytically important region the response time
is virtually independent of the specific activity on the electrode.
The behavior in the limit of a zero order reaction was also studied.
In this case, the enzyme concentration has no influence whatsoever
on the transient behavior, which is governed entirely by diffusion,
but it does influence the steady-state potential.

A kinetic analysis of a urea electrode was performed by Ollis
and Carter [240]. The purpose of the paper was to analyze the membrane concentration profiles in enzyme electrodes in order to predict
the design parameters for substrate and inhibitor determinations.
The complete kinetics of the soluble and immobilized urease system
were considered and compared with experimental data. An earlier
enzyme electrode analysis of Callmann [241] was used by substituting
urease kinetics for those of glucose oxidase. The availability of
experimental urease electrode data offers a much stronger test of
the electrode model, and the experimental data of the Guilbault
[53] urea electrode fit the theoretical predictions offered.

Pederson and Chotani [242] presented a theoretical model of
diffusion and reaction in an anisotropic enzyme membrane with partial emphasis on the application of such membranes in enzyme electrodes. The dynamic response of systems having linear kinetics
(which is the prime practical operational area for enzyme electrodes
in analysis) was investigated via an analytical solution of the gov-
erning differential equations. The response of the electrode was
presented as a function of a single dimensionless group μ, which is
the membrane modulus. The authors believed that in light of the
successful use of anisotropic enzyme electrodes for glucose detection [139,140], their model could be of value in optimizing the
electrode performance as well as for the design of other electrode
systems employing amperometric detection. The theoretical results
could be of interest to those working on model membrane systems
where heterogeneous enzymatic reactions take place.

IV. PROBES FOR ANALYSIS OF ENZYMES

In addition to all the work described in Sec. III, devoted to the
development of electrode probes for substrates, both organic and
inorganic, considerable effort has been devoted to the development
of electrodes for the assay of the enzymes themselves. A list of
these is presented in Table 5. Probes for the monitoring of over
25 different enzymes have been described. However, almost all the
work was performed with soluble rather than immobilized substrates.

    To develop a truly self-contained, enzyme-sensing "substrate
probe" would require an insolubilized substrate, something difficult
to achieve because of the small size of the substrate molecule.
Moreover, it is more difficult to maintain such a probe because
(1) the substrate, unlike the enzyme, is used up and must be replenished; (2) the enzyme must be monitored using a kinetic, not equilibrium, method.

    Nevertheless, several papers have been devoted to the concept
of a true substrate probe. In one Guilbault and Iwase [257] bound
the substrate acetylcholine to an anion-exchange resin by electrostatic charge attraction. The pH change upon enzymatic reaction is

then monitored and related to the cholinesterase content of the
solution. The substrate membrane had to be rejuvenated after every
20-25 assays by treating the anion exchanger with fresh substrate.
The system was based on the earlier work of Guilbault and Gibson
[99], who observed that this enzyme-catalyzed reaction could result
in a linear change of pH with time; 40 samples/hr could be processed
in an automated system. Crochet and Montalvo [256] described a system for assay of cholinesterase in which the substrate, acetylcholine, was continuously pumped in a thin layer over the surface of
a pH electrode. A pseudolinear curve was obtained from 10-70 U/ml
of cholinesterase and assays could be performed in 1.5-4.5 min.

Gebauer et al. [255] described the use of enzyme electrodes
for urea, AMP, and adenosine for kinetic assays of arginase, 3',5'-
cyclic nucleotide phosphodiesterase and 5'-nucleotidase, respectively.
The initial rate of potential change after addition of enzyme is
directly proportional to the enzyme activity present. Arginase
assays were found to be reproducible to a relative precision of 13%
or better, and applicable to the direct measurement of enzyme content of beef liver homogenates. The enzyme electrodes for urea,
AMP, and adenosine were prepared by placing a thin layer of urease,
AMP deaminase, or adenosine deaminase between the gas membrane of
an Orion NH_3 electrode.

Booker and Haslem [251] described an immobilized enzyme electrode for the determination of arginase. The coupled reactions:

$$\text{L-arginine} \xrightarrow{\text{arginase}} \text{L-ornithine + urea} \tag{56}$$

$$\text{Urea} \xrightarrow[\text{urease}]{\text{immobilized}} 2NH_4^+ + HCO_3^- \tag{57}$$

were used, monitored by a cation-selective electrode responsive to
NH_4^+ covered with a layer of urease enzyme. A precision of 3%, with
a range of analysis of 1.6-16 U of arginase, and an analysis time
of <10 min was observed. The change of potential with time is measured and the amount of arginase determined from a calibration curve.

Yoshino et al. [218] described a rapid measurement procedure
for amylase based on use of a glucose sensor. The system is based

TABLE 5 Electrode Probes for Enzymes

Enzyme	Substrate	Sensor	Ref.
O-Acetylserine sulfydrylase	O-Acetyl-L-serine	S^{2-}	245
Alcohol oxidase	Ethanol	C (NADH)	246,247
Amine oxidase	Primary amine	C [H_2O_2, $Fe(CN)_6^{3-}$]	248
L-Amino acid oxidase	L-Phenylalanine	Pt (O_2)	249
		Gas (O_2)	250
Amylase	Starch	Glucose	218
		Pt (H_2O_2)	273
Arginase	L-Arginine	NH_4^+	251
		NH_3	252
		Urea	255
Asparaginase	L-Asparagine	Glass (NH_4^+)	253,254
Cholinesterase	Acetylcholine	ACh^+	83
		H^+	99,256,257
	Acetylthiocholine	S^{2-}	24
Chymotrypsin	Diphenylcarbamylfluoride	F^-	258
Citrate synthase	Coenzyme A	Hg	259
3',5'-Cyclic nucleotide phosphodiesterase	AMP	AMP	255

Enzyme	Substrate	Species	Ref.
Glucose oxidase	Glucose	Pt (H_2O_2)	260,261
β-Glucosidase	Amygdalin	CN^-	261,262
		Ag $[CN]_2^-$	263
	Phenyl-β-glucopyranoside	O_2 (phenol)	264
β-Glucuronidase	Phenyl-β-D-glucuroniside	O_2 (phenol)	264
Gluconolactonase	Gluconic acid	H^+	265
Glutaminase	Glutamine	NH_4^+	266
Hydrogenase	H_2	C (Me viologen)	267
Lactate dehydrogenase	Lactate	C	268
		Pt	129
Lysozyme	*Micrococcus lysodeikticus*	$TMPA^+$	269
Malate dehydrogenase	Malate	C	268
5'-Nucleotidase	Adenosine	Adenosine	255
Phosphatase (alkaline)	Phenol phosphate	Gas (O_2)	270
		O_2 (phenol)	264
Protease	Casein	L-Leucine	271
Rhodanase	$S_2O_3^{2-}$	CN^-	261
Urease	Urea	NH_4^+	272
		NH_3	252

on the conversion of maltopentose into maltose and maltotriose,
followed by conversion with α-glucosidase of maltose to D-glucose.
The glucose is sensed by the immobilized glucose oxidase membrane
electrode, based on the H_2O_2 produced. Thus a current flows which
is proportional to the α-amylase activity. The C.V. is 5.3% at
727 U/liter and 4.4% at 1030 U/liter. From 30 to 33 samples/hr can
be processed, and only 20 μl of blood is required.

Barbarino et al. [273] developed a similar system for assay of
α-amylase utilizing an immobilized substrate (starch) and three or
more discrete immobilized enzymes (one column of glucose oxidase,
catalase, glucoamylase, or maltase and one of glucose oxidase). The
resulting H_2O_2 is detected by a three-electrode amperometric cell.
All immobilized reagents were placed on particulate, porous alumina
to allow rapid and constant flow rate. From 5 to 200 kU/liter of
amylase can be assayed in only a 50-μl sample.

Arwin and Lundstrom [274] described some different ways to use
adsorption of molecules on electrodes to measure enzymatic activity.
The method is based on use of a synthetic substrate (S-2160), which
has the ability to adsorb on metal electrodes inserted into a buffer.
The molecules adsorb in such a way that the electrodes are effectively screened from the ions in the buffer. The adsorption gives
rises to a large change in electrode capacitance, which is easily
measured with a capacitance bridge. The method has been applied
so far to reactions where the substrate is hydrolyzed by an enzyme:

$$\text{Substrate} + H_2O \longrightarrow \text{Part 1} + \text{Part 2} \tag{58}$$
$$\text{(S-2160)}$$

where

S-2160 = Bz-Phe-Val-Arg-NH-⟨○⟩-NO_2

Montalvo [272] designed an electrode for urease by continually
passing a layer of soluble urea between the tip of a NH_4^+ cation
electrode and a dialysis membrane. Urea diffuses through the membrane and is hydrolyzed by urease in the solution. The ammonium
ion diffuses back through the membrane to the cation electrode,
where it is sensed.

V. ELECTRODE PROBES UTILIZING WHOLE CELLS:
    MICROBIAL OR TISSUE ENZYMES

A. General Discussion

One of the newest areas in biological electrode probes has been the
application of whole cell microorganisms or tissue cells to the surface of an I.S.E. to form a bioselective sensor. Divies [307] in
1975 used bacteria with an electrochemical assay for ethanol, but
the first potentiometric microbial probe was reported by Divies
[308] in 1976.

Several excellent reviews of this area have been written by
Suzuki et al. (Biosensors in Japan) [279], Kobos [280], and Rechnitz
[281]. A rather complete coverage of research through 1982 is presented herein.

The use of microbes for electrode probes offers three main
advantages:

1. Purified enzymes are not necessary; the tissue slice or
    whole cell could be used directly without extensive puri-
    fication and separation steps.

2. The electrode can be regenerated by immersion in nutrient
    broth. The microbe is essentially "living" and can be fed
    and kept alive for long periods.

3. The whole cell can contain many enzymes and several cofac-
    tors that can catalyze extensive transformations that could
    be difficult, if not possible, to effect with single immo-
    bilized enzymes. In addition, the cofactors necessary for
    enzymatic reaction are held in a natural immobilized state.

Disadvantages include:

1. Poor selectivity can result because the bacteria or microbe
    contains several enzymes that can convert many different
    substrates in addition to the one desired.

2. Poor response times are often observed because the enzymes
    in the microbe are present at low concentrations and the
    electrode membrane is very thick subject to slow diffusion
    processes.

As an example of these advantages and disadvantages, consider first the aspartate electrode of Rechnitz and Kobos [310]. This bacterial sensor was stable for 10 days, using the ability of the bacterial colonies to regenerate themselves in appropriate growth media; the purified aspartate ammonia lyase was stable for only 3 days. Thus, the growth and replenishment of the inactive cells must be the explanation for the increased lifetime of the probe. However, typically 10-20 mg (10^9 cells) are placed on each electrode for good response; if only 1-2 mg of bacteria is used, very long response times of 20 min are encountered.

The arginine electrode [308,309] produced illustrates the lack of selectivity of the bacterial probe; not only arginine, but glutamine and asparagine also respond. This is typical of a bacterial probe because frequently many enzymes are present in the cell. *E. coli* has several decarboxylases present in the cell that could catalyze the decarboxylation of numerous amino acids. Purification can lead to great specificity for only one amino acid [108]. Some bacterial electrodes, on the other hand, are highly selective. Consider the glutamine electrode, which is based on the activity of glutamine deaminase present in the bacteria [230,303]; no interference was observed from alanine, arginine, asparagine, aspartic acid, histidine, or several other amino acids or amines tested.

An excellent example of the complex reaction sequences that can be catalyzed by the many enzymes and cofactors present in the microbe is the research of Kobos and Pyon [302], who described an electrode for nitrilotriacetic acid (NTA). The bacterial cells carry out a four-step reaction sequence converting the NTA to ammonia. The greatest activity was obtained with bacterial cells harvested in the exponential growth phase; good sensitivity was obtained, but again there were considerable interferences with the probe that could severely limit its usefulness.

A fairly complete list of all bacterial- and tissue-based electrodes is given in Table 6. Over 30 different electrode probes using whole cells have been described. The characteristics of some selected microbial sensors are presented in Table 7.

Essentially the sensors fall into two categories: (1) Those
that are based on an uptake of O_2 in the respiratory process (I of
Fig. 11).   In this case the microbial sensor is constructed by placing the immobilized microorganisms directly onto O_2 electrode.   (2)
The amperometric or potentiometric determination of electroactive
products liberated in the enzymatic reaction (II of Fig. 11).   Such
base electrodes could be pH, NH_3, H_2S, CO_2, or lactate, onto which
is placed a layer of the tissue or bacterial cells.

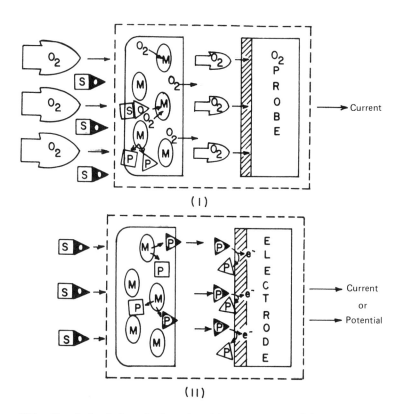

FIG. 11 Principle of the microbial sensor.   (I) Amperometric determination of respiration activity.   (II) Amperometric or potentiometric determination of metabolites (electroactive substances).
⑤▶, substrate; ℗ ℗▷, products (electrochemically inactive); ℗▷,
product (electrochemically active); Ⓜ, immobilized microorganism.

TABLE 6 Electrode Probes Using Microbial and Tissue Cells

Substrate	Species	Probe	Ref.
Acetic acid	T. brassicae	O_2	279
Adenosine	Mouse mucosal tissue	NH_3	315
Alcohols	T. brassicae	O_2	282,307
Ammonia	Nitrosomonas sp	O_2	283
Antidiuretic hormone	Toad bladder	Na^+	314
Arginine	Bacterial cells	NH_3	308,309
Aspartic acid	B. cadaveris	NH_3	310
BOD	Microbes (Clostridium butyricum, etc.)	O_2	284-286
	Yeast	O_2	287
Cephalosporin	C. freundii	pH	279,300
Cholesterol	Nocardia erythropolis	O_2	288
L-Cysteine	Proteus morgani	H_2S	289
Ethanol	Hansenula anomala	O_2	287
Formic acid	C. butyricum	Fuel cell	279
Glucose	P. fluorescens	O_2	279
	Hansenula anomala	O_2	287
Glucose, fructose, sucrose	Brevibacterium lactofermentum	O_2	290
L-Glutamic acid	E. coli	CO_2	292
	Yellow squash	CO_2	293
L-Glutamine	Porcine kidney	NH_3	230,291,303, 311,313

Histidine	Sarcina flava	NH$_3$	230,303,312
L-Lactate	Pseudomonas sp	NH$_3$	294
	Hansenula anomala	O$_2$	287
Lysozyme	Micrococcus lysodeikticus	TMPA$^+$	295
Microbes (assimilation)	Microorganisms	O$_2$	296
NAD	N. crassa + E. coli	NH$_3$	297
NADH	Mitochondria	O$_2$	298
Nicotinic acid	Lactobacillus arabinosus	pH	299
Nitrate	Azotobacter vinelandii	NH$_3$	301
Nitrilotriacetic acid	Pseudomonas sp	NH$_3$	302
Nystatin	S. cervisiae	O$_2$	279
Phenol	Trichosporon cutaneum	O$_2$	162
Phenylalanine	L. mesenteroides	Lactate	304
Pyruvate	Hansenula anomala	O$_2$	287
L-Serine	C. acidiurici	NH$_3$	305
Steroids	Nocardia opaca	C	306
Sucrose	Hansenula anomala	O$_2$	287
Sugars	Brevibacterium lactofermentum	O$_2$	290
	Dental plaque	pH	316
Vitamin B	L. fermenti	Fuel cell	279
	Yeast	O$_2$	333

TABLE 7 Characteristics of Microbial Sensors

Sensor	Species	Electrode	Response time (min)	Stability (days)	Range (ng/liter)
Acetic acid	*T. brassicae*	O_2	15	30	10-100
Ammonia	*N. europea*	O_2	8	15	0.2-1.2
Arginine	*S. faecium*	NH_3	10	21	2-200
Aspartate	*B. cadaveris*	NH_3	20	10	0-20
BOD	*T. cutaneum*	O_2	20	17	0-60
Cephalosporin	*C. freundii*	pH	10	7	60-500
Cholesterol	*N. erythropolis*	O_2	1	30	3-40
Ethanol	*T. brassicae*	O_2	10 steady state 6 pulse method	30	3-22.5
Formic acid	*C. butyricum*	Fuel cell	20	20	10-1000
Glucose	*P. fluorescens*	O_2	10	14	2-20
Glutamic acid	*E. coli*	CO_2	5	15	8-800
Glutamine	Porcine kidney	NH_3	5-7	30	3-300
Nicotinic acid	*L. arabinosus*	pH	60	30	0.05-5
Nitrate	*A. vinelandii*	NH_3	10	30	0.6-50
Nystatin	*S. cervisiae*	O_2	60	20	0.5-80
Phenol	*T. cutaneum*	O_2	0.25	5	0-15
Phenylalanine	*L. Mesenteroides*	Lactate	90	20	1-500
Sugars	*B. lactofermentum*	O_2	10	20	20-200
Vitamin B_1	*L. fermenti*	Fuel cell	360	60	1-500
	Yeast	O_2	3	5	0.01-0.5

B. Applications and Types

1. TISSUE MEMBRANES

The first electrode to effectively demonstrate the concept of use of tissue slices to form a bioprobe was in 1978 (311), a crude combination of beef liver tissue and separate urease to yield an arginine electrode:

$$\text{L-Arginine} \xrightarrow{\text{liver tissue}} \text{urea + ornithine} \tag{59}$$

$$\text{Urea} \xrightarrow{\text{urease}} CO_2 + NH_3 \tag{60}$$

In 1979, Rechnitz et al. [313] described the first complete tissue electrode using a slice from the cortex of porcine kidney, held at the surface of an ammonia gas electrode by a dialysis membrane. The kidney tissue catalyzes reactions (59) and (60) rather selectively (no response from L-alanine, L-arginine, L-histidine, L-glutamine acid, L-asparagine, L-aspartic acid, glycine, urea, creatinine, L-serine, or L-valine). The slope was 50 mV/decade over 6×10^{-5} to 6.7×10^{-3} M with a detection limit of 2×10^{-5} M. The response time was 5-7 min. Using sodium azide as a preservative (0.02%) assures lifetimes of >30 days.

Arnold and Rechnitz [291] applied this tissue-based electrode to the analysis of glutamine in cerebrospinal fluid. Good precision and accuracy over the range 10^{-4} to 10^{-2} M was obtained, with a detection limit of 2.9×10^{-5} M. The authors cited advantages of speed, convenience, and low cost by eliminating the need for fresh enzyme each measurement.

Mascini and Rechnitz [303] used porcine kidney tissue and *Sarcina flava* bacterial cells as biocatalysts for assay of glutamine, again with an NH_3 gas membrane electrode. Conversion to ammonia is 100% for $<10^{-4}$ M glutamine. Tubular reactors were used in a flow stream with detection of NH_3 liberated with the electrode.

A comparison of bacterial, mitochondrial, tissue, and enzyme biocatalysts for glutamine-selective membrane electrodes was made by Arnold and Rechnitz [230]. All the biocatalysts were immobilized at ammonia gas-sensing potentiometric electrodes and were compared in terms of electrode properties and operating requirements. *Sarcina flava* strain, porcine kidney cortex slices and kidney mitochondrial fractions were all employed. The electrode using tissue slices was found to have advantages of low cost, superior mechanical and time stability, and ease of preparation.

A plant tissue-based bioselective membrane electrode for L-glutamate was described by Kuriyama and Rechnitz [293]. The probe was constructed by placing a slice of yellow squash tissue at a CO_2 gas sensor, and represented the first successful use of intact plant materials as biocatalysts in the construction of bioselective poten-tiometric membrane electrodes. A range of 2×10^{-4} to 1.3×10^{-2} M was observed, and the probe had excellent selectivity characteristics on over 25 possible interferences tested. Good reproducibility and a lifetime of 7 days was observed. Pyridoxal-5'-phosphate was used as cofactor.

Updike and Treichel [314] described an electrode with response to antidiuretic hormone (ADH) by stretching a toad bladder tissue over the surface of a sodium ion-sensing glass electrode. The ADH effects an enhanced transport of Na^+ across the toad bladder, which is moni-tored by the Na^+ electrode. The response time is only 10 sec, but there are serious interferences. Other hormones, such as aldosterone, thyroxin, and angiotensin, as well as AMP, K^+, and Ca^{2+}, also effect the Na^+ transport, so the concept has only limited practical useful-ness.

The use of a mouse small intestine mucosal tissue selectively enhances the selectivity of an adenosine-sensing electrode, as de-scribed by Arnold and Rechnitz [315]. The response to phosphate nucleotides, due to alkaline phosphatase activity present, is sup-pressed by glycerol phosphate or an inhibitor like L-phenylalanine. The adenosine deaminase activity in the tissue then can act selec-tively on adenosine; the NH_3 liberated is monitored.

2.   *BACTERIAL ELECTRODES*

a.   *General Discussion*

The cells of microorganisms contain all the enzyme systems and cofactors necessary to effect vital transformations of substrates

to products. Stability and regeneration are characteristic of these
systems. In the cell the conversion of substrates occurs under the
action of multienzyme complexes and cofactors already present and is
easily performed. For these reasons considerable interest has developed, most of it within the last 5 years, on the combination of these
whole cells with electrode processes in analytical systems.

Historically, the first work in this area dates back to Divies
in 1975 and 1976 [307,308] on ethanol and arginine electrodes. A
layer of bacterial cells was placed at the surface of a gas-sensing
potentiometric probe. Most of the subsequent research on bacterial
probes centers on the use of selected cultural strains. A listing
of the many electrodes developed and their properties appears in
Tables 6 and 7.

b. *Amines and Amino Acids*

We have already mentioned the bacterial [230,303] and mitochondrial [230] glutamine-selective electrodes developed by Rechnitz
et al. An ammonia electrode was used as the base sensor. The bacterial cells of *Sarcina flava* are held at the surface of the NH_3
probe by means of a dialysis membrane. Using only 10-15 μl, 10^8-
10^9 living bacteria are present. The bacterial cells are easily
obtained by culture under sterile conditions, harvesting and washing,
followed by compacting from suspension by gentle centrifugation. The
cells are spread manually on the surface of the electrode. The electrode is quite selective for L-glutamine in the presence of other
amino acids. Lifetimes of >20 days are easily obtained.

Suzuki and his group have pioneered in the development of microbial electrodes. In one paper [292] a sensor for glutamic acid was
described, based on immobilized *E. coli* (which contain glutamate
decarboxylase) placed on a CO_2 electrode. Continuous introduction
of sample solution into a flow system incorporating the sensor gives
a steady potential in 5 min. Linear calibration plots are obtained
from 100 to 800 mg/liter. The sensor was observed to be stable,
selective, and reproducible, and was used for assays in fermentation
broths. In a second paper [304], Suzuki et al. described an electrode for L-phenylalanine using immobilized *Leuconostoc mesenteroides*

and a lactate electrode. The immobilized microorganisms can grow
and are active for a month in a polymer gel matrix [313]. A lactate
electrode is used to follow the reaction.

$$\text{L-Phenylalanine} \xrightarrow{\textit{L. mesenteroides}} \text{L-lactate} \qquad\qquad (61)$$

Linearity is observed from 10^{-4} to 5×10^{-2} g/liter.

Buck et al. [294] described a *Pseudomonas* bacterial electrode
for the assay of L-histidine. The bacterium is placed on an ammonia
gas-sensing electrode, where it converts one molecule of histidine
to two molecules of NH_3. The response of the electrode is linear
between 10^{-4} and 3×10^{-3} M histidine with a slope of 48-53 mV/decade.
The sensor has a lifetime of 3 weeks. A sterilizing filter is used
to immobilize the bacteria. Both the slope and the response of the
probe become more ideal with time. Urocanic acid and urea are major
interferences; arginine, tryptophan, and glycine are minor interferences. Lysine and glucose did not interfere, but ammonia and some
nitrogen-containing buffers are also interferences. In contrast to
these observations, see the work of Buck et al. cited [185] in which
purified histidine ammonia lyase is used. Only ammonia interferes
(as expected because the base sensor is an ammonia electrode). A
L-serine electrode was constructed using intact cells of *Clostridium*
acidiurici immobilized on an NH_3 sensor [305]. This is an example
of an anaerobic system.

The L-aspartate electrode [310] developed by Kobos and Rechnitz
has been described above. It utilizes immobilized *Bacterium cada-*
veris, whose biocatalytic activity could be regenerated by placing
the spent electrode backed into the same nutrient growth medium used
for culturing the strain. Fresh cells are grown in situ. The strain
Proteus morganii placed at a hydrogen sulfide gas electrode [289]
forms the basis for the L-cysteine electrode described by Jansen and
Rechnitz.

    c. *Carbohydrates and Vitamins*

A microbial sensor consisting of immobilized whole cells of
Brevibacterium lactofermentum and an oxygen electrode was prepared

by Suzuki and co-workers [290] for the determination of sugars
(glucose, fructose, and sucrose) in a fermentation broth for glutamic acid production. Total assimilable sugars were determined
from the extent of oxygen consumption by the immobilized microorganisms. The response time is about 10 min by the steady-state
and 1 min by the pulse method. A linear relationship was found for
glucose (up to 1 mM), fructose (up to 1 mM), and sucrose (<0.8 mM).
A reproducibility of 2% was observed, and the sensor could be used
for more than 960 assays or 10 days.

Kulys and Kadziauskiene [287] described a sensor for D-glucose,
L-lactate, pyruvate, sucrose, and ethanol based on the cells from
the yeast *Hansenula anomala*; the rate of conversion of metabolites
is determined by an O_2 electrode. In cells grown on a lactate media
as carbon source, an increased sensitivity to lactate is observed,
with lower response to pyruvate and glucose, and no response to
sucrose and ethanol. In a general growth media the concentration
ranges assayable are D-glucose (0.01-0.9 mM), L-lactate (0.01-0.2
mM), pyruvate (0.04-0.1 mM), sucrose (0.1-4 mM), and ethanol (1-
10 mM).

Suzuki et al. [279] also described an electrode for glucose in
the 2-20 mg/liter range using *Pseudomonas fluorescens* and an oxygen
electrode. The response time is about 10 min, and stability is good.
A vitamin B_1- (thiamine) sensitive electrode has been devised by combining an oxygen electrode with a yeast-containing membrane [333].
The assembly was used for assaying thiamine at concentrations down
to 10^{-11} g/liter. The analytical procedure developed should allow
the measurement of 10-20 samples/hr. The performance of the yeast
electrode was improved when alginate membranes reinforced with a
nylon network were used. An apparatus for preparing such membranes
was described together with a magnetic membrane holder facilitating
handling of membranes in combination with electrodes.

d. *Alcohols and Steroids*

A microbial electrode consisting of immobilized microorganisms,
a gas-permeable Teflon membrane, and an oxygen electrode was prepared

for the continuous determination of methyl and ethyl alcohols by
Suzuki et al. [282]. Immobilized *Trichosporon brassicae* was employed
for a microbial sensor for ethyl alcohol. When a sample solution
containing this substrate was injected into the microbial electrode
system, the current decreased markedly until a steady state was
reached. The response time was within 10 min by the steady-state
method and 6 min by the pulse method. A linear relationship between
decrease of the current and the concentration of ethyl alcohols is
observed up to 22.5 mg/liter with a reproducibility of ±3%. The
selectivity was good, and the microbial sensor was applied to a fermentation broth of yeasts with excellent results (correlation coefficient .98). The electrode could be used for 30 days and 2100 assays.
A microbial electrode sensor using immobilized bacterium AJ3993 for
methanol (O_2 electrode, linear range 2-22.5 mg/liter, response time
2 min, and stability 10 days) was also described.

An ethanol sensor using *Hansenula anomala* [287] has also been
described.

Neujahr and Kjellan [162] used a Clark oxygen electrode coated
with a thin paste of induced *Trichosporon cutaneum* to determine
phenol in the range of 0-15 mg/liter. The assay takes only 15 sec
after addition of sample, and the bioprobe is stable for 100 assays
or 5 days.

A microbial membrane electrode for steroid assay was described
by Wollenburger et al. [306]. The sensor for corticoids, gestogens,
and androgens is prepared by attaching a layer of *Nocardia opaca*
cells immobilized in polyacrylamide gel to the surface of a rotating
C electrode. Similarly, a microbial electrode using immobilized
Nocardia erythropolis [288] has been described for the determination
of cholesterol. The relationship between the measuring signal and
the cholesterol concentration was linear from 0.015-0.13 mM, with a
reproducibility of 2-7%. To shorten the measuring time the derivative of the current-time curve was monitored. The response time was
35-70 sec, and the sensor was stable for more than 4 weeks.

e. Inorganic Compounds

A novel potentiometric sensor for nitrate was developed by Kobos et al. [301] by coupling the bacterium *Azotobacter vinelandi* with an ammonia gas electrode. Nitrate is reduced to NH_3 by the two-step process involving the enzymes nitrate and nitrite reductase contained in the bacterial cells [Eqs. (62) and (63)]:

$$NO_3^- + NADH \xrightarrow[\text{reductase}]{\text{nitrate}} NO_2^- + NAD + H_2O \tag{62}$$

$$NO_2^- + 3NADH \xrightarrow[\text{reductase}]{\text{nitrite}} NH_3 + 3NAD + 2H_2O \tag{63}$$

Also present in the cell are several cofactors (including NADH) that enable this highly complex reduction to occur. Linearity is from 10^{-5} to 8×10^{-4} M with a slope of 45-50 mV/decade. The precision and accuracy is 3-4%. In this case the use of the whole cells simplifies the preparation of the bioselective electrode.

Hikuma et al. [283] described an ammonia electrode with immobilized nitrifying bacteria [Eq. (64)]:

$$2NH_3 + 3O_2 \xrightarrow[\text{spp}]{\textit{Nitrosomonas}} 2HNO_2 + 2H_2O \tag{64}$$

Nitrosomonas europaea was immobilized between two membranes, a porous acetylcellulose and the Teflon membrane of the O_2 electrode. From 0.2 to 1.2 mg/liter of NH_3 could be assayed. Urea and monomethylamine interfere at high concentrations.

f. Cofactors and Acids

Reichel and Rechnitz [297] proposed the use of a mixture of intact bacterial cells and an enzyme extract together with an ammonia electrode to construct a hybrid NAD electrode. The electrode measures NAD by means of the reaction sequence [Eqs. (65) and (66)] that produces ammonia in stoichiometric amounts.

$$NAD + H_2O \xrightarrow{\text{NAD nucleotidase}} \text{nicotinamide + adenosine diphos-}$$
$$\text{phate ribose}$$
$$\tag{65}$$

$$\text{Nicotinamide} + H_2O \xrightarrow[\text{deaminase}]{\text{nicotinamide}} \text{nicotinic acid} + NH_3 \tag{66}$$

Calibration plots are linear from 2.5×10^{-4} to 2.5×10^{-3} M NAD
with a slope of 49 mV/decade. The lifetime is about 1 week. Good
responses were obtained from *E. coli, Micrococcus lysodeikticus,*
S. flava, and *S. lutea. E. coli* provided the best response, and is
especially good mixed with *N. crassa* in the hybrid electrode.

Suzuki et al. [298] proposed an electrode for NADH using mitochondria and an O_2 electrode.

A very good demonstration of use of microbial cells that can
mediate complex reaction sequences in the construction of biosensing
probes was given by Kobos and Pyon [302]. A potentiometric sensor
was developed for nitrilotriacetic acid (NTA) in which a strain of
Pseudomonas was coupled to an ammonia gas-sensing electrode. The
bacterial cells carry out a four-step reaction sequence to produce
the measured species, NH_3, from NTA. The resulting microbial electrode has a linear response range of 1×10^{-4} to 7×10^{-4} M NTA,
with a sub-Nernstian slope of 35-40 mV/decade. The electrode is
stable for up to a month. Nitrite at concentrations above 10^{-5} M
and cupric ion at amounts above 10^{-4} M inhibit the electrode response.
Fluoride also inhibited the response. Other serious interferences
include glycine, serine, and urea, which gave responses similar to
that of NTA.

Suzuki et al. described several probes for organic acids:
T. brassicae with an O_2 probe for acetic acid (range 10-100 mg/liter)
[279], *Clostridium butyricum* with a fuel cell for formic acid (range
10-1000 mg/liter) [279], *L. arabinosus* with a pH electrode for nicotinic acid (0.05-5 mg/liter) [299], and pyruvate with *Hansenula*
anomala and an O_2 electrode [287]. The nicotinic acid electrode is
stable for 30 days, each assay requires 1 hour, and the relative
error is 5%.

g. BOD Sensors

Biochemical oxygen demand (BOD) is an important parameter showing the amount of metabolized organic compounds present in solution.
The microbiological assay takes 5 days.

The ability of microorganisms to consume O_2 in the presence of
organic compounds can lead to the development of new BOD determina-

tion methods. Suzuki and co-workers have been very active in this field.

In one paper [284] two different types of microbial electrodes were prepared. One, an electrode using a bacteria-collagen membrane and an O_2 electrode, when immersed in a solution containing glucose and glutamic acid as a model of wastewater showed a linear relationship between the steady-state current and the BOD of the solution. A stability of 10 days and a reproducibility of 7.5% were obtained. Also a biofuel cell utilizing immobilized C. butyricum and a Pt electrode was used to measure the BOD of wastewater. After 30-40 min a steady state was reached. A 10% reproducibility and 30 days stability was observed. In a second paper [285], a system for a rapid 5-day BOD assay by use of microorganisms and an O_2 electrode was described. The C.V. was 6% and linearity was observed at 50-400 ppm.

In a later manuscript [286], a new microbial electrode using immobilized C. butyricum was prepared for the BOD estimation of wastewaters. The current of the electrode decreased until a steady state was reached and a linear relationship of current and BOD was observed. Error was 7-10%, linear range 25-600 ppm BOD, and stability was 30 days.

Finally, a yeast BOD sensor was described by Kulys and Kadizianskiene [287]. The cells of H. anomala were used with an O_2 electrode. This electrode offers considerable promise because yeast microorganisms grow in media containing simple organic substances as C sources; hence their usefulness as BOD sensors is universal.

h. Other Assays

A rapid electrochemical method as an assimilation test for microorganisms was described by Suzuki et al. [296]. A microbial electrode consisting of the immobilized microorganisms to be tested and an O_2 electrode was used to study the assimilation characteristics of various microbes. Molds, yeasts, bacterias, activated sludges, and actinomycetes were tested with various substrates. Good correlations were observed between the electrochemical method and the conventional growth tests.

D'Orazio et al. [295] proposed a membrane electrode for measurement of lysozyme enzyme using living bacterial cells of the strain *M. lysodeikticus* as substrate. The cells are loaded with a marker ion which is released through the action of lysozyme upon the cell wall. The rate of ion release is monitored with a highly selective membrane electrode and is readily related to the concentration of enzyme present. The proposed method has excellent sensitivity and offers advantages of precision and convenience over previously used turbidometric methods.

VI.   IMMUNOELECTRODE PROBES

A.   General Discussion

Another possible application of biological probes would be the construction of sensor probes utilizing bound antibodies or antigens. The linkage of an enzyme directly to an antigen or antibody, or the direct binding of the antigen or antibody to a carrier, such as glass or collagen, can be effected as easily as the binding of an enzyme as described in Chapter 2. The enzyme-antigen or enzyme-antibody linkage, called EIA, enzyme immunoassay, has many advantages over radioimmunoassay (RIA): elimination of expensive counting equipment; no radioactive waste to deal with; reagents are cheap and stable. The technique of EIA, together with spectrophotometric assays, is widely used in many clinical analyzers (e.g., the Dupont ACA, Abbott 100, etc.) for analysis of drugs and metabolites. In this section the application of linked antigens and antibodies to electrode probes will be discussed.

B.   Linked Antibodies

One application of bound antibodies is the assay of antigens which react specifically with the antibody.

Yuan et al. [337] proposed a novel creatine kinase isoenzyme MB (CK-MB) electrode based on the principle that after immunoinhibition with goat-antihuman CK-M antibodies the residual activity of CK-B in serum is detected with a platinum electrode by coupling the NADH (generated from the hexokinase-glucose-phosphate dehydrogenase reactions) to the ferricyanide-diaphorase indicator reaction. The electrochemical oxidation of ferrocyanide ion is monitored at +0.36 V versus SCE. The whole assay took only 10 min and the linearity of a calibration plot of serum CK-MB enzyme activity extends up to 875 U/ liter. The C.V. and recovery value of this method were 3.0 and 97.8%, respectively. The results obtained were in excellent agreement with the Helena electrophoresis method for the isoenzyme.

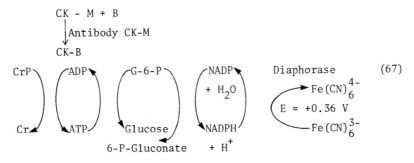

Yuan et al. [338] developed an immobilized immunostirrer for the determination of CK-MB isoenzyme in blood serum. The IgG antibodies are immobilized on alkylamine glass beads using glutaraldehyde as crosslinking agent, and the beads are packed into a rotating porous cell. After incubation with stirring, the CK-M isoenzymes in the blood serum sample are inhibited and are bound to the antibodies inside the stirrer. The residual CK-B isoenzyme activity is then determined electrochemically as described in Eq. (67). The binding capacity of the immunostirrer to CK-M isoenzyme was estimated to be 800 U/liter, with an efficiency of 97.8%. The within-day and day-to-day coefficients of variation were 5% and 4%, respectively, over a period of 52 days. An immunostirrer loaded with antibodies attached to cyanogen bromide-activated cellulose beads was also characterized, but the antibodies were not as stable as on glass beads.

An electrode-based enzyme immunoassay using urease conjugates was described by Meyerhoff and Rechnitz [ʋ39]. Urease conjugates were used for competitive binding enzyme immunoassays of a model

antigen, bovine serum albumin (BSA) and of cyclic AMP. Urease
activity bound to a double antibody solid phase is determined using
an ammonia gas sensing electrode. BSA (~10 ng/ml) and haptanes
(AMP $\sim 10^{-8}$ M) could be determined with good day-to-day reproducibility.

An enzyme immunoelectrode suitable for the assay of human serum
albumin and insulin was described by Mattiasson and Nilsson [317].
A radiometer oxygen electrode was covered with an antibody-containing
nylon net kept in place with an O ring. From 1 to 25 μg/nl of albumin
and 5 to 100 μg/nl of insulin can be assayed.

Urease conjugates were employed for competitive binding enzyme
immunoassays of a model antigen, BSA, and of acrylic AMP by Meyerhoff
and Rechnitz [318]. Cyclic AMP analogs coupled to urease are used to
determine their effect on the overall response characteristics of the
AMP assay. The use of urease as a label for EIA purposes is shown to
yield sensitive assays for both proteins (BSA ~ 10 ng/ml) and haptanes
(CAMP $< 10^{-8}$ M) with good day-to-day reproducibility.

Haga et al. [319] described an enzyme immunoelectrode for insulin
based on an antibody-bound membrane and an O_2 electrode. The assay
is based on the competitive immunochemical reaction of the antibody
with catalase-labeled and nonlabeled insulin, and separation of the
bound and nonbound insulin. The limit of detection was 40 nM.

Boitieux et al. described two procedures for assay of antigens
using bound antibody. In one [320], a sensitive and reliable procedure for assay of HB$_S$Ag in biological fluids was described. A gelatin membrane, onto which were immobilized anti-HB$_S$Ag-specific antibodies, was used in a sandwich procedure; a potentiometric determination of the antigen was effected using an iodide electrode.

In the second paper [321], an enzyme immunoassay procedure
developed from the biological model "hepatitis B surface-antigen/
antibody" was used in a sandwich procedure. The antibodies were
immobilized on gelatin membranes and the antigen concentration was
determined using an iodide electrode onto which was fixed the active
membrane. As little as 0.1 μg/liter of antigen HB$_S$Ag could be
assayed, and excellent correlation with the RIA procedure was found.

Aizawa et al. [322] developed a specific sensor for the tumor antigen (AFP), prepared from a membrane of immobilized antibody and an O_2 probe. Anti-AFP antibody is covalently immobilized on a membrane prepared from cellulose triacetate, 1,8-diamino-4-aminomethyl octane, and glutaraldehyde. The sensor was applied to an EIA based on the competitive antigen-antibody reaction with catalase-labeled antigen. After competitive binding of free and catalase-labeled AFP, the sensor is examined for catalase activity by amperometric measurement after addition of H_2O_2. AFP can be assayed in the range 10^{-8} to 10^{-11} g/ml.

Yamamoto et al. [323] investigated the antigen-antibody reaction of human choriogonadotropin potentiometrically, using a cyanogen bromide-treated electrode coated with the corresponding antibody. The potential of the modified electrode shifts in the positive direction after contact with a solution of the antigen. The change in potential is approximately proportional to the choriogonadotropin concentration. The technique, applied to samples of human urine, has shown a specific response to only choriogonadotropin.

Suzuki et al. [324,325] developed an immunosensor that used catalase-labeled IgG. The IgG was determined by competitive reaction with the membrane antibody and the assay of catalase activity. When peroxide is added, the change in current due to O_2 uptake is measured.

Yamamoto et al. [326] conducted extensive potentiometric investigations of antigen-antibody reactions using chemically modified electrodes. An immunoelectrode was made of titanium wire on which an antigen or an antibody was chemically fixed. The potential of the electrode sensitized with anti-hCG γ-globulin shifted in the positive direction in the presence of a small amount of hCG in solution.

On the other hand, the potential of a hCG-sensitized electrode ran in the negative direction on addition of anti-hCG. Similar changes were observed between trypsin and its inhibitor, aporotinin.

C. Bound Antigens

An enzyme immunoassay method using adenosine deaminase as the enzyme
label was described by Gebauer and Rechnitz [327]. Potentiometric
rate measurements were made with an ammonia gas-sensing electrode to
determine the activity of enzyme label bound to agarose bead-immobilized second antibody. The activity is related to the concentration
of either a model haptenic dinitrophenyl, DNP, or anti-DNP antibody.
The detection limit is 50 ng antibody.

    Aizawa and Suzuki [328] and Aizawa et al. [329] described an
immunosensor for determining specific protein. A liquid antigen
containing cardiolipin, phosphatidylcholine, and cholesterol was
immobilized to an acetyl cellulose membrane. The membrane-bound
antigen retained immunological reactivity to Wasserman antibody.
The asymmetrical potential developed was dependent on the concentration of the antibody.

    Suzuki [330] bound this antigen and developed a method for assay
of syphilis antibody in blood serum. The contact potential between
bond antigen and antibody was measured, with very low ΔmV (1-3 mV)
observed.

    Solsky and Rechnitz [331] described the preparation and proper-
ties of a membrane electrode with selective response for the antibodies of the antigenic determinant dinitrophenol (DNP). The DNP
ion carrier conjugates were incorporated into a polyvinylchloride
membrane matrix. The response mechanism of the antibody electrode
is postulated to involve a "selectivity shift" effect.

REFERENCES

1. A. H. Kadish and D. A. Hall, *Clin. Chem. 9*, 869 (1965).

2. Y. Makino and K. Koono, *Rinsho Byori 15*, 391 (1967) (Japan).

3. F. Cheng and G. Christian, *Anal. Chem. 49*, 1785 (1977).

4. A. Szabo, S. Szabo, and G. Velosy, *Z. Anal. Chem. 290*, 134
    (1978).

5. D. P. Nikolelis and H. Mottola, *Anal. Chem. 50*, 1665 (1978).

6. J. Dietschy, L. Weeks, and J. Delente, *Clin. Chim. Acta 73*,
    407 (1976).

7. Y. Kameno, N. Nakano, and S. Baba, *Clin. Chim. Acta 77*, 245 (1977).

8. A. Noma and K. Nakayama, *Clin. Chem. 22*, 336 (1976).

9. F. Cheng and G. Christian, *Anal. Chim. Acta 104*, 47 (1979).

10. C. Borrebaeck and B. Mattiasson, *Anal. Biochem. 107*, 446 (1980).

11. P. Alexander and P. Seegopaul, *Anal. Chim. Acta 121*, 61 (1980).

12. L. Meyerson, K. McMurtey, and V. Davis, *Anal. Biochem. 86*, 287 (1978).

13. S. Hassan and G. Rechnitz, *Anal. Chim. Acta 126*, 35 (1981).

14. C. Kiang, S. Kuan, and G. G. Guilbault, *Anal. Chem. 50*, 1319 (1978).

15. C. Tran-Minh and G. Broun, *C. R. Acad. Sci. Paris. Biochim. Anal. (Series D)*, 2215 *(April 2, 1973)*.

16. S. A. Katz, *Anal. Chem. 36*, 2500 (1964).

17. S. A. Katz and G. A. Rechnitz, *Z. Anal. Chem. 196*, 248 (1963).

18. G. G. Guilbault, J. Montalvo, and R. Smith, *Anal. Chem. 41*, 600 (1969).

19. G. G. Guilbault and F. Shu, *Anal. Chem. 44*, 2161 (1972).

20. R. Llenado and G. Rechnitz, *Anal. Chem. 46*, 1109 (1974).

21. H. Thompson and G. Rechnitz, *Anal. Chem. 46*, 246 (1974).

22. E. Hsiung, S. Kuan, and G. G. Guilbault, *Anal. Chim. Acta 90*, 45 (1977).

23. W. R. Hussein, L. H. Von Storp, and G. G. Guilbault, *Anal. Chim. Acta 61*, 89 (1972).

24. L. H. Von Storp and G. G. Guilbault, *Anal. Chim. Acta 62*, 425 (1972).

25. R. A. Llenado and G. A. Rechnitz, *Anal. Chem. 45*, 826 (1973).

26. M. Mascini and G. Palleschi, *Anal. Chim. Acta 100*, 215 (1978).

27. D. Papastathopoulos and G. Rechnitz, *Anal. Chem. 47*, 1792 (1975).

28. E. P. Diamandis and T. P. Hadjiioannou, *Clin. Chem. 27*, 455 (1981).

29. T. Hara, M. Imaki, and M. Toriyama, *Bull. Chem. Soc. (Japan) 54*, 1396 (1981).

30. M. Thompson, P. Worsfold, J. Holuk, and E. Stubley, *Anal. Chim. Acta 104*, 195 (1979).

31. L. Clark, C. Duggan, T. Grooms, L. Hart, and M. Moore, *Clin. Chem. 27*, 1982 (1981).

32. Y. Hahn and C. Olson, *Anal. Chem. 51*, 444 (1979).

33. V. Razumas, J. Kulys, and A. Malinauskas, *Anal. Chim. Acta 117*, 387 (1980).

34. T. Wallace and R. Coughlin, *Anal. Biochem. 80*, 133 (1977).

35. T. Wallace, M. Leh, and R. Coughlin, *Biotech. Bioeng. 19*, 901 (1977).

36. T. Matsunaga, I. Karube, and S. Suzuki, *Appl. Environ. Microb. 37*, 117 (1979).

37. T. Matsunaga, I. Karube, and S. Suzuki, *Anal. Chim. Acta 98*, 25 (1978).

38. A. Rey and M. Hanss, *Ann. Biol. Chim. 29*, 323 (1971).

39. M. Hanss and A. Rey, *Biochim. Biophys. Acta 227*, 630 (1971).

40. Product Bulletin, Beckman BUN and Creatinine Analyzers, Beckman Instrument Co., Fullerton, 1981.

41. A. Rey and M. Hanss, *Clin. Chim. Acta 30*, 207 (1970).

42. L. Clark and C. Lyons, *Ann. N.Y. Acad. Sci. 102*, 29 (1962).

43. S. J. Updike and G. P. Hicks, *Nature (London) 214*, 986 (1971).

44. G. G. Guilbault and G. J. Lubrano, *Anal. Chim. Acta 64*, 439 (1973).

45. G. G. Guilbault, *Handbook of Enzymatic Analysis*. Marcel Dekker, New York, 1977.

46. K. Mosbach, H. Nilsson, and A. Akerlund, *Biochim. Biophys. Acta 320*, 529 (1973).

47. G. J. Papariello, A. K. Mukerji, and C. M. Shearer, *Anal. Chem. 45*, 790 (1973).

48. G. G. Guilbault, R. McQueen, and S. Sadar, *Anal. Chim. Acta 45*, 1 (1969).

49. G. G. Guilbault and F. Shu, *Anal. Chem. 44*, 2161 (1972).

50. G. G. Guilbault and G. J. Lubrano, *Anal. Chim. Acta 69*, 183 (1974).

51. G. G. Guilbault and E. Hrabankova, *Anal. Lett. 3*, 53 (1970).

52. G. G. Guilbault and E. Hrabankova, *Anal. Chim. Acta 56*, 285 (1971).

53. G. G. Guilbault and J. G. Montalvo, *J. Am. Chem. Soc. 92*, 2533 (1970).

54. G. G. Guilbault and G. J. Lubrano, *Anal. Chim. Acta 64*, 439 (1973).

55. G. G. Guilbault and M. Nanjo, *Anal. Chem. 46*, 1769 (1974).

56. G. G. Guilbault and M. Nanjo, *Anal. Chim. Acta 73*, 367 (1974).

57. A. Johansson, J. Lundberg, B. Mattiasson, and K. Mosbach, *Biochim. Biophys. Acta 304*, 217 (1973).

58. G. G. Guilbault, G. Nagy, and S. S. Kuan, *Anal. Chim. Acta 67,* 195 (1973).

59. G. G. Guilbault and G. J. Lubrano, *Anal. Chim. Acta 69,* 189 (1974).

60. M. Mascini and A. Liberti, *Anal. Chim. Acta 68,* 177 (1974).

61. S. J. Updike and G. P. Hicks, *Nature 214,* 986 (1967).

62. H. Nilsson, A. Åkerlund and K. Mosbach, *Biochim. Biophys. Acta 320,* 529 (1973).

63. G. G. Guilbault and J. G. Montalvo, *J. Am. Chem. Soc. 91,* 2164 (1969).

64. G. G. Guilbault and E. Hrabankova, *Anal. Chim. Acta 52,* 287 (1970).

65. G. G. Guilbault and G. Nagy, *Anal. Chem. 45,* 417 (1973).

66. T. Anfalt, A. Granelli, and D. Jagner, *Anal. Lett. 6,* 969 (1973).

67. G. A. Rechnitz and R. Llenado, *Anal. Chem. 43,* 283 (1971).

68. G. A. Rechnitz and R. Llenado, *Anal. Chem. 43,* 1457 (1971).

69. G. G. Guilbault and G. J. Lubrano, *Anal. Chim. Acta 60,* 254 (1972).

70. W. J. Blaedel, T. R. Kissel, and R. C. Bogaslaski, *Anal. Chem. 44,* 2030 (1972).

71. L. Mell and J. Maloy, *Anal. Chem. 47,* 299 (1975).

72. G. G. Guilbault and F. Shu, *Anal. Chim. Acta 56,* 333 (1971).

73. G. G. Guilbault and M. Tarp, *Anal. Chim. Acta 73,* 355 (1974).

74. J. G. Montalvo and G. G. Guilbault, *Anal. Chem. 41,* 1897 (1969).

75. L. C. Clark, U.S. Patent 3,529,455 (1970).

76. G. Nagy, L. H. von Storp, and G. G. Guilbault, *Anal. Chim. Acta 66,* 443 (1973).

77. G. G. Guilbault and M. Nanjo, *Anal. Chim. Acta 78,* 69 (1975).

78. H. Kiang, S. S. Kuan, and G. G. Guilbault, *Anal. Chim. Acta 80,* 209 (1975).

79. H. Kiang, S. S. Kuan, and G. G. Guilbault, *Anal. Chem. 50,* 1323 (1978).

80. G. G. Guilbault and J. Montalvo, *Anal. Lett. 2,* 283 (1969).

81. J. Ruzicka and E. H. Hansen, *Anal. Chim. Acta 69,* 129 (1974).

82. P. W. Alexander and J. P. Joseph, *Anal. Chim. Acta 131,* 103 (1981).

83. G. Baum, *Anal. Biochem. 39,* 65 (1971); *42,* 487 (1971).

84. G. Baum and M. Lynn, *Anal. Chim. Acta 65,* 385 (1973); *Clin. Chim. Acta 36,* 406 (1972).

85. G. Baum, *Anal. Lett. 3,* 105 (1970).

86. R. Kobos and G. Rechnitz, *Arch. Biochim. Biophys. 175,* 11 (1976).

87. R. Kobos and G. Rechnitz, *Biochem. Biophys. Res. Commun. 71,* 762 (1976).

88. D. Papastathopoulos and G. Rechnitz, *Anal. Chem. 48,* 862 (1976).

89. G. Hjemdal-Monsen, D. Papasthatopoulos, and G. Rechnitz, *Anal. Chim. Acta 88,* 253 (1977).

90. F. Scheller and D. Pfeiffer, *Anal. Chim. Acta 117,* 383 (1980).

91. S. Suzuki, F. Takahashi, J. Satoh, and N. Sonobe, *Bull. Chem. Soc. (Japan) 48,* 3246 (1975).

92. M. Smith and C. Olson, *Anal. Chem. 47,* 1074 (1975).

93. G. G. Guilbault and M. Nanjo, *Anal. Chim. Acta 75,* 169 (1975).

94. G. G. Guilbault and G. J. Lubrano, *Anal. Chim. Acta 69,* 189 (1974).

95. R. Llenado and G. Rechnitz, *Anal. Chem. 46,* 1109 (1974).

96. G. Johannson, K. Edstrom, and L. Ogren, *Anal. Chim. Acta 85,* 55 (1976).

97. G. G. Guilbault and E. Hrabankova, *Anal. Chem. 42,* 1779 (1970).

98. C. Hsiung, S. Kuan, and G. G. Guilbault, *Anal. Chim. Acta 90,* 45 (1977).

99. G. G. Guilbault and K. Gibson, *Anal. Chim. Acta 76,* 245 (1975).

100. M. Mascini and G. Palleschi, *Ann. Chim. (Rome) 69,* 249 (1979).

101. G. Guilbault and G. Nagy, *Anal. Lett. 6,* 301 (1973).

102. C. Calvot, A. Berjonneau, C. Gellf, and D. Thomas, *FEBS Lett. 59,* 258 (1975).

103. R. Wawro and G. Rechnitz, *J. Membrane Sci. 1,* 143 (1976).

104. G. G. Guilbault and F. Shu, *Anal. Chim. Acta 56,* 333 (1971).

105. P. Davis and K. Mosbach, *Biochim. Biophys. Acta 370,* 329 (1974).

106. B. K. Ahn, S. K. Wolfson, and S. J. Yao, *Bioelectrochem. Bioenerg. 2,* 142 (1975).

107. T. Ngo, *Int. J. Biochem. 6,* 371 (1975).

108. W. White and G. Guilbault, *Anal. Chem. 50,* 1481 (1978).

109. D. Skogberg and T. Richardson, *J. Am. Assoc. Cereal Chem. 56,* 147 (1979).

110. K. W. Fung, S. S. Kuan, H. Y. Sung, and G. G. Guilbault, *Anal. Chem. 51,* 2319 (1979).

111. J. Havas and G. G. Guilbault, *Anal. Chem. 54,* 1991 (1982).

112. A. Kumar and G. Christian, *Clin. Chem. 21,* 325 (1975).

113. K. Matsumoto, K. Yamada, and Y. Osajima, *Anal. Chem. 53,* 1974 (1981).

114. G. G. Guilbault, E. K. Bauman, D. N. Kramer, and L. H. Goodson, *Anal. Chem. 37,* 1378 (1965).

115. H. Y. Neujahr, *Biotechnol. Bioeng. 22,* 913 (1980).

116. J. Varga and H. Neujahr, *Evr. J. Biochem. 12,* 427 (1970).

117. N. Huang, S. S. Kuan, and G. Guilbault, *Clin. Chem. 21,* 1605 (1975).

118. J. Dietschy, L. Wechs, and J. Delenk, *Clin. Chim. Acta 73,* 407 (1976).

119. J. Kamerro, N. Nakamo, and S. Baba, *Clin. Chim. Acta 77,* 245 (1977).

120. A. Kumar and G. Christian, *Clin. Chim. Acta 77,* 101 (1977).

121. A. Nonna and K. Nakajama, *Clin. Chem. 22,* 336 (1976).

122. I. Satoh, I. Karube, and S. Suzuki, *Biotechnol. Bioeng. 19,* 1095 (1977).

123. H. Thompson and G. Rechnitz, *Anal. Chem. 46,* 246 (1974).

124. B. Chen, S. Kuan, and G. G. Guilbault, *Anal. Lett. 13,* 1607 (1980).

125. W. Blaedel and R. Engstrom, *Anal. Chem. 52,* 1691 (1980).

126. F. Cheng and G. Christian, *Anal. Chim. Acta 104,* 47 (1979).

127. R. Ianniello and A. Yacynych, *Anal. Chem. 53,* 2090 (1981).

128. C. Bourdillon, J.-P. Bougeois, and D. Thomas, *Biotechnol. Bioeng. 21,* 1877 (1979); *J. Am. Chem. Soc. 102,* 4231 (1980).

129. D. L. Williams, A. R. Doig, and A. Korosi, *Anal. Chem. 42,* 118 (1970).

130. L. Gorton and K. Bhatti, *Anal. Chim. Acta 105,* 43 (1979).

131. W. Mindt, P. Racine, and P. Schlapfer, *Ber Bunsenges Physk. Chem. 77,* 805 (1973).

132. W. Mindt, P. Racine, and P. Schlapfer, DDR Patentshcrift 100, 556; INT KIG 01 n 27/30, KI 42L 3/06 (Sept. 20, 1973).

133. S. Bessmann and R. Schultz, 2nd Int. Conf. Med. Biol. Eng. Dresden (1973).

134. G. G. Guilbault and G. Lubrano, *Anal. Chim. Acta 97,* 229 (1978).

135. P. R. Coulet, J. H. Julliard, and D. C. Gautheron, *Biotechnol. Bioeng. 16,* 1055 (1974).

136. D. R. Thevenot, P. R. Coulet, R. Sternberg, and D. C. Gautheron, in *Enzyme Engineering,* Vol. 4. Plenum Press, New York, 1978, p. 221.

137. P. R. Coulet and D. Gautheron, *J. Chromatogr. 215,* 65 (1981).

138. D. R. Thevenot, P. R. Coulet, R. Sternberg, and D. C. Gautheron, *Bioelectrochem. Biotechnol. 5,* 541, 548 (1978).

139. P. R. Coulet, R. Sternberg, and D. C. Gautheron, *Biochem. Biophys. Acta 612,* 317 (1980).

140. D. R. Thevenot, P. R. Coulet, R. Sternberg, J. Laurent, and D. C. Gautheron, *Anal. Chem. 51,* 96 (1979).

141. D. N. Gray and M. H. Keyes, *Anal. Chem. 49,* 1067A (1977).

142. W. Blaedel and C. Olson, U.S. Patent 3367849 (Feb. 6, 1968).

143. D. Nikolelis, D. S. Papastathopoulos, and T. P. Hadjiioannou, *Anal. Chim. Acta 126,* 43 (1981).

144. E. Gulberg and G. Christian, *Fresenius' Z. Anal. Chem. 305,* 29 (1981).

145. H. Durliat, M. Comtat, J. Maheng, and A. Baudras, *J. Electroanal. Chem. 66,* 73 (1975) (French).

146. C. Guillot, D. Vanuxem, and C. Grimaud, *Pathol. Biol. Paris 24,* 431 (1976) (French).

147. L. Thomas and G. Christian, *Anal. Chim. Acta 78,* 271 (1975).

148. P. Racine, H.-J. Klenke, and K. Kochsieck, *Z. Klin. Chem. Klin. Biochem. 13,* 533 (1975).

149. H. Durliat, M. Comtat, and A. Baudra, *Clin. Chem. 22,* 1802 (1976).

150. H. Durliat, M. Comtat, J. Maheni, and A. Baudra, *Anal. Chim. Acta 85,* 31 (1976).

151. H. Durliat, M. Comtat, J. Maheni, and A. Baudra, *J. Electroanal. Chem. 66,* 73 (1976).

152. C. Guillot, D. Vanuxem, and C. Grinaud, *Pathol. Biol. Paris 24,* 431 (1976).

153. M. Cordonier, F. Lawny, D. Chapot, and D. Thomas, *FEBS Lett. 59,* 263 (1975).

154. F. Cheng and G. Christian, *Anal. Chim. Acta 102,* 124 (1977).

155. C. Borrebaeck and B. Mattiasson, in *Proteins of Biological Fluids,* 27th Colloquin (H. Peeters, ed.), Pergamon Press, New York, 1979.

156. C. Kiang, S. Kuan, and G. G. Guilbault, *Anal. Chem. 50,* 1319 (1978).

157. S. Yao, S. Wolfson, and J. Tokarsky, *Bioelectrochem. Bioenerg. 2,* 348 (1975).

158. L. F. Cullin, J. F. Rusling, A. Schleifer, and G. J. Papariello, *Anal. Chem. 46,* 1955 (1974).

159. C. J. Olliff, R. T. Williams, and J. M. Wright, *J. Pharm. Pharmacol. 30*, 45p (1978).

160. A. Aizawa, J. Karube, and S. Suzuki, *Anal. Chim. Acta 69*, 431 (1974).

161. J. Kyzlink and L. Macholan, *Vodni Hospod, Part B, 29*, 240 (1979).

162. H. Neujahr and K. Kjellen, *Biotechnol. Bioeng. 21*, 671 (1979).

163. K. Kjellen and H. Neujahr, *Biotechnol. Bioeng. 22*, 299 (1980).

164. I. Satoh, J. Karube, and S. Suzuki, *Biotechnol. Bioeng. 18*, 269 (1976).

165. G. G. Guilbault and M. Nanjo, unpublished results, University of New Orleans (1976).

166. P. W. Alexander and P. Seegopaul, *Anal. Chim. Acta 125*, 55 (1981).

167. I. Satoh, I. Karube, and S. Suzuki, *Biotechnol. Bioeng. 18*, 269 (1976).

168. R. Llenado and G. Rechnitz, *Anal. Chem. 44*, 1366 (1972).

169. M. Mascini and G. G. Guilbault, *Anal. Chem. 49*, 795 (1977).

170. G. Johannson and L. Olgren, *Anal. Chim. Acta 84*, 23 (1976).

171. G. G. Guilbault, R. M. Smith, and J. Montalvo, *Anal. Chem. 41*, 600 (1969).

172. T. Kawashima, A. Arima, N. Hatakeyama, N. Tominaga, and M. Ando, *Nippon Kagaku Kaishi (1980)*, 1542 (Japanese).

173. G. G. Guilbault and T. Cserfalvi, *Anal. Lett. 9*, 277 (1976).

174. C. Tran-Minh and J. Breaux, *Anal. Chem. 51*, 91 (1979).

175. C. Tran-Minh and J. Breaux, *C. R. Acad. Sci. Paris Biochim. Anal. (Series C)*, p. 191 *(September 18, 1978)*.

176. E. Fogt, L. Dodd, E. Jenning, and A. Clemens, *Clin. Chem. 24*, 1366 (1978).

177. S.-O. Enfors, *Enzyme Microbiol. Technol. 3*, 29 (1981).

178. T. Tsuchida and K. Yoda, *Enzyme Microbiol. Technol. 3*, 326 (1981).

179. J. Bachner, *Sci. Pharm. 48*, 156 (1980) (German).

180. M. Koyama, Y. Sato, M. Aizawa, and S. Suzuki, *Anal. Chim. Acta 116*, 307 (1980).

181. I. M. Jensen and G. Rechnitz, *J. Membrane Sci. 5*, 117 (1979).

182. A.-M. Berjonneau, D. Thomas, and G. Broun, *Pathol.-Biol. 22*, 497 (1974).

183. C. Calvot, A.-M. Berjonneau, G. Gellf, and D. Thomas, *FEBS Lett. 59*, 258 (1975).

184. L. Macholan, *Coll. Czech. Chem. Commun. 43,* 1811 (1978).

185. R. Walters, P. Johnson, and R. Buck, *Anal. Chem. 52,* 1684 (1980).

186. F. Cheng and G. D. Christian, *Clin. Chem. 24,* 621 (1978).

187. H. Neujahr, *Biotechnol. Bioeng. 22,* 913 (1980); *Appl. Biochem. Biotechnol. 7,* 107 (1982).

188. A. Torstensson, G. Johansson, M. Mansson, P. Larsson, and K. Mosbach, *Anal. Lett. 13,* 837 (1980).

189. A. Chen and C. Liu, *Proc. Biochem. (January 1979),* p. 12.

190. J. G. Schiller, A. K. Chen, and C. C. Liu, *Anal. Biochem. 85,* 25 (1976).

191. G. Guilbault and T. Cserfalvi, *Anal. Chim. Acta 84,* 259 (1976).

192. P. Davies and K. Mosbach, *Biochim. Biophys. Acta 370,* 329 (1974).

193. R. Sternberg, A. Apoteker, and D. R. Thevenot, in *Electroanalysis in Hygiene, Environmental, Clinical and Pharmaceutical Chemistry* (W. F. Smyth, ed.), Elsevier, Amsterdam, 1980, p. 461.

194. P. R. Coulet and C. Bertrand, *Anal. Lett. 12,* 581 (1979).

195. C. Bertrand, P. R. Coulet, and D. C. Gautheron, *Anal. Chim. Acta 126,* 23 (1981).

196. F. Scheller, M. Jänchen, D. Pfeiffer, J. Seyer, and K. Müller, *Z. Med. Labor Diag. 18,* 312 (1977).

197. H. Weise, F. Scheller, and K. Siegler, *Die Nahrung 25,* 127 (1981).

198. J. J. Kulys, M. V. Pesliakiene, and A. S. Samalius, *Bioelectrochem. Bioenerg. 8,* 81 (1981).

199. C. Tran-Minh and G. Broun, *Anal. Chem. 47,* 1359 (1975).

200. S. Gondo, M. Morishita, and T. Osaki, *Biotechnol. Bioeng. 22,* 1287 (1980).

201. J.-R. Mor and R. Guarnaccia, *Anal. Biochem. 79,* 319 (1977).

202. L. B. Wingard, C. C. Liu, and N. L. Nagda, *Biotechnol. Bioeng. 13,* 629 (1971).

203. L. B. Wingard, J. G. Schiller, S. K. Wolfson, C. C. Liu, A. L. Drash, and S. J. Yao, *J. Biomed. Materials Res. 13,* 921 (1979).

204. J.-L. Romette, B. Froment, and D. Thomas, *Clin. Chim. Acta 95,* 249 (1979).

205. P. Taylor, E. Kmetec, and J. Johnson, *Anal. Chem. 49,* 789 (1977).

206. R. Ianniello and A. Yacynych, *Anal. Chim. Acta 131,* 123 (1981).

207. F. W. Janssen and H. W. Ruelius, *Biochim. Biophys. Acta 151*, 330 (1968).

208. F. W. Janssen, R. M. Kerwin, and H. W. Ruelius, *Biochim. Biophys. Res. Commun. 20*, 630 (1965).

209. T. Yao and S. Musha, *Anal. Chim. Acta 110*, 203 (1979).

210. L. Macholan and L. Schanel, *Coll. Czech. Chem. Commun. 42*, 3667 (1977).

211. H. Durliat, T. M. Comtat, and J. Mahenc, *Anal. Chim. Acta 106*, 131 (1979).

212. T. Shinbo, M. Sugiura, and N. Kamo, *Anal. Chem. 51*, 100 (1979).

213. J. J. Kulys and G. S. Svirmickas, *Anal. Chim. Acta 117*, 115 (1980).

214. A. Malinauskas and J. Kulys, *Anal. Chim. Acta 98*, 31 (1978).

215. R. K. Kobos and T. A. Ramsey, *Anal. Chim. Acta 121*, 111 (1980).

216. J. G. Schindler and M. Gülich, *Biomed. Technik 26*, 43 (1981).

217. R. A. Kamin and G. S. Wilson, *Anal. Chem. 52*, 1198 (1980).

218. F. Yoshino, H. Osawa, and K. Harada, Abstracts of 33rd National Meeting of the American Association of Clinical Chemistry, Kansas City, Missouri, July (1981) in *Clin. Chem. 27*, 1098 (1981).

219. L. C. Clark, C. Emory, C. Glueck, and M. Campbell, in *Enzyme Engineering*, Vol. 3 (E. K. Pye and H. H. Weetall, eds.). Plenum Press, New York, 1978, pp. 409-425.

220. S. Enfors and H. Nilsson, *Enzyme Microb. Technol. 1*, 260 (1979).

221. J. W. Hewetson, T. H. Jong, and P. P. Gray, *Biotechnol. Bioeng. Symp. 9*, 125 (1979).

222. E. Diamandis and T. Hadjiioannou, *Clin. Chem. 27*, 455 (1981).

223. D. Pfeiffer, F. Scheller, M. Janchen, and K. Bertermann, *Biochimie 62*, 587 (1980).

224. I. Deng and C. Enke, *Anal. Chem. 52*, 1937 (1980).

225. A. Malinauskas and J. Kulys, *Biotechnol. Bioeng. 21*, 513 (1979).

226. T. L. Riechel and G. A. Rechnitz, *Biochem. Biophys. Res. Commun. 74*, 1377 (1977).

227. C. Tran-Minh and J. Breaux, *C. R. Acad. Sci. Paris (Comp Rendu) 286* (Jan. 23, 1978), Series C, p. 115.

228. P. Durand, A. David, and D. Thomas, *Biochim. Biophys. Acta 527*, 277 (1978).

229. M. Arnold and G. Rechnitz, *Anal. Chim. Acta 113*, 351 (1980).

230. M. Arnold and G. Rechnitz, *Anal. Chem. 52*, 1170 (1980).

231. H. Imai and Y. Kawauchi, *Bunseki Kagaku 30,* 94 (1981).

232. C. Borrebaeck and B. Mattiasson, *Anal. Biochem. 107,* 446 (1980).

233. L. Ogren and G. Johansson, *Anal. Chim. Acta 96,* 1 (1978).

234. B. Mattiasson, H. Nilsson, and B. Olsson, *J. Appl. Biochem. 1,* 377 (1979).

235. W. J. Blaedel and J. Wang, *Anal. Chem. 52,* 1426 (1980).

236. W. J. Blaedel and J. Wang, *Anal. Chem. 52,* 1697 (1980).

237. H. F. Haneka and G. Rechnitz, *Anal. Chem. 53,* 1586 (1981).

238. L. Mell and J. Maloy, *Anal. Chem. 48,* 1597 (1976).

239. P. Carr, *Anal. Chem. 49,* 799 (1977).

240. D. F. Ollis and R. Carter, *Enzyme Engineering,* Vol. 2 (E. K. Pye and L. B. Wingard, eds.), Plenum Press, New York, 1974, p. 271.

241. W. A. Callanan, M.S. thesis, University of Pennsylvania, 1972.

242. H. Pederson and G. Chotani, *Appl. Biochem. Biotechnol. 6,* 309 (1981).

243. Hoffman La Roche and Co., British Patent 1329520 (1973).

244. I. Karube, I. Satoh, Y. Araki, S. Suzuki, and H. Yamada, unpublished results, 1983.

245. T. Ngo and P. Shargool, *Anal. Biochem. 54,* 247 (1973).

246. L. Thomas and G. Christian, *Anal. Chim. Acta 78,* 271 (1976).

247. L. Thomas and G. Christian, *Anal. Chim. Acta 82,* 265 (1976).

248. W. Mason and L. Olsen, *Anal. Chem. 42,* 488 (1970).

249. D. Luppa and H. Aurich, *Enzyme 12,* 688 (1971).

250. L. Thomas and G. Christian, *Anal. Chim. Acta 89,* 83 (1977).

251. H. Booker and J. Haslem, *Anal. Chem. 46,* 1054 (1974).

252. N. Larsen, E. Hansen, and G. Guilbault, *Anal. Chim. Acta 79,* 9 (1975).

253. D. Fergerson, J. Boyd, and A. Phillips, *Anal. Biochem. 62,* 81 (1974).

254. S. Lovett, *Anal. Biochem. 64,* 110 (1975).

255. C. Gebauer, M. Meyerhoff, and G. Rechnitz, *Anal. Biochem. 95,* 479 (1979).

256. K. Crochet and J. Montalvo, *Anal. Chim. Acta 66,* 259 (1973).

257. G. Guilbault and A. Iwase, *Anal. Chim. Acta 85,* 295 (1976).

258. B. Erlanger and R. Sack, *Anal. Biochem. 33,* 318 (1970).

259. R. Weitzmann, *Biochem. Soc. Trans. 4,* 724 (1976).

260. I. Grigorov, J. Ignatov, and G. Spassov, *Enzymologia 42*, 377 (1972).

261. R. Llenado and G. Rechnitz, *Anal. Chem. 45*, 826 (1973).

262. R. Llenado and G. Rechnitz, *Anal. Chem. 44*, 468 (1972).

263. G. Guilbault and D. Kramer, *Anal. Biochem. 18*, 313 (1967).

264. L. Macholan, *Coll. Czech. Cehm. Commun. 44*, 3033 (1979).

265. M. Dumontier and M. Hass, *Biochimie 56*, 1291 (1974).

266. Y. Huang, *Anal. Biochem. 61*, 464 (1974).

267. T. Yagi, M. Goto, K. Nakamo, K. Kimura, and H. Inokuchi, *J. Biochem. (Tokyo) 78*, 443 (1975).

268. T. Yagi, M. Goto, K. Nakamo, K. Kimura, and H. Inokuchi, *J. Biochem. (Tokyo) 78*, 1347 (1975).

269. P. D'Orazio, M. Meyerhoff, and G. Rechnitz, *Anal. Chem. 50*, 1531 (1978).

270. A. Kumar and G. Christian, *Anal. Chem. 48*, 1283 (1976).

271. P. Chien and L. Michael, *Anal. Biochem. 68*, 626 (1975).

272. J. Montalvo, *Anal. Chem. 41*, 2093 (1969); *Anal. Biochem. 38*, 359 (1970).

273. R. C. Barbino, D. N. Gray, and M. H. Keyes, *Clin. Chem. 24*, 1393 (1978).

274. H. Arwin and I. Lundstrom, *FEBS Lett. 109*, 252 (1980).

275. J. Mahenc and H. Aussaresses, *Compt. Rend. 28*, 357 (1979).

276. S. J. Updike, M. C. Shults, and M. Bushby, *J. Lab. Clin. Med. 93*, 518 (1979).

277. Z. Toul and L. Macholan, *Coll. Czech. Chem. Commun. 40*, 2208 (1975).

278. F. R. Shu and G. S. Wilson, *Anal. Chem. 48*, 1679 (1976).

279. S. Suzuki, I. Satoh, and I. Karube, *Appl. Biochem. Biotechnol. 7*, 147 (1982).

280. R. Kobos, in *Ion-Selective Electrodes in Analytical Chemistry*, Vol. 2 (H. Freiser, ed.), Plenum Press, New York, 1980, pp. 69-84.

281. G. Rechnitz, *Science 214*, October 1981, p. 287.

282. M. Hikuma, T. Kubo, T. Yasuda, I. Karube, and S. Suzuki, *Biotechnol. Bioeng. 21*, 1845 (1979).

283. M. Hikuma, T. Kubo, T. Yasuda, I. Karube, and S. Suzuki, *Anal. Chem. 52*, 1020 (1980).

284. I. Karube, T. Matsunaga, S. Mitsuda, and S. Suzuki, *Biotechnol. Bioeng. 19*, 1535 (1977).

285. I. Karube, S. Mitsuda, T. Matsunaga, and S. Suzuki, *J. Ferment.*
     Technol. 55, 243 (1977).

286. I. Karube, T. Matsunaga, and S. Suzuki, *J. Solid Phase Biochem.*
     2, 97 (1977).

287. J. Kulys and K. Kadziauskiene, *Biotechnol. Bioeng. 22*, 221
     (1980).

288. U. Wollenberger, F. Scheller, and P. Atrat, *Anal. Lett. 13*,
     825 (1980).

289. M. A. Jensen and G. A. Rechnitz, *Anal. Chim. Acta 101*, 125
     (1978).

290. M. Hikuma, H. Obana, T. Yasuda, I. Karube, and S. Suzuki,
     Enzyme Microb. Technol. 2, 234 (1980).

291. M. A. Arnold and G. A. Rechnitz, *Anal. Chim. Acta 113*, 351
     (1980).

292. M. Hikuma, H. Obana, T. Yasuda, I. Karube, and S. Suzuki, *Anal.*
     Chim. Acta 116, 61 (1980).

293. S. Kuriyama and G. Rechnitz, *Anal. Chim. Acta 131*, 91 (1981).

294. R. Walters, B. Moriarty, and R. Buck, *Anal. Chem. 52*, 1680
     (1980).

295. P. D'Orazio, M. Meyerhoff, and G. Rechnitz, *Anal. Chem. 50*,
     1531 (1978).

296. M. Hikuma, H. Suzuki, T. Yasuda, I. Karube, and S. Suzuki,
     Eur. J. Appl. Microbiol. Biotechnol. 9, 305 (1980).

297. T. Riechel and G. Rechnitz, *J. Membrane Sci. 4*, 243 (1978).

298. M. Aizawa, M. Wada, and S. Suzuki, *Biotechnol. Bioeng. 22*,
     1769 (1980).

299. T. Matsunaga, I. Karube, and S. Suzuki, *Anal. Chim. Acta 99*,
     233 (1978).

300. K. Matsumoto, H. Seijo, T. Watanabe, I. Karube, I. Satoh, and
     S. Suzuki, *Anal. Chim. Acta 105*, 429 (1979).

301. R. K. Kobos, D. J. Rice, and D. S. Flournoy, *Anal. Chem. 51*,
     1122 (1979).

302. R. K. Kobos and H. Y. Pyon, *Biotechnol. Bioeng. 23*, 627 (1981).

303. M. Mascini and G. Rechnitz, *Anal. Chim. Acta 116*, 169 (1980).

304. T. Matsunaga, I. Karube, N. Teraoka, and S. Suzuki, *Anal. Chim.*
     Acta 127, 245 (1981).

305. C. L. Di Paolantonio, M. Arnold, and G. Rechnitz, *Anal. Chim.*
     Acta 128, 121 (1981).

306. U. Wollenberger, F. Scheller, and P. Atrat, *Anal. Lett. Part B*
     13, 1201 (1980).

307. C. Divies, *Ann. Microb. (Paris)* 126A, 175 (1975).

308. C. Divies, *Chem. Eng. News* 54(44), 23 (1976).

309. G. A. Rechnitz, R. K. Kobos, T. L. Reichel, and C. R. Gebauer, *Anal. Chim. Acta* 94, 357 (1977).

310. R. K. Kobos and G. A. Rechnitz, *Anal. Lett.* 10, 751 (1977).

311. J. W. Ross, J. H. Riseman, and J. A. Krueger, *Chem. Eng. News* 56, 16 (1978).

312. G. A. Rechnitz, T. L. Riechel, R. K. Kobos, and M. E. Meyer-hoff, *Science* 199, 440 (1978).

313. G. A. Rechnitz, M. A. Arnold, and M. E. Meyerhoff, *Nature* 278, 466 (1979); U.S. Patent 4,216,065 (August 1980); I. Karube, T. Matsumaga, S. Tsura, and S. Suzuki, *Biochim. Biophys Acta* 444, 338 (1976).

314. S. Updike and I. Treichel, *Anal. Chem.* 51, 1643 (1979).

315. M. A. Arnold and G. A. Rechnitz, *Anal. Chem.* 53, 515 (1981).

316. S. R. Grobler and G. A. Rechnitz, *Talanta* 27, 283 (1980).

317. B. Mattiasson and H. Nilsson, *FEBS Lett.* 78, 251 (1977).

318. M. Meyerhoff and G. Rechnitz, *Anal. Biochem.* 95, 483 (1979).

319. M. Haga, H. Itagaki, and T. Okano, *Nippon Kagaku Kaishi* 10, 1549 (1980).

320. J. Boitieux, G. Desmet, and D. Thomas, *Biologie Prospective,* 4e Colloque de Pont-a-Mousson, 1978.

321. J. Boitieux, G. Desmet, and D. Thomas, *Clin. Chem.* 25, 318 (1979).

322. M. Aizawa, A. Morioka, and S. Suzuki, *Anal. Chim. Acta* 115, 61 (1980).

323. N. Yamamoto, Y. Nagasawa, S. Shuto, H. Tsubomura, M. Sawal, and H. Okumura, *Clin. Chem.* 26, 1569 (1980).

324. M. Aizawa, A. Morioka, and S. Suzuki, *J. Membrane Sci.* 4, 221 (1978).

325. M. Aizawa, A. Morioka, H. Matsuoka, S. Suzuki, Y. Nagamura, R. Shinohara, and I. Ishiguro, *J. Solid Phase Biochem.* 1, 319 (1977).

326. N. Yamamoto, Y. Nagasawa, M. Sawai, T. Sudo, and H. Tsubomura, *J. Immol. Meth.* 22, 309 (1978).

327. C. Gebauer and G. Rechnitz, *Anal. Lett.* 14, 97 (1981).

328. M. Aizawa and S. Suzuki, *Chem. Lett.,* pp. 779-782 (1977).

329. M. Aizawa, S. Kato, and S. Suzuki, *J. Membrane Sci.* 2, 125 (1977).

330. S. Suzuki, *J. Solid Phase Biochem.* 4, 25 (1979).

331. R. L. Solsky and G. Rechnitz, *Anal. Chim. Acta 123*, 135 (1981).

332. P. M. Kovach and M. E. Meyerhoff, *Anal. Chem. 54*, 217 (1982).

333. B. Mattiasson, P. Larsson, L. Lindahl, and P. Sablin, *Enzyme Microb. Technol. 4*, 153 (1982).

334. F. Cheng and G. Christian, *Clin. Chim. Acta 91*, 295 (1979).

335. H. Jaegfeldt, A. Torstensson, and G. Johansson, *Anal. Chim. Acta 97*, 221 (1978).

336. L. C. Clark, C. Duggan, T. Grooms, L. Hart, and M. Moore, *Clin. Chem. 27*, 1978 (1981).

337. C. Yuan, S. Kuan, and G. Guilbault, *Anal. Chem. 53*, 190 (1981).

338. C. Yuan, S. Kuan, and G. Guilbault, *Anal. Chim. Acta 124*, 169 (1981).

339. M. Meyerhoff and G. Rechnitz, *Anal. Biochem. 95*, 483 (1979).

340. F. Mizutani, K. Sasaki, and Y. Shimura, *Anal. Chem. 55*, 35 (1983).

341. T. Tsuchida and K. Yoda, *Clin. Chem. 29*(1), 51 (1983).

4

Enzyme Reactors and Membranes

I. GENERAL DISCUSSION

One of the first reported analytical uses of immobilized enzymes for analysis utilized a column approach in which the substrate was passed over a tube of insolubilized biocatalyst followed by fluorometric detection [1]. Such columns, called immobilized enzyme reactors (IMER), are one of the most versatile techniques because of their ability to interface with virtually any detector and with most flow systems. Shortly thereafter, Hornby and co-workers [2] developed another type of reactor, the nylon tube-bound enzyme approach, which today is the most widely used example of immobilized enzyme technology. A third type of reactor, the "membrane" and helicoid reactors, was developed in 1976 by Coulet and his co-workers.

In this chapter all of these techniques will be discussed, as applied with spectrophotometric, thermal, electrochemical, and luminescence detectors. The theory of enzyme reactors will not be discussed because an excellent treatment of the subject was written by Carr and Bowers [44] and by Johansson [45]. In the latter Johansson showed that, from a practical point of view, the theory simplifies if complete conversion (100%) rather than incomplete (<100%) is attained. In the former case, neither flow rate, temperature, pH, ionic strength, nor small additions of inhibitors or activators has any effect on the analytical response of an enzyme reactor.

At first it may appear that a system that relies on complete conversion of substrate will be slow since the reaction requires time to go to completion. But this is not true, as shown by Johansson; the time as variable has been replaced by the flow rate. If the dead volume is kept small by completely filling the reactor with enzyme particles on a rigid support, the reaction will be complete and fast within the reactor, and compression of the packing by the flow will be prevented. The second factor that affects the time of attainment of a steady state is channeling. This factor is not important if the percentage conversion of the reactor is high.

In those assays of inhibitors, which depend on a reproducible amount of enzyme, a very small reactor must be packed carefully.

One disadvantage of the fixed or packed bed reactor is a high-pressure drop, which can eventually clog the flow line.

An alternate approach is that of the open tubular heterogeneous enzyme reactor (the OTHER), which is prepared by immobilizing enzyme onto the inner surface of a tube (originally nylon, but now using many other types of plastic). The OTHER is to the IMER as the capillary tube was to the packed column in gas chromatography. Advantages of the OTHER include (1) low-pressure drop, permitting lower flow rates, (2) compatibility with flow analyzers, such as the Technicon Auto-Analyzer or others, and (3) high throughput of number-of-samples. The only disadvantage is the possible long length of enzyme tube required to effect complete conversion of substrate.

The first immobilization studies using nylon tubes were conducted by Hornby and Sundaram [2]. Most of the research in this area has been conducted by the groups of Hornby, Sundaram, Ngo, and Horvath, and has resulted in the development of commercial tubes sold formerly by Miles and Technicon, and now by Carla Erba in Milan.

Although nylon tubing, which has been extruded from high molecular weight nylon, has very few free amino groups, the numerous secondary amide linkages can be easily cleaved to free amino groups by treatment with 3.7 M HCl for 30 min at 35°C. These free amino groups, plus numerous free carboxyl groups, are readily available for binding to enzymes, using glutaraldehyde treatment (see Chapter 2). Alternatively, bisimidate (dimethyl adipimidate) has been used for coupling. The glutaraldehyde-coupled OTHER has a net negative charge, whereas the dimethyl adipimidate-coupled OTHER possesses a net positive charge on its inner surface.

Because it is so easy to carry out chemical modification on nylon, tubes of this substance have been used extensively as the support for immobilization of enzymes. However, the amide bonds of nylon are very susceptible to acid or alkaline hydrolysis, and nylon tubes can undergo complete dissolution in half-strength mineral acids within a few hours.

For this reason a number of synthetic polymers have been tested in an attempt to obtain a better substrate for enzyme coupling. The best of these appears to be high molecular weight polyethylene, which is more resistant to acid and alkaline hydrolysis. It is chemically more inert, has better mechanical strength, and possesses flexibility. For example, treatment of polyethylene with concentrated H_2SO_4, concentrated HCl, or 30% NaOH at 50°C has little or no effect.

To activate polypropylene, carboxylic functional groups are generated on the inner surface with a 1:2:1 mixture of $CrO_3/H_2O/H_2SO_4$ at 70°C for 30 min, followed by treatment of the tube with 70% HNO_3 at 50°C for 30 min.

The preparation of nylon and polyethylene tubes has been described in detail by Ngo [48].

Coulet and his group at the Laboratory of Biology and Membrane Technology, University Claude Bernard, Lyon [38-41] investigated a totally different type of reactor from the two described above. Enzyme membranes are used to form the reactor, either stacked [38-40] or in a helicoid [41]. In the helicoid approach plastic bands with knobs regularly distributed maintain a constant spacing between the active films, and glucoamylase was used to produce glucose from maltose or soluble starch solutions. Tracer studies have shown that this reactor behaves as a continuous stirred tank reactor. The catalytic support kept its activity for 18 days in continuous operation at 40°C, and the activity was 80% after 17 months storage at 4°C.

Using the same enzyme, a thin-layer flow reactor consisting of
stacks of several 0.15 x 0.16 m collagen membranes maintained by
spacers was constructed [40]. Two Altuglas blocks with inlet and
outlet pipes ensured the water tightness of the system. This reactor
was of the plug-flow type and a large enzyme area was used per volume
of reaction fluid.

A bench scale polymembrane reactor was constructed by Coulet
et al. [39] using an artificial kidney module (RP5 model from Rhone-
Poulenc, Paris) modified by substituting the dialysis membranes with
glucohydrolase collagen membranes prepared in their labs. The modified module consisted of 11 compartments with two enzymatic 32.5 x
11.8 cm membranes in each separated by grooved holding plates. The
enzyme membranes were prepared by soaking the activated membranes in
a two-enzyme solution containing amyloglucosidase and α-amylase.
Good results were obtained for starch conversion.

This "membrane" reactor should prove a valuable type for enzymatic conversion in biotechnology.

II. ANALYTICAL USES OF IMMOBILIZED ENZYME
REACTORS AND NYLON TUBE BOUND ENZYMES

A. Electrochemical Detectors

The combination of an enzyme column with an ion-selective (NH_3, I^-,
NH_4^+, pH) or redox (Pt, Au, O_2) electrode has been quite widely used
as indicated in Table 1. The principles of a detector are similar
to those of an enzyme electrode, except that a column or tube of
enzyme is used rather than a thin layer. Advantages of this technology are:

1. Complete conversion of substrate, which results in a
    greater yield of electroactive product to be detected.
2. Greater stability of the enzyme. Because much more enzyme
    is used than in an electrode, considerable activity can be
    lost with time and usage, and yet the system is still quite
    useful. Nylon tubular enzymes have been used for 10,000-
    20,000 assays, without much loss of activity.

3. Adaptability to continuous analysis on an AutoAnalyzer,
    such as the Technicon. Thus 60 or more assays per hour
    can be performed.

The advantages of using an electrode system are (1) operability
in turbid solutions, (2) no need for color developing reagents,
(3) excellent selectivity, and (4) low cost.

1. AMINO ACIDS

Johansson et al. [3] described an analysis of L-amino acids
such as L-leucine using L-amino acid oxidase immobilized on porous
glass. If oxygen for the reaction is provided only from that dissolved, an upper limit of only 5×10^{-4} M amino acid was found.
Thus catalase (1/3) was immobilized to porous glass and mixed with
the immobilized amino acid oxidase (2/3). Hydrogen peroxide is
added to the buffer as a source of O_2. An NH_3 sensor was used and
10^{-3} to 3×10^{-5} M L-leucine can be assayed using a 3.2 x 45 mm
column.

Ngo has described OTHERs for the reagentless determination of
L-asparagine and L-arginine via the combined use of asparaginase
or arginase plus urease and an NH_4^+ ion electrode [4,5].

2. CARBOHYDRATES

The first of such "electrochemical" systems using immobilized
enzymes was described by Updike and Hicks [6] for the determination
of glucose in blood. The oxygen electrode was used to monitor the
O_2 uptake in an apparatus shown in Fig. 1. Immobilized glucose oxidase in polyacrylamide was placed in a chromatographic column, and
samples containing glucose to be analyzed were pumped over the column
at the rate of 0.4 ml/min with a peristaltic pump. Weibel and co-
workers [7,49], for example, have described a prototype analytical
device, now marketed by Leeds and Northrup [10,11], for the assay
of glucose in blood using glucose oxidase [7] or glucose oxidase
plus peroxidase [49] bound chemically to glass beads. The conversion of glucose is monitored by the uptake of dissolved oxygen in
the sample stream using a Clark O_2 electrode mounted into a miniature

TABLE 1   Immobilized Enzyme Reactors

Substrate	Enzyme	Type	Electrode	Ref.
	A. Electrochemical Detectors			
Alcohol	Alcohol dehydrogenase	Column	Pt	25
	Alcohol oxidase	Tubular	O_2	34
L-Amino acids	Amino acid oxidase/catalase	Column	NH_3	3
L-Arginine	Arginase/urease	Nylon tube	NH_4^+	4, 5
Amylase	Starch/glucoamylase/maltase/glucose oxidase	Column	$Pt(H_2O_2)$	50
L-Asparagine	Asparaginase	Nylon tube	NH_4^+	4, 5
Glucose	Glucose oxidase	Column	O_2	6-11, 27, 47
			Au	36
	Glucose oxidase/catalase	Column	O_2	12, 60
	Glucose oxidase/peroxidase	Column	O_2	49
	Glucose oxidase/catalase	Column	Conductance	13
	Glucose oxidase	Column	$Pt(H_2O_2)$	14
Galactose	Galactose oxidase	Column	$Pt(H_2O_2)$	15
Guanase	Xanthine oxidase	Tubular	I^-	35
Lactose	B-galactsidase/glucose oxidase	Column	H_2O_2	16
			I^-	37
Lecithin	Phospholipase D/choline oxidase	Column	O_2	17
L-Leucine	L-amino acid oxidase/catalase	Column	NH_3	3
Mercury(II)	Urease (inhibition)	Column	NH_3	18

Analyte	Enzyme	Configuration	Detection	References
Maltose	Glucoamylase	Membrane column	Glucose	38, 39, 40
		Helicoid	Glucose	38, 41
Neutral lipids	Lipoprotein lipase	Column	pH	19
Nitrate	Nitrate reductase/nitrite reductase	Column	NH_3	20
Nitrite	Nitrite reductase	Batch	NH_3	51
Organophosphorus pesticides	Cholinesterase	Enzyme pad	S^{2-}	21, 22, 26
Penicillin	Penicillinase	Column	pH	23, 24
Starch	Glucoamylase	Membrane reactor	Glucose	38-40
		Helicoid	Glucose	38, 41
Sucrose	Invertase/mutarotase/glucose oxidase	Column	H_2O_2	28
Urea	Urease	Column	NH_3	29-32
			pH stat	46
Uric acid	Uricase	Column	O_2	33
Zinc(II)	Cholesterase (inhibition)	Pad	S^{2-}	22

B. UV/Visible Absorbance

Analyte	Enzyme	Configuration	Detection	References
Alcohol	Alcohol dehydrogenase	Column	UV	90
L-Aspartic acid	Aspartate amino transferase/MDH	Column	UV	52
Asparagine	Asparaginase	Column	Visible	92, 95
CPK	Hexokinase/glucose-6-PDH	Column	UV	53
		Nylon tube	UV	54
Cholesterol	Cholesterol oxidase	Nylon tube	UV	101
Creatine	CK/PK/LDH	Nylon tube	UV	55

TABLE 1 (Continued)

Substrate	Enzyme	Type	Electrode	Ref.
Creatinine	Creatinine amidohydrolase/CK/PK/LDH	Nylon tube	UV	55
	Creatine amidohydrolase	Nylon tube	Visible	56
Glucose	Glucose oxidase	Column	Visible	57, 58, 93
		Fiber	Visible	95
		Tubular	Visible	54, 58, 59-62, 89, 94
	Glucose dehydrogenase	Tubular	UV	63
		Sandwich	Visible	88
	Hexokinase/G6PDH	Tubular	UV	64, 65
		Tubular	UV	66, 67
Galactose	Galactose dehydrogenase	Column	UV	68
GOT	Malate DH	Tubular	UV	69
Hydroxysteroids	3α-hydroxysteroid dehydrogenase	Tubular	Visible	100
Lactate	LDH/alanine transaminase	Tubular	UV	70
Lactose	β-Galactosidase/glucose oxidase	Tubular	Visible	71
Maltose	Amyloglucosidase/glucose oxidase	Tubular	Visible	71
Nitrate	Nitrate reductase	Column	Visible	72
Oxalate	Oxalate oxidase	Column	Visible	166
Penicillin	Penicillinase	Column	Visible	92
		Fiber	Visible	95
Phosphate	Alkaline phosphatase	Column	Visible	73

Phosphoenolpyruvate	Pyruvate kinase/LDH	Column	UV	74
Pyruvate	Lactate dehydrogenase	Tubular	UV	75
		Column	UV	74
		Tubular	UV	70
Starch	Amyloglucosidase	Helicoil membrane reactor	Visible	40, 41
Sucrose	Invertase	Column	Visible	76, 97
	Invertase/glucose oxidase	Tubular	Visible	71
Sulfate	Aryl sulfatase	Column	Visible	73
Testosterone	3β,17β-Hydroxysteroid dehydrogenase	Column	UV	77
Triglycerides	Glycerol kinase	Tubular	UV	54
	Glycerol dehydrogenase	Tubular	UV	78, 99
Tryptophan	Tryptophanase	Column	UV	79
	Tryptophanase/LDH	Column	UV	79
Urea	Urease	Column	Visible	80, 92
		Tubular	Visible	58, 61, 62, 81, 95
		Pipette	Visible	91
		Fiber	Visible	95
Uric acid	Uricase	Column	Visible	82, 93
		Tubular	Visible	61, 62, 86, 87, 98
		Sandwich	Visible	88

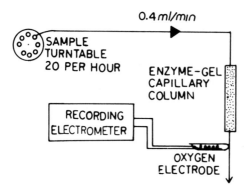

FIG. 1   Instrumentation system used by Updike and Hicks to monitor glucose in blood.   (From Ref. 6.)

vortex mixer (Fig. 2).   Complete conversion of glucose was found in less than 60 sec.   In this system only 0.1 ml of sample is needed, which is injected and diluted.   The technique suffers from disadvantages of large peak widths and O_2 depletion.   The latter problem was solved by using a column of glucose oxidase/peroxidase coimmobilized on Corning glass.   The peroxidase recycles the O_2:

$$H_2O_2 \xrightarrow{\text{peroxidase}} H_2O + \tfrac{1}{2}O_2 \tag{1}$$

Plasma and whole blood could be assayed.

A similar system was used by Dritschilo and Weibel [33] for the assay of uric acid in urine and serum using uric acid oxidase.   Satisfactory reproducibility and accuracy for clinical work was demonstrated.

Kunz and Stasny [12] reported an analytical system for glucose in serum using immobilized glucose oxidase packed in a column that was part of a continuous flow system.   Oxygen depletion was measured with an O_2 electrode and the resulting data were stored and evaluated using a computer.   The column could be used for 1000 assays.   Using 9 μl of sample and appropriate dilution, the width of the detection peak was decreased from the 90 sec observed in the Weibel system [7] to less than 30 sec.   The slope of the response curve, the peak area,

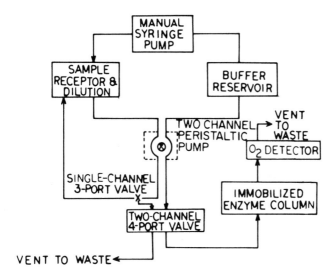

FIG. 2 A block schematic of the glucose monitor showing the principal components. (From Ref. 7.)

and the height of the peak are all proportional to the glucose concentration.

Bergmeyer and Hagen [47] described a permanently circulating system for the determination of glucose that used a chemically bound enzyme with a flow-through oxygen electrode as detector and a reservoir for the buffer solution and all necessary reagents except the enzyme. The system is pictured in Fig. 3. The result of 60 samples an hour is obtained within 1 min and up to 10,000 measurements can be made on one enzyme column. Similarly, the system could be used for absorbance measurements using a flow-through cuvette.

Gorton and Bhatti [36] demonstrated an assay of glucose using immobilized glucose oxidase and an in-line dialyzer to remove proteins from the samples. The normal substrate, oxygen, was replaced by benzoquinone, and the quinone/hydroquinone ratio was monitored with gold electrodes. The reaction was designed to give complete conversion of β-D-glucose to gluconic acid, but some of the advantages of use of an enzyme reactor with complete conversion (discussed above) are lost because the dialyzed fraction is dependent

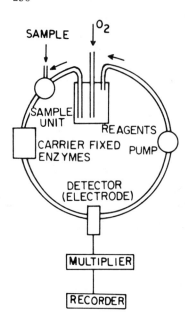

FIG. 3 Permanently circulating system for the determination of
glucose with immobilized enzymes. (From Ref. 47.)

on the flow rate and temperature of the dialyzer. A second Au
electrode is used to correct for interferences. A linear range of
4×10^{-5} to 10^{-2} M is obtained.

Schindler et al. [27] used a flow-through glucose analyzer consisting of a column of glucose oxidase and an O_2 sensor. The latter
was constructed using a leak-current-proof welded membrane.

Watson et al. [14] described a prototype instrument for glucose
that uses a column of glucose oxidase bound to an inorganic support,
and a Pt electrode to sense the H_2O_2 liberated in the enzymatic reaction. The column allows highly reproducible results (1-2% C.V.) over
the range 0-200 mg% glucose in blood.

Messing [13] measured the conductance produced by the gluconic
acid liberated on a glucose oxidase column, and Campbell et al. [8]
used a nylon tube activated by alkylation using dimethyl sulfate with
glucose oxidase attached using lysine, hexamethylene diamine, and
polyethylene imine spacers. The activities of these different tubes

[8] were compared by incorporation into standard Technicon automated analysis systems. The decrease in dissolved O_2 concentration using a flow-through oxygen electrode served as an indicator of the glucose concentration.

Dahodwala et al. [15] described a flow system for assay of galactose similar to those above except for the use of galactose oxidase instead of glucose oxidase. A Pt electrode is used to sense the H_2O_2 liberated.

Similarly, systems for lactose (using immobilized β-galactosidase and glucose oxidase) [16] and sucrose (immobilized invertase mutarotase and glucose oxidase) [28] have been described and are marketed by Leeds and Northrup.

Coulet and his group at University Claude Bernard [38-41] investigated the helicoid (two rolled strips of enzymatic collagen film) and the enzyme membrane (a thin-layer flow reactor composed of stacks of several collagen membranes maintained by spacers) reactors for analysis. By coimmobilizing α-amylase and glucoamylase by random coupling, the hydrolysis of maltodextrins to glucose at high flow rates in recycling experiments was effected. The extent of conversion was monitored with a glucose electrode (glucose oxidase collagen membrane on a Pt anode).

Barabino et al. [50] described a reactor system for the analysis of α-amylase activity. The solution is passed over a column packed with immobilized starch. The resulting oligosaccharides are successively exposed to a column or columns containing immobilized glucose oxidase, catalase, glucoamylase or maltase, and glucose oxidase. The resulting hydrogen peroxide is detected by a three-electrode amperometric cell. All immobilized reagents were insolubilized on a particulate, porous alumina to allow rapid and constant flow rate. From 5 to 200 kU/liter of α-amylase activity could be measured with a 50-μl sample size.

A continuous flow system for assay of lactose in a liquid mixture of various sugars was developed by Volesky and Emond [37]. Lactase and glucose oxidase were immobilized on a phenol-formaldehyde resin. The peroxide produced reacts with iodide, whose concentration is monitored with an iodide membrane electrode.

$$\text{Lactose} \xrightarrow{\text{lactase}} \text{glucose} + \text{galactose} \tag{2}$$

$$\text{Glucose} \xrightarrow{\text{oxidase}} H_2O_2 + \text{gluconic acid} \tag{3}$$

$$H_2O_2 + I^- \xrightarrow{MoO_4^-} I_3^- \tag{4}$$

A multichannel pump fed two independently operating streams, thus eliminating the glucose background. The response time delay was shortened to 15 min to allow lactose control in a fermentation system.

Bouin et al. [96] constructed dual catalysts of varying glucose oxidase and catalase activities by immobilizing the enzymes onto silanized nickel, silica, or alumina with glutaraldehyde. The efficiency of the catalysts were compared by measuring the O_2 uptake with a flow-through electrode of the Clark type. Dual catalysts were compared to mixed catalysts (each enzyme immobilized to separate particles) and to soluble, homogeneous systems at the same equal total activity. The dual catalysts were found to be superior to either of the others.

3. ALCOHOLS AND PENICILLIN

Gulberg and Christian [34] used immobilized alcohol oxidase for the determination of blood alcohol. The alcohol oxidase catalyzed the aerobic oxidation of ethanol and the oxygen concentration was monitored with an oxygen electrode in a flow cell. The enzyme was attached covalently via glutaraldehyde to the inside walls of nylon tubing and by adsorption into three separate controlled-pore glass support materials: SiO_2, Al_2O_3, and TiO_2. The supports were packed into 10-cm lengths of 3-mm i.d. glass tubing or into 30-cm lengths of 5-mm i.d. tubing. A comparison of methods showed that immobilization on silanized glass beads resulted in the highest activity and greatest stability. From 1-10 mg% ethanol could be assayed with good precision.

An electrochemical system for oxidation of reduced NAD directly and after reduction in an enzyme reactor was described by Johansson and his group [25]. NADH was coulometrically oxidized at a rotating Pt gauze electrode at 0.7 V versus SCE at pH 9. The NAD formed was reduced enzymatically in a reactor containing immobilized alcohol dehydrogenase. The overall conversion was 99.3% current efficient. A continuous method for assay of alcohol and removal of aldehyde by dialysis was tested and described.

An immobilized enzyme-based flowing stream reactor-analyzer for measurement of penicillin in fermentation broths was described by Rusling et al. [24]. The enzyme hydrolysis liberates H^+, which is measured with a glass electrode. A covalently bound penicillinase-glass derivative was used in a microcolumn. The potentiometric response is proportional to penicillin in the range 8×10^{-5} to 5×10^{-4} M with a reproducibility of 2%. The new penicillin analyzer was shown to be applicable to the rapid assay of penicillin in fermentation broths. Using a dialysis system cleanup, the results of the penicillin determination on fermentation broths have been found to be in good agreement with results by an established hydoxamic acid procedure colorimetrically. A similar flow system with a column of immobilized penicillinase and a pH electrode detector was described by Marconi et al. [23].

4. UREA, URIC ACID, AND LIPOPROTEINS

Watson and Keyes [29] were the first to describe a dedicated instrument system for the assay of blood urea nitrogen (BUN) using a column of immobilized enzymes. Urease is immobilized on a porous alumina support (Fig. 4) and a gas detecting electrode (Fig. 5) is used to sense the NH_3 resulting from the enzymatic hydrolysis:

$$\text{Urea} \xrightarrow{\text{urease}} NH_4^+ \xrightarrow{OH^-} NH_3 \qquad (5)$$

A comparison of patient samples (plasma and serum) showed no statistical difference when compared with the diacetyl monoxine AutoAnalyzer method. Excellent reproducibility (1-2% C.V.), accuracy, sensitivity,

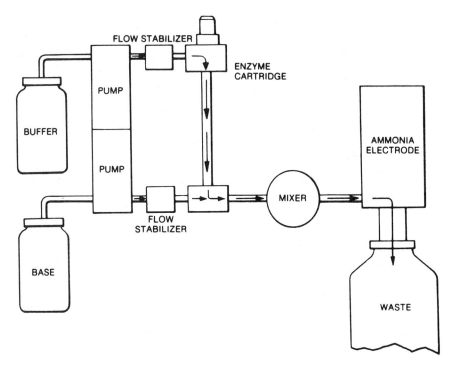

FIG. 4   Schematic diagram of a potentiometric immobilized enzyme
instrument. (Reprinted from B. Watson and M. H. Keyes, *Anal. Lett.*
9, 713 (1976), by courtesy of Marcel Dekker, Inc., New York.)

and specificity are obtained. This instrument also minimizes the
disadvantage of the generally slow response of the NH_3 gas electrode
(23-150 sec) by not waiting for a baseline to be reestablished; a
throughput rate of 45-60 samples/hr has been reported. A commercial
unit, the Kimble BUN analyzer, based on this principle is now sold by
Technicon. Hanson and Bretz evaluated the Kimble BUN analyzer [31]
and found excellent results.

A similar system was described by Johansson and Ogren [30] in
which a buffer carries the sample over a column of glutaraldehyde-
bound enzyme on glass beads. The effluent is mixed with NaOH and
fed to an ammonia gas electrode. At flow rates of 40 ml/hr the
reaction had a slope of 57 mV/decade for 5×10^{-5} to 3×10^{-2} M

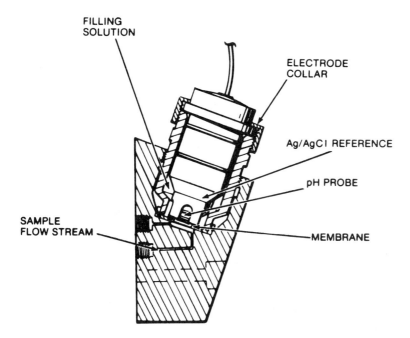

FIG. 5 Details of ammonia electrode. (Reprinted from Watson and
Keyes, *Anal. Lett.* *9*, 713 (1976), by courtesy of Marcel Dekker, Inc.,
New York.)

urea. Only eight samples/hr were reported analyzable. The experimental arrangement is shown in Fig. 6. The reactor was 45 mm long
and the i.d. was 3.4 mm, so that the total volume was 400 µl.

A coulometric flow analyzer for urea was described by Adams and
Carr [46]. A fast electrochemical pH stat was used to titrate the
basic NH_3 formed in the enzymatic reaction. Linearity was achieved
from 1 to 15 mM with 5% precision.

Lovett [32] used a polyacrylamide column of urease and a potentiometric detector for the NH_3 liberated. Only limited enzyme stability was obtained.

Dritschilo and Weibel [33] described a reaction for uric acid
that uses an O_2 electrode to sense the uptake of O_2 in the uricase-
uric acid reaction.

$$\text{Uric acid} + O_2 \xrightarrow{\text{uricase}} H_2O_2 \qquad\qquad (6)$$

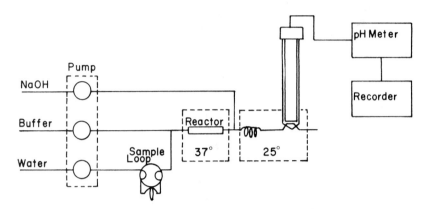

FIG. 6 Experimental arrangement for urea determination. (From Ref. 30.)

A system for lecithin was described [17] that utilizes a cyano-gen bromide-octyl Sepharose 4B column of phospholipase D and choline oxidase and an O_2 detector electrode. Linearity was observed up to 3 g/liter, and better than 70% stability is retained after 9 days. Ca^{2+} and Triton X-100 are needed to assume complete solubilization of reactants and solution.

Neutral lipids were assayed with a polystyrene-γ-aminopropyl triethoxysilane-glutaraldehyde packed column of lipoprotein lipase [19]. A pH electrode was used to measure the acids produced an hydrolysis:

$$\text{Lipids} \xrightarrow{\text{lipoprotein}\atop\text{lipase}} H^+ + \text{alcohols} \tag{7}$$

5. ASSAY OF INHIBITORS

Ogren and Johansson [18] showed that the design considerations for an enzyme reactor become completely different when an inhibitor is to be assayed. A linear relationship of percent inhibition to concentration of inhibitor is obtained only if the ratio of products/ substrate (P/S) is small. Also the sensitivity and linearity increase when the ratio of K_m/S decreases.

The determination of Hg(II) [18] via inhibition of immobilized urease has been described. A buffer containing urea is pumped through

the urease reactor containing 14 µl of enzyme-glass. Only 3% of
the urea is converted to NH_3, which is sensed by the gas electrode.
A valve is then switched so that a 5-ml sample containing Hg(II) is
passed through the reactor; urease is inhibited, and from a plot of
percent inhibition versus Hg(II) concentration (linear from 0 to 0.75
nmoles) the concentration is assayed. The reaction was regenerated
by adding thioacetamide and EDTA between samples. Only silver and
copper interfere.

In 1965 the first application of an immobilized enzyme for detection of atmospheric pollutants was described by Guilbault and
co-workers [21]. Horse serum cholinesterase, immobilized by physical
entrapment in a starch gel on the surface of a urethane foam, catalyzes the hydrolysis of butyrylthiocholine iodide (E $\approx$ 0.6 V) to
thiocholine (E $\approx$ 0.2 V).

$$\text{Butyrylthiocholine iodide} \xrightarrow{\text{cholinesterase}} \text{thiol + alcohol} \quad (8)$$
$$E = 0.6 \text{ V} \qquad\qquad\qquad\qquad E = 0.2 \text{ V} \quad (2)$$

The pad containing the bound enzyme is held between two Pt electrodes;
a cathode on top, anode at the bottom, and a stream of substrate and
air is passed through the pad and electrodes. In the absence of inhibitor the thiol is produced and a low potential is observed; in the
presence of inhibitor the hydrolysis is blocked and the potential
rises to 0.6 V.

A system for the detection of pesticides in air and water has
been described by Goodson et al. [22,26]. The cholinesterase is
adsorbed on aluminum hydroxide gel during the precipitation of the
aluminum hydroxide. The adsorbed enzyme is suspended in a starch
slurry and applied to a polyurethane foam. For water assay a separate line applies sample to the pad. In a second cycle the air and
substrate buffer solution are passed through, and the presence of
inhibitors is noted by a change in the potential of the system. A
device is marketed by Midwest Research on this system, called CAM
(Cholinesterase Antagonist Monitor), and is commercially available
for air and water monitoring. With automatic pad changer, both
pesticides and zinc in the ppb range can be assayed [22].

6. GENERAL ASSAYS

Reactors for assay of nitrate and nitrite have been described.
The nitrate reactor uses a column of coimmobilized nitrate and nitrite
reductase and an NH_3 electrode. The effluent from the column is collected and the NH_3 produced analyzed by an air gap electrode [23].
At 10^{-4} M a 5- to 7-min response is observed. The detection limit
is 2×10^{-5} M nitrate.

$$NO_3^- \xrightarrow{\text{reductase}} NO_2^- \qquad\qquad\qquad (9)$$

$$NO_2^- \xrightarrow{\text{reductase}} NH_3 \qquad\qquad\qquad (10)$$

The nitrite reactor [51] is a batch-type system in which the
nitrite reductase-bovine serum albumin copolymer effects a reduction
of nitrite to ammonia in the presence of methyl viologen as donor.
After 5 minutes incubation, NaOH is added and the NH_3 assayed with
an air gap electrode. Linearity is 0.1-100 mM.

Blaedel and Wang [42] described a mixed immobilized enzyme-
porous reactor that utilizes a flow through electrochemical cell.
A Sepharose-bound enzyme is packed into a retriculated vitreous
carbon (RVC) disc. The effects of various experimental parameters
on the response are described. A rapid 15-sec response is obtained
due to intimate contact between enzyme and the electrode surface.
Substrate measurements could be made at the micromolar concentration
level.

In a subsequent paper [43] the characteristics of this rotated
porous flow-through electrode were evaluated. Pulsed rotating voltammetry was used to discriminate against the major background currents.
Using dopamine, ascorbic acid, and ferrocyanide as test systems,
detection limits at nanomolar levels were obtained.

Immobilized xanthine oxidase was used in a flow system for the
determination of guanase (guanine deaminase) by Gulberg and Christian
[35]. The xanthine produced from guanine in the enzyme reaction is
passed through a column of xanthine oxidase immobilized on silica.
The resulting H_2O_2 is reacted with KI in the presence of molybdate

at pH 5, and the liberated I_3^- is measured amperometrically. A simple assay takes 15 min.

B. UV and Visible Detectors

1. *CARBOHYDRATES*

The first spectrophotometric studies using immobilized enzymes were described by Hicks and Updike [57] using the system illustrated in Fig. 7. A column was packed with the enzyme gel prepared as described above, and substrate was passed over the gel at a rate of 0.8 ml/min. The column and tubing preceding the column were thermostatted at constant temperature. A portion of the column effluent (S_1) (0.2 ml/min) is mixed with a stream of color reagent to detect the reaction products in the effluent stream. After passing through a short delay line to permit the color reaction to develop, the reaction stream passes to a photometer cell. Simultaneously, the color reagent is mixed with buffer or a standard control serum (S_2) and passed through a second delay line into a photometer cell to serve as a reagent blank. The difference in absorbance between the two

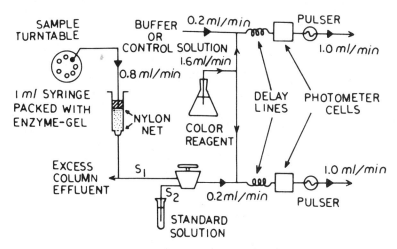

FIG. 7 Diagram of instrumentation used by Hicks and Updike for spectrophotometric studies of immobilized enzymes. (From Ref. 57.)

cells is a measure of the product concentration in the column effluent stream. Standard solutions are introduced at S_2 to permit calibration. Any dehydrogenase or oxidase enzyme system could be monitored by the coupled reactions:

a. Dehydrogenase system:

$$\text{Substrate} + \text{NAD} \xrightarrow{\text{dehydrogenase}} \text{product} + \text{NADH} \qquad (11)$$

$$\text{NADH} + \text{dye}_{(ox)} \xrightarrow{\text{PMS}} \text{NAD} + \text{dye}_{(red)} \qquad (12)$$
$$\text{(blue)} \qquad\qquad\qquad \text{(colorless)}$$

b. Oxidase systems:

$$\text{Substrate} + O_2 \xrightarrow{\text{oxidase}} \text{product} + H_2O_2 \qquad (13)$$

$$H_2O_2 + \text{dye}_{(red)} \xrightarrow{\text{peroxidase}} H_2O + \text{dye}_{(ox)} \qquad (14)$$
$$\text{(colorless)} \qquad\qquad\qquad \text{(blue)}$$

Hornby et al. [59] and Inman and Hornby [58] used urease and
glucose oxidase covalently attached to nylon tubes in a system for
the automated analysis of glucose and urea. Glucose was determined
spectrophotometrically by the indicator reaction:

$$H_2O_2 + I^- \longrightarrow I_3^- \qquad (15)$$
$$\text{(abs. at 349 nm)}$$

and urea by the ammonia formation. The system could be used to
assay 60 samples/hr. The paper cited [59] was the first example of
covalent binding to nylon tubes, a technique that has produced some
of the most stable immobilized enzymes.

Five thousand assays could be performed on one nylon tube with
no detectable loss of activity. These nylon tubes with attached
enzymes were once sold commercially by Miles (Elkhart, Indiana)
under the name Catalinks.

Campbell et al. [8] used a nylon tube activated by alkylation
with dimethyl sulfate for the immobilization of glucose oxidase.
The activities of all of the different nylon tube-glucose oxidase
derivatives were compared by their incorporation into Technicon

automated analysis systems. Activities were measured either spectrophotometrically by the hydrogen peroxide-iodide reaction or polarographically by following the decrease in the dissolved oxygen concentration using an O_2 electrode. Up to 10,000 glucose determinations
were made with no significant decrease in the activity of the enzyme.

Bisse and Vonderschmitt [64] immobilized glucose dehydrogenase
onto a nylon tube and used this preparation for the automated analysis
of glucose. The NADH produced is measured in a flow system at 340 nm.
Linearity was observed from 10 to 800 mg/100 ml of glucose.

Campbell et al. [88] physically entrapped glucose oxidase between
two dialysis membranes and placed the resulting sandwich reactor in a
standard Technicon dialyzer unit. An automated analysis of glucose
in the 2.5-50 mg% range was effected, with good stability observed
over 10 days of operation. A uricase reactor was stable for 12 days
or 1000 assays with a C.V. of 3.6%.

Leung et al. [89] used glucose oxidase nylon tubes (Catalinks,
Miles Laboratories) to specifically oxidize glucose to H_2O_2. Peroxidase was used to couple the H_2O_2 to 4-aminophenazone and 3,5-
dichloro-2-hydroxybenzenesulfonate to form a chromogen at 505 nm.
Very good reproducibility was obtained.

Immobilized glucose oxidase nylon tube reactors (OTHERs) have
been sold commercially by Miles (Catalink) and by Technicon (Autozyme).
These tubes have been withdrawn from the marketplace despite excellent
reported results [54,67,89,94]. Carla Erba (Milan) now sells enzyme-
linked OTHER tubes for glucose, urea, uric acid, and creatinine under
the trade name Clinibond. At least 10,000 assays are guaranteed by
the manufacturer. In one paper by Chirillo et al. [62] the use of
these reactors for assay of glucose, urea, and uric acid was described. The reactors consist of plastic cartridges containing
spirally bound nylon tubes (1 mm i.d.) on the inner walls of which
the enzymes have been bound by covalent binding. The tubes come in
different lengths to fit different instruments. The enzyme reactor
tubes are adapted to continuous flow analyzers (Technicon AAII, SMA
12/60, and SMAC) used in routine laboratory assays as shown in Fig.
8. The reactors were substituted for the free-enzyme reagents in

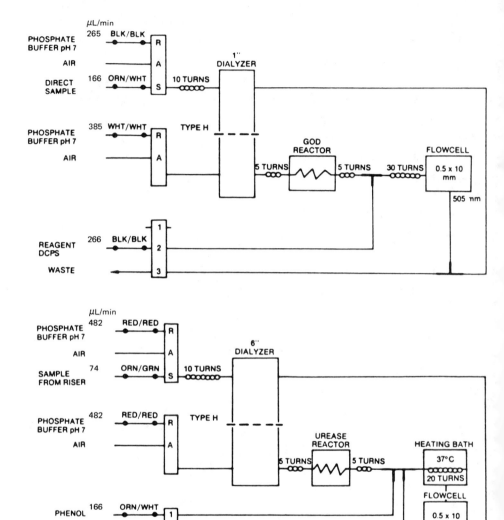

FIG. 8 SMAC analyzer flow diagram for analysis of (top) glucose with 25-cm glucose oxidase reactor, (bottom) urea nitrogen with 25-cm urease reactor, and (upper right) uric acid with 25-cm uricase reactor. DCPS, 2,4-dichlorophenol sulfonate; GOD reactor, glucose oxidase reactor. (From Ref. 62.)

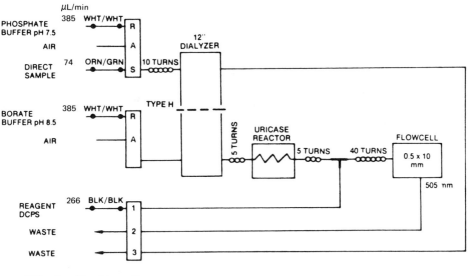

FIG. 8   (Continued)

Fig. 8.   The reactors were substituted for the free-enzyme reagents
in the respective channels of the SMA 12/60 and SMAC, without modi-
fying the parameters of the remaining channels.   Typical tracings are
shown in Figs. 9 and 10, and in Table 2 is presented data on the precision of the Clinibond reactors, used in assay of urea glucose and
uric acid on three automated analyzers.   Excellent results are obtained.

Werner et al. [61] immobilized glucose oxidase, uricase, and
urease on the interior surface of activated polyamide tubing and used
these tubes for continuous flow analysis of glucose, uric acid, and
urea with conventional systems and with hybrid microsystems in which

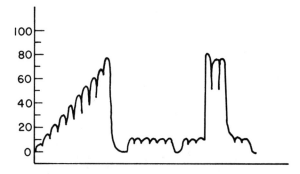

FIG. 9   Tracing obtained with the glucose oxidase reactor for the determination of glucose (SMAC analyzer, 150 samples/hr).   (From Ref. 62.)

modules of different manufacture were combined.   The glucose oxidase tubes were more stable than those of urease or uricase and all tubes could be used for thousands of assays.

Endo et al. [93] bound glucose oxidase to alkylamine glass and prepared a microcolumn with immobilized enzyme for use in an Auto-Analyzer I continuous flow system.   The sensitivity and wash charac-teristics of a column of 1.5 mm i.d. and 40 mm long were satisfactory at an assay speed of 50 samples/hr.   The immobilized enzymes were stable for 2 months or 2000 assays.   Good correlation with accepted methods was observed.

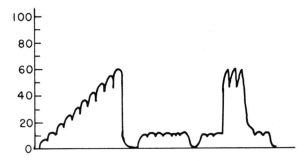

FIG. 10   Tracing obtained with the urease reactor for the determina-tion of urea nitrogen (SMAC analyzer, 150 samples/hr).   (From Ref. 62.)

TABLE 2 Precision of Glucose Oxidase, Urease, and Uricase Reactors in Analysis of Low, Medium, and High Amounts of Glucose, Urea Nitrogen, and Uric Acid with AutoAnalyzer II, SMA 12/60, and SMAC*

	AutoAnalyzer II			SMAC 12/60			SMAC		
	Low	Medium	High	Low	Medium	High	Low	Medium	High
Glucose									
Mean value (mg/liter)	775	2005	4716	769	2135	4953	850	2250	4980
SD (mg/liter)	6.8	21.2	32.3	11.2	19.9	41.7	7.4	17.0	19.2
CV (%)	0.88	1.06	0.68	1.46	0.93	0.84	0.87	0.75	0.38
Urea nitrogen									
Mean value (mg/liter)	235	751	1283	156	506	1022	220	660	1360
SD (mg/liter)	5.0	9.1	15.0	4.7	5.6	11.0	5.4	8.8	31.5
CV (%)	2.13	1.21	1.18	3.06	1.11	1.08	2.49	1.34	2.30
Uric acid									
Mean value (mg/liter)	49.5	91.6	128.3	42.5	84.2	121	45.3	91.7	135.4
SD (mg/liter)	1.3	1.2	1.2	0.5	0.7	0.9	0.6	1.8	3.3
CV (%)	2.63	1.35	1.00	1.19	0.84	0.75	1.27	2.04	2.44

*Average of 48 determinations with SMAC and 30 each with AutoAnalyzer II and SMA. The assays were performed on different days.
Source: Ref. 62.

Joseph et al. [63] described a stopped-flow clinical analyzer that utilizes a reaction loop containing immobilized enzyme (Fig. 11). The concept of the analyzer was tested by determining glucose with immobilized glucose oxidase. The system cost less than $500 to build, and by separating the enzymatic reaction from the follow-up spectrophotometric indicator reaction, it could be used in either fixed time, kinetic, or equilibrium modes. Good precision and accuracy were obtained.

Gorton and Ogren [60] also described a flow injection system for glucose and urea using a reactor to give 100% substrate conversion. The hydrogen peroxide formed is converted to a colored complex with 4-aminophenazone and N,N-dimethylaniline. The coupling is catalyzed by a reactor containing immobilized peroxidase. The colored

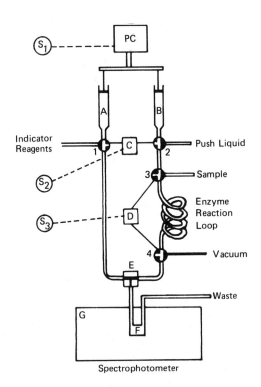

FIG. 11 Diagram of stopped-flow clinical analyzer. (From Ref. 63.)

complex is measured in a flow-through spectrophotometric cell. Urea
is converted to ammonia in a reactor with immobilized urease and detected with an ammonia gas membrane electrode. Proteins and other
interfering species from serum samples are removed in an on-line
dialyzer. Calibration curves are linear for glucose in the range
1.6×10^{-4} to 1.6×10^{-2} M and for urea in the range 10^{-4} to 10^{-1} M.
The samples are 25 µl for glucose determination and 100 µl for urea
determination. Linear ranges can be changed by varying the sample
sizes. The effects of the dialyzer, enzyme reactors, and detection
on dispersion were evaluated.

Marconi et al. [95] described the potential applications of a
silicon tube containing fiber-entrapped enzymes coupled with the
Technicon AutoAnalyzer. The enzyme fibers were found to eliminate
any problems of nonuniform distribution of air bubbles that could
result in overlapping of samples. Assays for glucose, urea, asparagine, penicillin, and sucrose were demonstrated. The automated analysis of glucose, for example, was based on the following reactions:

$$\text{Glucose} + O_2 \xrightarrow{\text{glucose oxidase}} \text{gluconate} + \text{peroxide} \qquad (16)$$

$$\text{Peroxide} + \text{ferrocyanide} \xrightarrow{\text{peroxidase}} \text{ferricyanide} \qquad (17)$$

$$\text{Ferricyanide} + \text{phenol} + \text{4-aminoantipyrene} \longrightarrow \text{quinone} \qquad (18)$$
$$\text{(abs. at 525 nm)}$$

For sucrose, fiber-entrapped invertase effected 100% hydrolysis in
30 min; a linearity of 1-5 mg/ml was observed.

A UV method for assay of galactose using immobilized galactose
dehydrogenase in a column followed by UV detection of the NADH liberated was described by Fleischmann and Schery [68]. Inman and
Hornby [71] detailed the preparation of some immobilized linked
enzyme systems and their use in the automated determination of
disaccharides. In a linked enzyme system, comprising invertase
immobilized within a nylon tube acting in series with glucose oxidase likewise immobilized, an automated assay of sucrose was developed. Similarly, mixtures of glucose oxidase and invertase within
a nylon tube were used for sucrose analysis. The linked enzyme

system of β-galactosidase immobilized within a nylon tube acting in
series with glucose oxidase was used for assay of lactose. Both the
linked system and mixture system containing amyloglucosidase immobilized with glucose oxidase were used for the automated determination
of maltose. Ranges were 0.5-30 mM for sucrose, 1-30 mM for maltose,
and 0.1-10 mM for lactose.

Onyezili and Onitiri [97] immobilized invertase on O-alkylated
nylon tubes modified with amine arms and activated with glutaraldehyde. The performance of the immobilized invertase derivative was
dependent on the nylon tube pretreatment. A method for assay of
sucrose using these coils was proposed.

Coulet and co-workers described the use of a thin-layer flow
reactor consisting in a piling up of collagen membranes [40] (Fig.
12) and a helicoidal reactor with amyloglucosidase collagen mem-

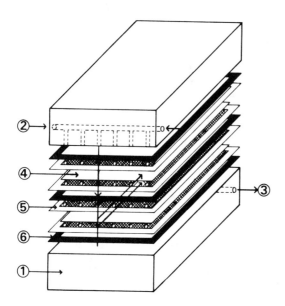

FIG. 12 Scheme of the thin-layer flow reactor. 1, Altuglas block
(50 x 220 x 265 mm); 2, feeding pipe; 3, draining-off pipe; 4, enzymatic collagen membrane (150 x 160 mm); 5, retentate spacer (165 x
195, 150 x 160 mm, external and internal dimensions); 6, plastic
frame (174 x 210, 153 x 185 mm, external and internal dimensions).
(From Ref. 40.)

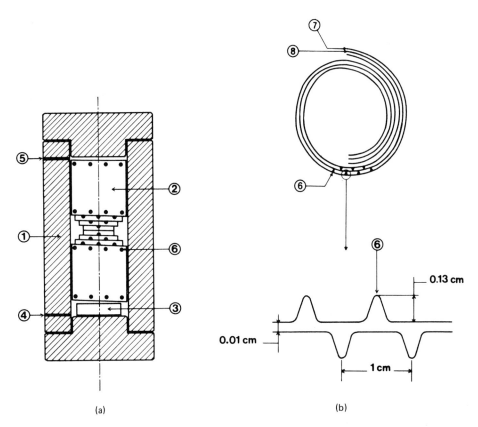

FIG. 13 (a) Reactor design: 1, reactor body; 2, helicoid; 3, magnetic bar; 4, inlet; 5, outlet. (b) Helicoid section view and plastic band details: 6, spike; 7, plastic band; 8, enzymatic collagen
membrane. (From Ref. 41.)

branes [41] (Fig. 13) for use in producing glucose from maltose or
soluble starch solutions. The glucose liberated is measured by the
glucose oxidase-peroxidase-ABTS method (which has greater sensitivity
than o-dianisidine). Analytical methods for assay of soluble starch
were possible.

2. *AMINO ACIDS AND NITROGEN-CONTAINING COMPOUNDS*

Ikeda and Fukui [79] described the application of immobilized
tryptophanase and tryptophanase/lactate dehydrogenase systems for
the assay of tryptophan. The enzymes were bound to cyanogen bromide-

activated Sepharose 4B. The tryptophan was passed over the column
and measured by the absorbance change of NADH at 340 nm. From 5 to
50 mmoles of tryptophan can be assayed.

$$\text{Tryptophan} \longrightarrow \text{indole} + \text{pyruvate} + NH_3 \tag{19}$$

$$\text{Pyruvate} + \text{NADH} \longrightarrow \text{NAD} + \text{lactate} \tag{20}$$

Ikeda et al. [52] proposed the application of immobilized
aspartate aminotransferase (AAT) and immobilized aspartate amino-
transferase-malate dehydrogenase (MDH) for the assay of L-aspartic
acid. From 8 to 50 nmoles of L-aspartate could be assayed.

$$\text{L-aspartate} \xrightarrow{\text{AAT}} \text{2-oxoglutarate} \tag{21}$$

$$\text{NADH} + \text{2-oxoglutarate} \xrightarrow{\text{MDH}} \text{NAD} + \text{malate} \tag{22}$$

The enzymes were immobilized on cyanogen bromide-activated
Sepharose 4B.

Many column reactor methods have been described for the assay
of urea. Immobilized urease covalently bound to the inner surface
of nylon tubes, followed by a spectrophotometric measurement of the
ammonia produced was proposed by Coliss and Knox [84], Filippusson
et al. [82], and Reisel and Katchalski [80]. Glutaraldehyde was used
for coupling and an automated flow system was used. A 3-min assay
time was achieved [84], and assay of both serum and urine could be
effected [80].

Inman and Hornby [58] studied three different immobilized derivatives of urease in the analysis of urea (nylon tube-supported
enzyme, nylon powder-supported enzyme in the form of small packed
beds, and nylon membrane-supported enzyme). Up to 5000 urea samples
could be assayed with the nylon tube urease without incurring any
loss of activity. The nylon tube was also the most sensitive (linear
range 0.01-0.1 mM), the membrane the least sensitive (0.2-2.0 mM);
the powder was intermediate (0.05-0.4 mM range).

Marconi et al. [95] used fiber-entrapped enzymes in silicon
tubing for the assay of urea (urease) and asparagine (asparaginase).
The NH_3 produced was titrated with a pH stat. Up to 20 samples/hr
could be assayed.

Werner et al. [61] immobilized urease on the interior surface
of an activated polyamide tubing and found the resulting tubes were
stable for at least 6 months and several thousand assays. The NH_3
liberated was assayed with phenol-nitroprusside reagent. The calibration curve was linear up to 250 mg% urea.

In an evaluation of the Clinibond enzyme reactor for urea,
Chirillo et al. [62] also used the phenol-nitroprusside reagents, and
with the SMAC analyzer 150 samples/hr could be assayed. The C.V.
ranged from 1 to 3% depending on the analyzer used (AutoAnalyzer II,
SMA 12/60, SMAC) and range of sample.

Sundaram et al. [85] described an immobilized enzyme nylon tube
reactor for routine assay of urea and citrulline in serum. The reactor was used as part of a Technicon AutoAnalyzer I, and the results
obtained for assay of urea were compared with those of the soluble
urease and diacetyl monoxine procedures. Good reliability, reproducibility, and low cost were claimed. The reactor was useful for
at least 2000 tests or 4 months. Linearity is 10-150 mg%.

In a subsequent paper Sundaram and Jayaraman [91] described the
preparation, properties, principles of operation, and use of a new
device, the immobilized enzyme pipette or inpette, for routine assay
of urea. Clinical trials conducted at a medium-size hospital gave
very reliable and reproducible results with very high precision.
The phenol or hypochlorite method was used for colorimetric assay of
NH_3 liberated. A coiled nylon tube was used as the tip for the enzymatic reactor. Over 9 months of use showed 92% stability at 25°C,
80% at 35°C.

Two totally different reactors have been proposed for assay of
creatinine. The first, described by Sundaram and Igloi [55], used
creatininase and creatine kinase (CK) linked to the pyruvate kinase
(PK)-lactate dehydrogenase (LDH) system for the assay of creatinine
and creatine. The four enzymes were bound to a nylon tube; the assay
of NADH is measured spectrophotometrically.

$$\text{Creatinine} \xrightarrow{\text{amidohydrolase}} \text{creatine} \xrightarrow[\text{ATP}]{\text{CK}} \text{ADP} \xrightarrow[\text{pK}]{\text{PEP}} \text{pyruvate}$$

(23)

Pyruvate + NADH $\xrightarrow{\text{LDH}}$ lactate + NAD $\qquad$ (24)

In a better approach Cambiaghi et al. [56] bound creatinine desimidase (EC 3.5.4.21) on the interior surfaces of activated polyamide tubing to form a Clinibond reactor suitable for the analysis of creatinine in continuous flow systems (Fig. 14) with colorimetric assay of the ammonia produced. Each reactor could be used for 20,000 assays (25% activity lost). The samples are premixed with glutamate dehydrogenase (GDH) and NADH to minimize interference from endogenous ammonia. The creatinine dialyzed into the reactor where NH_3 is produced.

NH_3 (endogenous) + α-oxoglutamate + NADH $\xrightarrow{\text{GDH}}$ NAD + glutamate
$\qquad$ (25)

Creatinine + H_2O $\xrightarrow{\text{creatininase}}$ N-methylhydantoin + NH_3 $\quad$ (26)

NH_3 + phenol + hypochlorite $\longrightarrow$ indophenol $\qquad$ (27)

The specificity of the enzyme is excellent, no response being observed with creatine, urea, arginine, guanine, or cytosine. The stability is quite good at up to 50°C, being excellent at 5°C for 12 months (Table 3).

Sundaram et al. [87] described a nylon tube reactor with immobilized uricase for use in assay of acid in a flow through Technicon AutoAnalyzer II. Reliable and reproducible results are obtained with the reactor which is good for 3 months or 4000 assays.

Uric acid $\xrightarrow{\text{uricase}}$ H_2O_2 $\xrightarrow[\text{dye}]{\text{peroxidase}}$ color $\qquad$ (28)
$\qquad\qquad\qquad\qquad\qquad\qquad$ (abs. at 505 nm)

From 1 to 12 mg% uric acid is assayable.

The Clinibond-type polyamide tubing reactor with uricase was used by Werner et al. [61] for the assay of uric acid. The peroxide produced is reacted with 2,4-dichlorophenolsulfonate (Trinder's reagent). At 40 samples/hr linearity is obtained from 10 to 100 mg/liter.

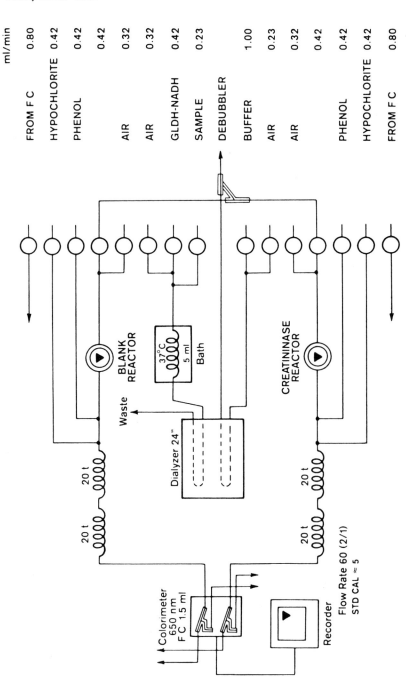

FIG. 14 Flow diagram for automated analysis of creatinine using 1-m immobilized creatinase nylon tube reactor in Technicon AAII. (From Ref. 56.)

TABLE 3   Stability of Immobilized Creatinase

Time	Relative activity		
	4°C	35°C	50°C
Zero	100	100	100
1 Week	100	100	100
2 Weeks	100	100	100
1 Month	100	98	93
2 Months	98	95	90
3 Months	100	89	
6 Months	100	75	
12 Months	100		

Source: Ref. 56.

Chirillo et al. [62] used the Clinibond tube reactors in Technicon AAII, SMA 12/60, and SMAC systems for assay of uric acid in serum.   Linearity was observed to 120 mg/liter (12 mg%).   Ascorbic acid and reduced glutathione interfere at concentrations >15 and 200 mg/liter, respectively.

Endo et al. [93] immobilized uricase from *Candida utilis* on alkylamine glass beads and prepared a column for assay of uric acid. At 50 samples/hr good sensitivity and wash characteristics were observed.   The enzyme was stable for at least 2 months or 2000 assays. A correlation coefficient of .998 to the soluble enzyme-peroxidase-dye method was obtained.

Salleh and Ledingham [98] immobilized uricase onto glutaraldehyde-activated nylon tubes and incorporated these into a continuous flow analyzer.   The immobilized enzyme shows good storage and operational stability and the assay system compares favorably with established methods.   A Technicon coil for uricase was used by Leon et al. [86] in the SMA and SMAC systems for assay of uric acid, and a uricase sandwich procedure was described by Campbell et al. [88].

3. TRIGLYCERIDES, CHOLESTEROL, AND STEROIDS

Leon et al. [54] used small-bore (Autozyme) tubes with immobilized enzymes at the inner wall in SMAC analyzers for the assay of triglycerides. A tube with glycerol kinase is used in the Bucolo assay.

$$\text{Triglycerides} \xrightarrow{\text{lipase}} \text{glycerol + fatty acids} \tag{29}$$

$$\text{Glycerol + ATP} \xrightarrow{\text{kinase}} \text{glycerol-1-phosphate + ADP} \tag{30}$$

$$\text{ADP + PEP} \xrightarrow{\text{PK}} \text{pyruvate + ATP} \tag{31}$$

$$\text{Pyruvate + NADH} \xrightarrow{\text{LDH}} \text{lactate + NAD} \tag{32}$$

All the other enzymes and substrates were soluble and added to the solution; the change in absorbance of NADH is measured. A 1-mm i.d. 15-cm tube is used at 150 assays/hr. At least 3000 assays could be performed. An attempt to immobilize all enzymes was unsuccessful.

Hinsch and Sundaram [78] used an immobilized glycerol dehydrogenase nylon tube reactor in a Technicon AutoAnalyzer II system for the routine assay of glycerol in serum and beverages. The reactor was stable for at least 3500 tests and for several months.

$$\text{Glycerol + NAD} \xrightarrow[\text{dehydrogenase}]{\text{glycerol}} \text{NADH} \tag{33}$$

The system could be used for assay of triglycerides, as described in a subsequent paper [99]. The combination of the glycerol dehydrogenase reactor (stable for 2000 assays) with the lipolytic enzymes added in solution yields a highly reliable and reproducible assay which correlates well with the commonly used fully enzymatic triglyceride determination (.993 coefficient). Using the new method, the cost of analysis could be reduced by one-third. A rate of 60 samples/hr was used.

The 3α-hydroxysteroid dehydrogenase from *Pseudomonas testosteroni* was coimmobilized with diaphorase on cellulose beads and used to prepare a continuous flow microreactor by Bovara et al. [100]. Thus it was possible to determine urinary and serum 3-OH steroids with spectrophotometric monitoring in the visible region. The increased

stability of the enzymes following immobilization and the mechanical
characteristics of the matrix made the reactor usable for at least
20 days. The assay speed was 10-15 samples/hr. The results obtained
with immobilized enzymes followed closely those obtained with free
3α-hydroxysteroid dehydrogenase, and the method is sensitive, reliable,
and economical.

Cremonesi and Bovara [77] used a column of 3β,17β-hydroxysteroid
dehydrogenase for the assay of testosterone. The UV absorbance of
NADH was measured, and the device was used as an HPLC detector.

Ogren et al. [101] described an enzyme reactor for use as a post-
column detector after the HPLC separation of a mixture of cholesterol
and some autooxidation products. The reactor contained cholesterol
oxidase immobilized on controlled-pore glass. The UV absorbance of
the product cholest-4-ene-3-one at 241 nm is measured as an indicator
of the enzymatic reaction. Linearity was observed from 10 to 80 μM.

4. ALCOHOLS AND ACIDS

May and Landgraff [90] described a cofactor-recycling method in
liquid membrane systems that uses a hydrocarbon-based liquid surfac-
tant membrane to effectively retain NAD and NADH. The activity of
an immobilized alcohol dehydrogenase NAD system and of a coupled
cofactor recycling system involving alcohol dehydrogenase/diaphorase/
ferricyanide were examined by noting the extent of ethanol consumption. The system could be used for assay of ethanol.

Newirth et al. [74] used lactate dehydrogenase (LDH) and pyruvate kinase (PK) immobilized in packed-bed reactors to analyze for
both pyruvate (PYR) and phosphoenolpyruvate (PEP) through the disappearance of NADH monitored at 340 nm.

$$\text{PEP} + \text{ADP} \xrightarrow{\text{PK}} \text{PYR} + \text{ATP} \tag{34}$$

$$\text{PYR} + \text{NADH} \xrightarrow{\text{LDH}} \text{lactate} + \text{NAD} \tag{35}$$

Packed-bed reactors containing PK and/or LDH were also capable of
monitoring continuously varying concentrations of ADP, PEP, and PYR.
Preparations of immobilized LDH and PK exhibited enhanced stability

in the presence of mercaptoethanol and NADH, or EDTA, respectively.
75% of the original activity remained after 4 months at 4°C.

A continuous flow clinical analyzer with an immobilized nylon
tube reactor was described by Sundaram and Hinsch [70] for the routine assay of lactate and pyruvate in serum. These reactors are incorporated into the flow system of a modified Technicon AutoAnalyzer.

For pyruvate: pyruvate + NADH $\xrightarrow{\text{LDH}}$ lactate + NAD   (36)

For lactate: lactate + NAD $\xrightarrow{\text{LDH}}$ pyruvate + NADH   (37)

alanine + 2-oxoglutarate $\xleftarrow[\text{glutamate}]{\text{ALT}}$ (38)

In assay of lactate the transaminase (ALT) is added to trap the pyruvate formed. The change in absorbance of NADH is measured. From 1
to 10 μM lactate and from 50 to 1000 μM pyruvate can be assayed.
About one-third activity is left in 22 days.

L-aspartic acid was assayed by Ikeda et al. [52] using bound
aspartate aminotransferase and malate dehydrogenase via cyanogen-
bromide on Sephadex. The change in absorbance of NADH was monitored
at 340 nm. The sample was introduced into the reactor, left 5 min,
eluted, and the absorbance measured.

5. *INORGANIC IONS AND INHIBITORS*

Immobilized enzyme reactors have also been used for the detection and quantitation of inorganic ions.

For example, nitrate in environmental water samples can be
measured using nitrate reductase:

$$NO_3^- + 2MV^{+\bullet} + 2H^+ \xrightarrow{\text{enzyme}} NO_2^- + 2MV^{2+} + H_2O \qquad (39)$$

where MV^{2+} and $MV^{+\bullet}$ are the oxidized and reduced forms of methyl
viologen, respectively. The absorbance of methyl viologen is measured at 543 nm [72]. In Fig. 15 is shown a schematic drawing of
the continuous flow device. The range of assay was 50 ppb to 5 ppm.

Weetall and Jacobsen [73] developed methods for phosphate and
sulfate using immobilized alkaline phosphatase and aryl sulfatase:

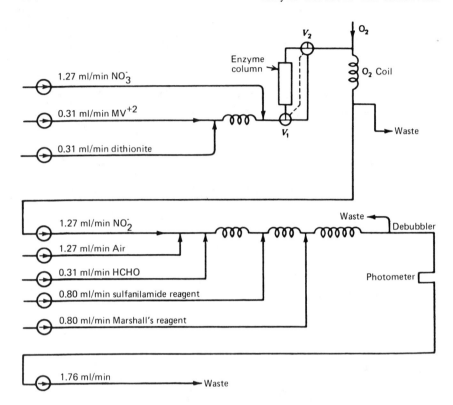

FIG. 15 Schematic diagram of an IMER system for the trace level
determination of nitrate. (From Ref. 72.)

$$\text{p-Nitrophenyl phosphate} + H_2O \rightleftharpoons \text{p-nitrophenol} + HPO_4^{2-} \quad (40)$$
$$\text{(colorless)} \qquad\qquad\qquad \text{(yellow)}$$

$$\text{p-Nitrophenyl sulfate} + H_2O \rightleftharpoons \text{p-nitrophenol} + HSO_4^- \quad (41)$$
$$\text{(colorless)} \qquad\qquad\qquad \text{(yellow)}$$

The competitive inhibitory effect of the inorganic anion causes a
shift in the equilibrium and hence a change in color which is monitored by an apparatus as shown in Fig. 16. Because of the mode of
operation, the system is relatively unaffected by column flow rate,
temperature, or even a relatively wide range of molar concentrations
of the substrate. Operating the system with a relatively high salt

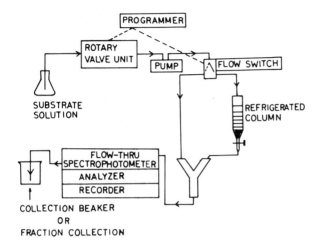

FIG. 16 Schematic representation of apparatus for continuous
monitoring of inorganic phosphate in solution. (From Ref. 73.)

concentration eliminates the nonspecific anion effects, and the
system responds only to the anion of interest.

C. Thermal Detectors

1. GENERAL DISCUSSION

In the two techniques described above, monitoring cannot be
effected unless there is (1) an electroactive or (2) an absorbing
species present. Since all enzyme reactions generate a change of
heat (Table 4), some (e.g., catalase) very large, others (e.g.,
urease) small, in theory a measurement system could be developed
for the assay of any substrate using a column of immobilized enzyme
and a thermistorlike probe to measure the heat of reaction. In those
reactions where the heat change is very small (i.e., ester hydrolysis),
suitable amplification steps could be introduced [102,103].

An important advantage of thermal analysis is that the enzymatic
reaction can be followed irrespective of the optical properties of
the sample (e.g., turbid or highly colored solution). Additionally,
there is no need to add coupling reactions to the main reaction which

TABLE 4   Molar Enthalpies of Some Enzyme-Catalyzed Reactions

Enzyme	EC number	Substrate	$-\Delta H$ (kJ/mole)
Catalase	1.11.1.6	Hydrogen peroxide	100.4
Cholesterol oxidase	1.1.3.6	Cholesterol	52.9
Glucose oxidase	1.1.3.4	Glucose	80
Hexokinase	2.7.1.1	Glucose	27.6
Lactate dehydrogenase	1.1.1.27	Sodium pyruvate	62.1
Trypsin	3.4.21.4	Benzoyl-L-arginineamide	27.8
Urease	3.5.1.5	Urea	6.6
Uricase	1.7.3.3	Urate	49.1

give rise to a color change or pH shift yet require expensive cofactors and additional reagents.

But despite an excellent potential, the application of calorimetry to biochemical analysis has not gained widespread usage, probably due to rather high instrument cost of commercially available microcalorimeters and the lack of readily available instrumentation for enzymatic studies. A new instrument incorporating the technology of Mosbach and Danielsson developed at the Chemical Center in Lund will soon be available from Bifork, Sweden.

One technique of the thermal analysis is that of direct-injection enthalpimetry (thermal titrimetry) [104,105] in which the sensor is placed directly in the reaction chamber (Fig. 17b). This technique can be described as "instantaneous thermometric titration," that is, a temperature pulse is formed and recorded in an adiabatic cell on injection of a reagent to the enzyme solution in the vessel followed by rapid mixing; the electrodes are called thermal enzyme probes (TEP). Thermal titrimetry is a very sensitive method and easier to calibrate than the thermistor device. However, the number of successive runs possible is limited in the case of thermal titrimetry because of the accumulation of reaction product(s) in the vessel. Further, it appears to be difficult to use immobilized enzymes with

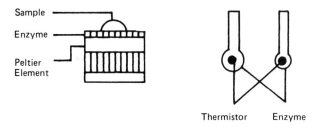

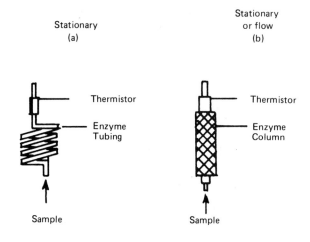

FIG. 17 Schematic of (a) small-volume calorimeter, (b) thermal
enzyme probes (TEP), and (c) enzyme thermistor or immobilized
enzyme flow enthalpimetric analyzer. (From Ref. 113.)

such a technique. In addition, the enzyme thermistor can also be
applied to continuous analysis of various processes, whereas this
is not possible with thermal titrimetry.

   An additional approach to thermal analysis has recently been
described [106] in which the enzyme is coupled onto the surface of
a small solid-state (Peltier) detector in a vesselless design (Fig.
17a). The detector surface, a thin sheet (0.009 in.) of aluminum,

is anodized prior to binding the enzyme. The reported sensitivity
of the enzymatic reaction studied, i.e., peroxidase was good, gave
reproducible signals with a low substrate concentration when 10 μl
of the solution was applied. However, further studies are necessary
to fully assess this method, in particular whether it can be applied
to continuous analyses. Another important aspect to be studied is
whether sufficient amounts of enzyme can be immobilized to the detector permitting analysis of larger amounts of substrate concentrations.
Both these aspects have previously been found to be limiting factors
in early enzyme thermistorlike devices [107].

Although useful, the low sensitivity of the TEP and Peltier
probes have made these systems unattractive. A solution to the problem is the use of flow-through systems, which allows the heat evolved
to be transported to or along the thermal probe, thus minimizing heat
losses. By packing the immobilized enzyme in columns surrounding the
probe, as exemplified by the enzyme thermistor [102,108] or the immobilized enzyme flow-enthalpimetric analyzer [109], increased sensitivity and continuous sampling is possible. Either end point or
kinetic measurements can be made. Such flow devices are indicated
in Fig. 17c. Recent modifications of the original TEP method now
utilize a laminar flow [110], making it a far more attractive technique.

Some excellent reviews on thermal bioanalyzers have been written
by Danielsson [111,112], Mosbach and Danielsson [113], Martin and
Marini [114], and Grime [115]. In this section the apparatus, methodology, and analytical applications of the use of a thermistor
device with immobilized enzymes will be described.

Finally, Schifreen et al. [116] studied the fundamental principles of operation of an immobilized enzyme flow analyzer, using a
high-speed thermoanalyzer (thermistor versus thermistor, ΔT measured)
with urease immobilized on porous glass by covalent attachment.
Models were developed to explain the dependence of peak height, area,
and half-width on sample concentration (0.1-2000 mM), sample volume
(40-2500 μl), flow rate (0.3-4.5 ml), and size of column (0.4-0.9 ml).

Up to 60 samples/hr can be processed in columns of 1000 U of enzyme
at flow rates of 2 ml/min.

2. INSTRUMENTATION

A small, glass-encapsulated thermistor is placed at the tip of
a stainless steel tube (2 mm o.d.) which is placed at the top of the
enzyme column (a small 0.2- to 1.0-ml plastic column) which contains
the immobilized enzyme preparation, usually covalently bound to small,
porous glass beads. The column is mounted in a Plexiglas housing sur-
rounded by airspace for thermal insulation in a holder designed for
rapid changing of columns (Fig. 18). Buffer is continuously pumped
through the enzyme thermistor using a peristaltic pump after passing
through a heat exchanger made of a piece of thin-walled stainless
steel tubing (40-50 cm long, 0.8 mm i.d.). The heat exchanger coil
sits in a water-filled plastic cup to minimize temperature fluctuations in the solution passing through the column. These temperature
fluctuations are on the order of $\pm 0.01°$ in the water baths normally
used and less than $\pm 10^{-4}°C$ around the thermistor.

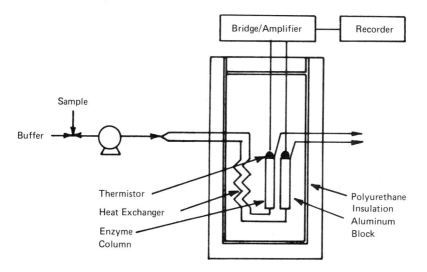

FIG. 18 Schematic representation of an enzyme thermistor system
(can be used as a single thermistor device or, if required, in the
split-flow arrangement including reference column indicated to the
right). (From Ref. 113.)

The temperature signal is registered with a potentiometric
recorder coupled to a commercially available Wheatstone bridge,
which delivers a 100-mV signal for a temperature change of 0.02° on
the most sensitive range. In order to increase baseline stability
and insensitivity to nonspecific heat production caused by dilution
or interactions with the enzyme support, a split-flow enzyme thermistor arrangement can be applied (Fig. 18) [117]. In this case two
columns are used: one containing the immobilized enzyme and the
other, the reference column, containing support material only, or,
if required, inactivated enzyme. Nonspecific effects are assumed
to be equal in the two columns, so that the differential temperature
signal recorded by the two thermistors at the outlets of the columns
represents the true signal originating from the enzymatic reaction.
Two pumps are used in this arrangement because it is important that
the flow through each column be identical.

Immersion of the device in a good water bath gives sufficient
accuracy for a great number of analyses. However, to allow use as
a convenient routine instrument, the water bath may be replaced with
a temperature-controlled metal block. This unit, which includes a
new Wheatstone bridge, now produces at maximum sensitivity a 100-mV
change in the recorder signal for a temperature change of 10^{-3}°C
[112], thus permitting measurement of temperature changes down to
10^{-5}°C. Electrical and other calibrations have shown that as much
as 80% of the total heat evolved in a semiadiabatic device can optimally be registered as a change in temperature. This implies that
for a given substrate present at a concentration of 1 mM and with a
molar enthalpy change for the enzymatic reaction of 80 kJ/mole, a
peak height corresponding to 10^{-2}°C or higher will be obtained, and
a temperature resolution of 10^{-4}°C gives 1% accuracy in the measurement. The enzyme thermistor can quickly (i.e., within 15 min) be
recharged with a new enzyme column when required. The complete
enzyme thermistor setup is shown in Fig. 19.

3. *APPLICATIONS*

Analytical applications of immobilized enzymes and heat change
measurements have been in four major areas: clinical analysis,

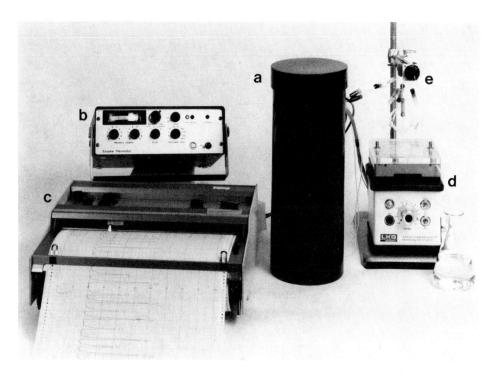

FIG. 19 Photograph of a complete enzyme thermistor setup from the
phototype series: recorder (c), pump (d), and enzyme thermistor (a)
connected to Wheatstone bridge (b).   (From Ref. 113.)

immunological analysis, fermentation analysis and process control,
and environmental analysis.   These are listed in Table 5.   A great
variety of substances have been determined, and in theory any substance that participates in any enzymatic reaction either as substrate, activator, or inhibitor can be measured by this technique.

a. Clinical Analysis

As usual most attention has been devoted to the assay of urea
and glucose, the two most important metabolites, and the range assayable lies well within that of interest clinically.

Using a column of glucose oxidase, Mattiasson et al. [117] were
able to assay 0.03- to 0.5-mM concentrations of glucose.   The range
could be extended to 5 mM by adding catalase.   Danielsson et al.
[131] used a packed bed (glass beads) of glucose oxidase with a

TABLE 5 Applications of Thermal Detectors

Substance	Immobilized biocatalyst	Range (mM)	Ref.
Albumin	Immobilized antibodies + enzyme-linked antigen	$10^{-7}-10^{-5}$	118, 119
Ascorbic acid	Ascorbate oxidase	0.05-0.6	120
ATP	Apyrase or hexokinase	0.001-0.008	121, 122
Benzoyl arginine ethyl ester	Trypsin	0.001-0.005	123, 124
Cellobiose	β-Glucosidase + glucose oxidase/catalase	0.05-5	125
Cephalosporin	Cephalosporinase	0.005-10	120
Cholesterol	Cholesterol oxidase	0.03-0.15	117, 120
Cholesterol esters	Cholesterol esterase/cholesterol oxidase	0.03-0.15	117, 120
Creatinine	Creatinine iminohydrolase	0.01-10	120
Cyanide	Rhodanase	0.02-1	126, 129
Ethanol	Alcohol oxidase	0.01-1	127
Galactose	Galactose oxidase	0.01-1	125
Gentamicin	Immobilized antibodies + enzyme-linked antigen	0.1 μg/ml (detection)	128
Glucose	Glucose oxidase	0.002-1.0	124
	Glucose oxidase/catalase	0.002-0.8	117, 130-132, 143
	Glucose oxidase/albumin/antibody	0.05-0.5	133
	Hexokinase	2.5-25	134

Analyte	Reagent	Range	Ref.
Heavy-metal ions (Hg²⁺, Cu²⁺, Ag⁺)	Urease	$0.075-1.0$ (Cu^{2+}), $0.01-0.1$ (Hg^{2+})	135
Insecticides (e.g., parathion)	Acetylcholinesterase	$5 \times 10^{-3} - 10^{-2}$	136
Insulin	Immobilized antibodies/enzyme-linked antigen	$0.1-1.0$ U/ml	137
Lactate	Lactate-2-monooxygenase	$0.1-1.0$	125
Lactose	Lactase/glucose oxidase/catalase	$0.05-10$	117
Oxalic acid	Oxalate oxidase	$0.005-0.5$	120, 125
	Oxalate decarboxylase	$0.1-3$	125
Penicillin G	Penicillinase	$0.01-500$	120, 124, 138
	Penicillinase-albumin-Ab	$0.1-10$	133
Peroxide	Catalase	$0.005-10$	120, 139, 143
	Catalase-albumin-Ab	$0.05-10$	133
Phenol	Phenol oxidase	$0.01-1$	120
	Tyrosinase-albumin-Ab	$0.1-1$	133
Sucrose	Yeast	$0.05-100$	140
	Invertase-albumin-Ab	$1-50$	133
Triglycerides	Lipase, lipoprotein	$0.1-5$	141
Tyrosine	Tyrosinase-albumin-Ab	$0.1-1$	133
Urea	Urease	$0.01-500$	110, 118, 124, 142-146
Uric acid	Uricase/catalase	$0.5-4$	117

dual-column/dual-thermistor split-flow system. A linear range of
0.01- to 0.45-mM glucose was observed, which could be extended to
0.90 mM by coimmobilization of catalase.

Bowers et al. [134] described an immobilized enzyme flow enthalpimetric analyzer for glucose by direct phosphorylation catalyzed by
hexokinase. The linearity observed was 50-500 mg/100 ml, and good
correlation was obtained with the o-toluidine and hexokinase procedures. The precision is 5% and only 120 μl of sample is used.

Mosbach et al. [124] placed the thermistor probe in close proximity to the site of reaction in order to develop maximum sensitivity.
The probe, placed directly into a microcolumn filled with enzyme,
allowed the assay of glucose (0.0625-1 μmole), urea (2.7-7 mM), penicillin (7.5-150 μmoles) and benzoyl-L-arginine ethyl ester (1-5 μmoles).

Johansson et al. [123] placed immobilized trypsin in polyacrylamide beads in a column with a thermistor probe. Upon reaction with
benzoyl arginine ethyl ester a heat change was produced proportional
to concentration (1-5 mM).

Bowers et al. [142] coupled a differential thermal detector with
an immobilized urease reaction to determine urea in serum. The measured temperature changes were proportional to serum urea concentration. No interference was noted from protein, bilirubin, or hemoglobin. Results correlate well with the diacetyl and indophenol
procedures, with a linear range of 5-70 ng/100 ml.

Fulton et al. [110] using a flow-through cell that provides
quiet laminar flow past a thermal enzyme probe (TEP) consisting of
two thermistors (one blank, one with immobilized urease) showed that
temperature differences of 2-3μ°C between the two probes could be
resolved. Thus, good sensitivity is obtained for as little as 10^{-4} M
urea.

Tran-Minh and Vallin [143] coated the sensitive part of a thermistor with an artificial enzyme membrane and used this device for the
assay of hydrogen peroxide, glucose, and urea. Excellent results
were cited.

Mattiasson et al. [117,120] described column reactors for cholesterol and creatinine. Using cholesterol esterase plus oxidase,

total cholesterol esters could be assayed, from 0.03 to 0.15 mM.
Creatinine was linear from 0.1 to 10 mM.

Satoh et al. [141] applied a split-flow type of enzyme thermistor for the determination of triglycerides. The device measured
the protonation heat produced when the triglyceride is passed through
a column containing triacyl glycerol lipase covalently bound to controlled-pore glass beads. The time required for reaction is less
than 5 min, and the calibration curve is linear to 5 mM. Triglycerides in serum could be assayed with excellent correlation to conventional spectrophotometric procedures.

Mosbach and Danielsson [121] described a heat sensor placed
directly in a bed of immobilized enzyme in development of a method
for ATP and BAEE:

$$ATP + H_2O \xrightarrow[\text{EC 3.6.1.5}]{\text{apyrase}} ADP + Pi \tag{42}$$

With immobilized apyrase 2-8 μmole of ATP could be assayed.

A very simple procedure was developed by the Lund group for
lactate [125]. Only a single enzyme, lactate oxidase, was used.
Linearity of ΔT to lactate concentration was found from 0.1 to
1.0 mM lactate, and both serum and fermentation broths could be
analyzed. Additionally, oxalate was assayed with both oxalate
decarboxylase [125] and oxalate oxidase [120,125] with good sensitivity (as little as 0.005 mM oxalate detectable).

b. Environmental Analysis

The enzyme thermistor probe has been used for the determination of specific components in wastewater samples. Specificity was
achieved by the enzymatic reaction, even in turbid and cloudy solutions. The researchers at Lund have shown the potential of this
technology in two areas: determination of the inhibitors of a certain
enzymatic reaction and determination of a substrate of an enzyme.

An example of the first is the analysis of heavy-metal ions
using immobilized urease [135]. The degree of inhibition was expressed as the ratio of the temperature peaks obtained before and

after introduction of inhibitor; 0.075-1.0 mM Cu^{2+} and 0.01-0.1 mM Hg^{2+} were assayed.

The second type of analysis demonstrated by Mattiasson et al. [126,129] was that of cyanide in standard solutions and in blast furnace water. The heat signal generated in the conversion of cyanide, catalyzed by the immobilized rhodanase and injectase, is measured. With injectase 0.020-0.6 mM CN^- can be assayed and with rhodanase 0.02-1.0 mM. After 200 hr operation, 80% stability was retained. The response time was 2-3 min.

   c. *Process Fermentation Control*

Of primary concern in this area is analysis of carbohydrates, such as sucrose, lactose, cellobiose, antibiotics such as penicillin and cephalosporins, and metabolites such as ethanol and lactate. Sucrose can be determined in the range of 0.05-100 mM in a very simple and reliable fashion using the enzyme invertase (EC 3.2.1.26) [140]. Other glucose-containing disaccharides such as lactose and cellobiose can be determined by a glucose oxidase/catalase thermistor after enzymatic hydrolysis of the disaccharide. This is preferably done with the hydrolytic enzyme in a precolumn, as was done with β-glucosidase in a recent study on cellobiose [125]. Enzyme thermistor assays for carbohydrates are sufficiently accurate to be used for control of biotechnological processes [140], as demonstrated for sucrose assay using immobilized yeast, or glucose with immobilized glucose oxidase. Also ethanol fermentation was monitored with an enzyme thermistor as a sucrose sensor [140].

Very good results have been obtained in studies on antibiotics. Penicillin can be measured in discrete samples or continuously with the aid of immobilized penicillinase (β-lactamase, EC 3.5.2.6) [138]. The enzyme has been covalently bound either to CPG or to the inside of nylon tubing. Both systems work well. CPG gives a somewhat higher sensitivity, but the nylon tubing is less susceptible to fouling (Fig. 20). Enzyme thermistor assay of penicillin in fermentation broths competes favorably with existing methods [120,124, 138]. Another promising enzyme thermistor application is the assay

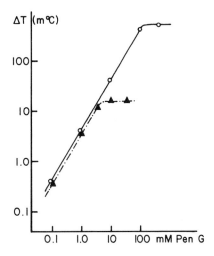

FIG. 20 Calibration curves (log-log scale) for penicillin G with
β-lactamase as the enzyme. Comparison of the linear ranges obtained
for CPG-bound (o) and for nylon tubing-coupled (▲) enzyme. (From
Ref. 138.)

of cephalosporins using cephalosporinase [120] as mentioned above.
It has the same advantages as the penicillin assay. Figure 21 shows
calibration curves for two different commercial preparations of
cephalosporin.

 d. *Immunological Analysis*

The same measuring device can also be applied with slight modifications to the growing area of immunochemical analysis (i.e.,
antigen/antibody determination). For this alternative procedure we
have suggested the name thermometric enzyme-linked immunosorbent
assay (TELISA) [118]. In principle, the column of the enzyme thermistor is filled with an immunosorbent such as antibodies immobilized
on Sepharose CL 4B. The antigen to be determined and an enzyme-
labeled antigen are introduced into the flow, whence the amount of
enzyme-bound antigen remaining bound to the column is a function of
the content of antigen. The less antigen present in the sample, the
more enzyme-labeled antigen will be found in the column, and thus
the more heat that will be evolved after the subsequent introduction

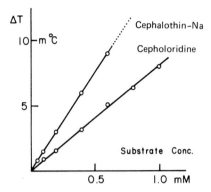

FIG. 21 Temperature response curves for two different commercial cephalosporins: cepholoridine and the sodium salt of cephalothin. Sample pulses (1-ml) were introduced with a flow rate of 0.8 ml/min. About 100 U of cephalosporinase was coupled to CPG. The buffer used was 0.1 M tris-HCl, pH 8.3. (From Ref. 120.)

of the enzyme's substrate into the flow stream. Excellent sensitivity down to 10^{-13} moles/liter has been found with the TELISA technique (Table 5).

In another broad manuscript Mattiasson [133] described a general enzyme thermistor using antigen-antibody reactions for the assay of hydrogen peroxide, penicillin, sucrose, glucose, phenol, and tyrosine. Anti-human serum albumin (HSA) is prepared and coupled with the enzymes catalase (hydrogen peroxide, range 5×10^{-5} to 10^{-2} M), penicillinase (penicillin G, range 10^{-4} to 10^{-2} M), tyrosinase (phenol, range 10^{-4} to 10^{-3} M and tyrosine, range 10^{-4} to 10^{-3} M), invertase (sucrose, 10^{-3} to 5×10^{-2} M linear range) and glucose oxidase (glucose, range 5×10^{-5} to 5×10^{-4} M).

A schematic presentation of an assay cycle is shown in Fig. 22.

D. Luminescence Detectors

   1. GENERAL DISCUSSION

   When a molecule absorbs radiation, it is excited to a higher electronic, vibrational, or rotational level. It then returns to the ground state via a radiationless transition (an absorbing molecule, such as NADH) or via an emission of radiation (a fluorescent

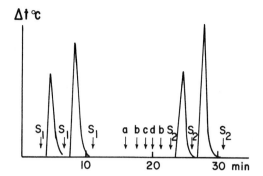

FIG. 22 Schematic presentation of an assay cycle. The arrows indicate changes in the perfusion medium, normally 0.1 M potassium phosphate buffer, pH 7.0 (flow rate 0.74 ml/min). The cycle starts with an enzyme-antigen (E_1, HSA) bound to the antibody- (anti-HSA) containing support material. At the arrows marked S_1 substrate for enzyme E_1 is introduced as pulses, and at the arrows marked S_2 substrate for enzyme E_2 is introduced. The heat signals obtained upon substrate pulses are represented by the peaks. At the arrow a, a 2-min pulse of 0.2 M glycine-HCl, pH 2.2, is introduced in order to split the complex and to wash the system. After a pulse of potassium phosphate buffer b, new enzyme-antigen complex (E_2, HSA) is introduced (c), followed by a 1-min pulse of 0.1 M potassium phosphate in 0.5 M KCl (d) and buffer (b). The system is then ready for new assays. (From Ref. 133.)

molecule like fluorescein). In the first case the molecule is said to be absorbing, in the latter luminescent.

A molecule can also be excited to a higher state by a chemical reaction occurring; it can then emit its excess radiation in returning to the ground state. Such a molecule is said to be chemiluminescent (excitation via a chemical reaction) or bioluminescent (excitation by a biological reaction).

In this section we shall consider those assays performed using immobilized enzymes and either fluorescent and chemiluminescent detections. A list of all these assays developed to 1982 are given in Table 6.

2. APPLICATIONS

The majority of applications in this area have centered about the assay of glucose and ATP.

TABLE 6 Analysis Using Luminescence Techniques

Substrate	Enzyme	Support	Type	Ref.
Androsterone	Luciferase/oxidoreductase/ steroid dehydrogenase	Rods	Chemi.	147
ATP	Luciferase	Rods	Chemi	148
		Films	Chemi	149, 150
		Suspension	Chemi.	151, 152
Glucose	Glucose oxidase	Column	Chemi.	153, 154, 155
	Glucose oxidase/glucose-6-phosphate dehydrogenase/luciferase	Column	Chemi.	158
Glucose-6-phosphate	Glucose-6-Pi dehydrogenase/luciferase	Column	Chemi.	147, 158
Lactic acid	Lactate dehydrogenase	Column	Fluor.	156
			Chemi.	157
NADH	Luciferase/FMN reductase	Column	Chemi.	147, 158
		Rods	Chemi.	159
NADPH	Luciferase/FMN reductase	Column	Chemi.	158
Organophosphorus insecticides	Cholinesterase	Column	Fluor.	160
Peroxide	Peroxidase	Column	Fluor.	161, 162
		Tubular	Chemi.	163
Pyruvate	Lactate dehydrogenase	Column	Chemi.	157
Testosterone	Luciferase/oxidoreductase/ steroid dehydrogenase	Rods	Chemi.	147
Uric acid	Uricase	Column	Chemi.	164
		Sandwich	Fluor.	165

In one of the first papers in this area Bostick and Hercules
[153] measured blood glucose using the enzymatic conversion of β-D-
glucose to D-gluconic acid and hydrogen peroxide in an immobilized
glucose oxidase column. The peroxide formed reacts with a mixed
luminol-ferricyanide reagent to produce a chemiluminescence, the
intensity of which is proportional to glucose over the range 10^{-8}
to 10^{-4} M. Good correlation with standard methods were obtained.
Peroxide in the range of 10^{-8} to 10^{-5} M can also be assayed. Ferricyanide was chosen as catalyst, although several metals (Cu^{2+}, Co^{2+},
Ni^{2+}) catalyze the reaction; these could be assayed based on their
catalysis. The small, Sepharose-bound glucose oxidase reaction is
placed in an arrangement as shown in Fig. 23. The precision is
about 1%.

Williams et al. [154] modified this system and assayed for
glucose in urine; good results were obtained. Auses et al. [155]
described a system similar to that of Bostick and Hercules at almost
the same time. Hornby [163] used a nylon tubular column of peroxidase placed directly on a photomultiplier tube in developing an assay
for peroxide liberated in enzymatic reactions.

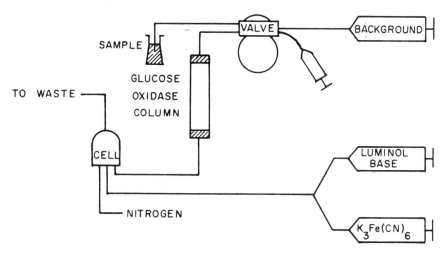

FIG. 23 Schematic diagram of apparatus designed for IMER-based
chemiluminescent determination of serum glucose. (From Ref. 153.)

Kamoun and Douay [165] used an enzyme membrane sandwich reactor for assay of uric acid. The peroxide produced is reacted with peroxidase and p-hydroxyphenylacetic acid to give a fluorescence. Gorus and Schram [164] measured the chemiluminescence produced by the peroxide luminol reaction after passing uric acid over a column of uricase.

Brolin et al. [157] entrapped lactate dehydrogenase in polyacrylamide microparticles and developed an assay for both lactate and pyruvate. The chemiluminescence of the NADH-FMN-luciferase reaction is measured. From 10 to 100 pmoles of lactate can be assayed.

Jablonski and DeLuca [158,159] described the use of immobilized bacterial luciferase and FMN reductase on glass for the assay of NADH, glucose phosphate, and NADPH. The enzymes, isolated from *Beneckea harveyi*, were covalently linked via diapotization to arylamine glass beads. These beads were either packed into a column [158] or were cemented onto a glass rod [159]. These immobilized enzymes are individually active and function to produce light via a coupled reaction utilizing NADH or NADPH:

$$NAD(P)H + FMN + H^+ \xrightarrow{\text{oxidoreductase}} NAD(P) + FMNH_2 \qquad (43)$$

$$FMNH_2 + RCHO + O_2 \xrightarrow{\text{luciferase}} FMN + RCOOH + h\nu \qquad (44)$$

A linearity is observed between intensity of light emitted as a function of NADH or NADPH concentration. Linearity is obtained in the range 1 pmole to 50 nmoles of NADH, and between 10 pmoles and 200 nmoles for NADPH [159]. Using the column as little as 0.2 pmole of NADH could be assayed [158]. Glucose-6-phosphate dehydrogenase was coimmobilized with these enzymes, and with this system it is possible to quantitate 1 pmole of glucose-6-phosphate. By coimmobilizing a fourth enzyme 20 pmoles/liter of glucose can be determined [158]. The bound enzymes are quite stable and reusable.

Many groups have devoted effort to developing a sensitive assay for ATP using the firefly reaction:

$$\text{Luciferin} + \text{luciferase} + \text{ATP} \xrightarrow{Mg^{2+}} h\nu \qquad (45)$$

Lee et al. [148] covalently linked firefly luciferase with glutaraldehyde to alkylamine glass beads which were then cemented to glass
rods. The immobilized enzyme has a lower pH optimum than the soluble
enzyme and emits light with a major peak at 615 nm (the soluble enzyme
emits light with a peak at 562 nm). The immobilized enzyme is stable
and can be used for multiple assay. The peak light intensity is linear with ATP in the range of 10^{-5} to 10^{-8} M. The luciferase rods
were used also in a coupled assay to measure the rate of ATP production catalyzed by creatine phosphokinase.

Ugarova et al. [149,150] immobilized firefly luciferase on
Sepharose 4B and cellophane films, and applied these for the detection of ATP at concentrations of 0.1 pM-1 mM as well as for activity
determination of enzymes catalyzing ATP synthesis (pyruvate kinase)
and ATP hydrolysis (ATPase). The same sample of luciferase immobilized on cellophane film was used 40 times in ATP quantification.
Error was 7%.

In subsequent papers, Brovko et al. [151,152] used a suspension
of immobilized firefly luciferase for the quantitative assay of ATP
and of the activity of those enzymes which synthesize or destroy ATP.

Ford and DeLuca [147] developed a new assay for picomole levels
of androsterone and testosterone using coimmobilized luciferase,
oxidoreductase, and steroid dehydrogenase.

Purified NADH, flavine mononucleotide oxidoreductase and luciferase from *Beneckea harveyi* have been covalently attached to Sepharose
4B. Compared with the arylamine glass bead-immobilized enzymes, the
Sepharose-immobilized preparations have a much higher activity. The
Sepharose-bound enzymes were used to determine quantitatively NADH
in the range of 1 pmole to 2 nmole. When glucose-6-phosphate dehydrogenase was coimmobilized with the oxidoreductase and luciferase on
Sepharose, it is possible to assay glucose 6-phosphate in the range
of 10 pmoles-20 nmoles. Immobilization of 3α-hydroxysteroid dehydrogenase or 3β-hydroxysteroid dehydrogenase with the light-emitting
system gives a preparation that can rapidly determine 0.8 pmole of
either androsterone or testosterone. In all examples cited, the
immobilized enzymes were found to be catalytically more efficient
than a comparable mixture of the soluble enzymes.

Guilbault and Kramer [160] prepared a column of immobilized horse serum cholinesterase in a flow system and developed an assay of organophosphorus pesticides. The fluorescence due to hydrolysis of a fluorescein or resorufin ester by the cholinesterase was monitored. When the enzyme was inhibited no hydrolysis occurred and the fluorescence was no longer liberated. As little as 1 ppb of parathion, systox, or malathion could be assayed.

REFERENCES

1. G. G. Guilbault and D. Kramer, *Anal. Chem. 37*, 1665 (1967).

2. W. Hornby, H. Filippusson, and A. MacDonald, *FEBS Lett. 9*, 8 (1970); W. Hornby and P. V. Sundaram, *FEBS Lett. 10*, 325 (1970).

3. G. Johansson, K. Edstrom, and L. Ögren, *Anal. Chim. Acta 85*, 55 (1976).

4. T. T. Ngo, *Can. J. Biochem. 54*, 62 (1976).

5. T. T. Ngo, *Int. J. Biochem. 6*, 633 (1975).

6. S. J. Updike and G. P. Hicks, *Science 158*, 270 (1967).

7. M. K. Weibel, W. Dritschilo, H. J. Bright, and A. Humphrey, *Anal. Biochem. 52*, 402 (1973).

8. J. Campbell, W. Hornby, and D. Morris, *Biochim. Biophys. Acta 384*, 307 (1975).

9. K. F. O'Driscoll, A. Kapoulas, A. M. Albisser, and R. Gander, in *Hydrogels for Medical and Related Applications* (J. D. Andrade, ed.). ACS Symposium Series, No. 31, 1976.

10. H. W. Levin and M. K. Weibel, *Food Eng. 47*, 58 (1975).

11. Leeds and Northrup Co., New Product Information Sheet C2-5111-TP.

12. H. J. Kunz and M. Stasny, *Clin. Chem. 20*, 1018 (1974).

13. R. A. Messing, *Biotechnol. Bioeng. 16*, 897 (1974).

14. B. Watson, D. N. Stiffel, and F. E. Semersky, *Anal. Chim. Acta 106*, 233 (1979).

15. S. K. Dahodwala, M. K. Weibel, and A. E. Humphrey, *Biotechnol. Bioeng. 18*, 1679 (1976).

16. Leeds and Northrup, New Product Information Sheet C2-5113-TP.

17. I. Karube, K. Hara, I. Satoh, and S. Suzuki, *Anal. Chim. Acta 106*, 243 (1979).

18. L. Ögren and G. Johansson, *Anal. Chim. Acta 96*, 1 (1978).

19. I. Satoh, I. Karube, S. Suzuki, and K. Aikawa, *Anal. Chim. Acta 106*, 369 (1979).

20. W. Marconi, F. Bartoli, S. Gulinelli, and F. Morisi, *Proc. Biochem. 9*, 22 (1974).

21. E. K. Bauman, G. G. Guilbault, D. N. Kramer, and L. H. Goodson, *Anal. Chem. 37*, 1378 (1965).

22. L. H. Goodson and W. B. Jacobs, Midwest Research Institute Report No. MRI 1173; *Enzyme Engineering,* Vol. 2 (E. K. Pye and L. B. Wingard, eds.). Plenum Press, New York, 1974, p. 383.

23. W. Marconi, F. Bartoli, S. Gulinelli, and F. Morisi, *Proc. Biochem. 9*, 22 (1974).

24. J. F. Rusling, G. H. Luttrell, L. F. Cullen, and G. J. Papariello, *Anal. Chem. 48*, 1211 (1976).

25. H. Jaegfeldt, A. Torstensson, and G. Johansson, *Anal. Chim. Acta 97*, 221 (1978).

26. L. Goodson, W. Jacobs, and A. Davis, *Anal. Biochem. 51*, 362 (1973).

27. J. G. Schindler, W. Riemann, D. Sailer, G. Berg, and W. Schäl, *J. Clin. Chem. Clin. Biochem. 15*, 709 (1977).

28. Leeds and Northrup Co., New Product Information Sheet C2-5112-TP.

29. B. Watson and M. H. Keyes, *Anal. Lett. 9*, 217 (1976).

30. G. Johansson and L. Ogren, *Anal. Chim. Acta 84*, 23 (1976).

31. D. J. Hanson and N. S. Bretz, *Clin. Chem. 23*, 477 (1977).

32. S. Lovett, *Anal. Biochem. 64*, 110 (1975).

33. W. Dritschilo and M. K. Weibel, *Biochem. Med. 11*, 242 (1974).

34. E. L. Gulberg and G. D. Christian, *Anal. Chim. Acta 123*, 125 (1981).

35. E. L. Gulberg and G. D. Christian, *Z. Anal. Chem. 305*, 29 (1981).

36. L. Gorton and K. M. Bhatti, *Anal. Chim. Acta 105*, 43 (1979).

37. B. Voleski and C. Emond, *Biotechnol. Bioeng. 21*, 1251 (1979).

38. P. R. Coulet and D. C. Gautheron, *J. Chromatogr. 215*, 65 (1981).

39. P. R. Coulet, F. Paul, D. Dupret, and D. C. Gautheron, *Enzyme Engineering,* Vol. 5 (H. H. Weetall and G. P. Royer, eds.). Plenum Press, New York, 1980, p. 231.

40. J. M. Brillouet, P. R. Coulet, and D. C. Gautheron, *Biotechnol. Bioeng. 18*, 1821 (1976).

41. J. M. Brillouet, P. R. Coulet, and D. C. Gautheron, *Biotechnol. Bioeng. 19*, 125 (1977).

42. W. Blaedel and J. Wang, *Anal. Chem. 52*, 1426 (1980).

43. W. Blaedel and J. Wang, *Anal. Chem. 52*, 1697 (1980).

44. P. W. Carr and L. D. Bowers, *Immobilized Enzymes in Analytical and Clinical Chemistry.* Interscience, New York, 1980, Ch. 7 and 8.

45. G. Johansson, *Appl. Biochem. Biotechnol. 7,* 99 (1982).

46. R. E. Adams and P. W. Carr, *Anal. Chem. 50,* 944 (1978).

47. H. U. Bergmeyer and A. Hagen, *Z. Anal. Chem. 261,* 333 (1972).

48. T. Ngo, *Int. J. Biochem. 11,* 459 (1980).

49. W. Dritschilo and M. Weibel, *Biochem. Med. 9,* 32 (1974).

50. R. C. Barabino, D. Gray, and M. Keyes, *Clin. Chem. 24,* 1393 (1978).

51. L. J. Forrester, D. M. Yourtrie, and H. D. Brown, *Anal. Lett. 7,* 599 (1974).

52. S. Ikeda, Y. Sumi, and S. Fukui, *FEBS Lett. 47,* 295 (1974).

53. M. S. Denton, W. D. Bostick, S. R. Dinsimore, and J. E. Mrochek, *Clin. Chem. 24,* 1408 (1978).

54. L. P. Leon, M. Sansur, L. R. Snyder, and C. Horvath, *Clin. Chem. 23,* 1556 (1977).

55. P. V. Sundaram and M. P. Igloi, *Clin. Chim. Acta 94,* 295 (1979).

56. S. Cambiaghi, D. Bassi, E. Murador, G. Aimo, A. Caropreso, and C. Rosso, *Progress in Clin. Biochem.,* Vol. 2, 1983.

57. G. P. Hicks and S. J. Updike, *Anal. Chem. 38,* 726 (1966).

58. D. J. Inman and W. E. Hornby, *Biochem. J. 129,* 255 (1972).

59. W. E. Hornby, H. Filippusson, and A. MacDonald, *FEBS Lett. 9,* 8 (1970).

60. L. Gorton and L. Ögren, *Anal. Chim. Acta 130,* 45 (1981).

61. M. Werner, R. J. Mohrbacher, C. J. Riendeau, E. Mruador, and S. Cambachi, *Clin. Chem. 25,* 20 (1979).

62. R. Chirillo, G. Caenaro, B. Pavan, and A. Pin, *Clin. Chem. 25,* 1744 (1979).

63. M. D. Joseph, D. Kasiprzak, and S. R. Crouch, *Clin. Chem. 23,* 1033 (1977).

64. E. Bissie and D. J. Vonderschmitt, *FEBS Lett. 81,* 326 (1977).

65. P. V. Sundaram, B. Blumenberg, and W. Hinsch, *Clin. Chem. 25,* 1436 (1979).

66. D. L. Morris, J. Campbell, and W. E. Hornby, *Biochem. J. 147,* 593 (1975).

67. C. C. Garber, D. Feldbruegge, R. C. Miller, and R. N. Carey, *Clin. Chem. 24,* 1186 (1978).

68. W. D. Fleischmann and H. Schery, *Microchim. Acta 2,* 443 (1976).

69. W. E. Hornby, J. Campbell, D. Inman, and D. L. Morris, in *Enzyme Engineering,* Vol. 2 (E. K. Pye and L. Wingand, eds.). Plenum Press, New York, 1974.

70. P. V. Sundaram and W. Hinsch, *Clin. Chem. 25,* 285 (1979).

71. D. J. Inman and W. E. Hornby, *Biochem. J. 137*, 25 (1974).

72. D. R. Serrir, P. W. Carr, and L. N. Klatt, *Anal. Chem. 48*, 954 (1976).

73. H. H. Weetall and M. A. Jacobsen, *Proc. IV IFS Ferment., Technol. Today 361*, 361 (1972).

74. T. L. Newirth, M. Diegelman, E. Pye, and R. Fallen, *Biotechnol. Bioeng. 15*, 1089 (1973).

75. P. V. Sundaram, *Solid Phase Biochem. 3*, 185 (1978).

76. J. W. Finley and A. C. Olsen, *Cereal Chem. 52*, 500 (1974).

77. P. Cremonesi and R. Bovara, *Biotechnol. Bioeng. 18*, 1487 (1976).

78. W. Hinsch and P. V. Sundaram, *Clin. Chim. Acta 104*, 87 (1980).

79. S. Ikeda and S. Fukui, *FEBS Lett. 41*, 216 (1974); *Biochem. Biophys. Res. Commun. 52*, 482 (1973).

80. E. Reisel and E. Katchalski, *J. Biol. Chem. 239*, 1521 (1964).

81. P. V. Sundaram and W. E. Hornby, *FEBS Lett. 10*, 325 (1971).

82. H. Filippusson, W. E. Hornby, and A. MacDonald, *FEBS Lett. 20*, 291 (1972).

83. D. R. James and B. Pring, *Clin. Chim. Acta, 62*, 435 (1975).

84. J. S. Coliss and J. M. Knox, *Med. Lab. Sci. 34*, 275 (1978).

85. P. V. Sundaram, M. P. Igloi, R. Wasserman, and W. Hinsch, *Clin. Chem. 24*, 234 (1978).

86. L. P. Leon, J. B. Smith, L. R. Snyder, and C. Horvath, *Clin. Chem. 24*, 1023 (1978).

87. P. V. Sundaram, M. P. Igloi, R. Wasserman, and W. Hinsch, *Clin. Chem. 24*, 1813 (1978).

88. J. Campbell, A. Chawla, and T. M. S. Chang, *Anal. Biochem. 83*, 330 (1977).

89. F. Y. Leung, P. Ward, and B. Janik, *Clin. Biochem. 10*, 4 (1977).

90. S. May and L. Landgraff, *Biochem. Biophys. Res. Commun. 68*, 786 (1976).

91. P. V. Sundaram and S. Jayaraman, *Clin. Chim. Acta 94*, 309 (1979).

92. F. Bartoli, S. Giovenco, O. Lostia, W. Marconi, F. Morisi, F. Pittalis, G. Prosperi, and G. Spotorno, *Pharmacol. Res. Commun. 9*, 521 (1977).

93. J. Endo, M. Tabata, S. Okada, and T. Murachi, *Clin. Chim. Acta 95*, 411 (1979).

94. L. P. Leon, S. Narayanan, R. Dellenbach, and C. Horvath, *Clin. Chem. 22*, 1017 (1976).

95. W. Marconi, F. Bartoli, S. Gulinelli, and F. Morisi, *Proc. Biochem.*, May 1974, p. 22.

96. J. C. Bouin, M. Atallah, and H. O. Hultin, *Biochim. Biophys. Acta 438*, 23 (1976).

97. F. N. Onyezili and A. C. Onitiri, *Anal. Biochem. 113*, 203 (1981).

98. A. B. Salleh and W. M. Ledingham, *Anal. Biochem. 116*, 40 (1981).

99. W. Hinsch, W. Ebersbach, and P. V. Sundaram, *Clin. Chim. Acta 104*, 95 (1980).

100. R. Bovara, G. Carrea, P. Cremonesi, and G. Mazzola, *Anal. Biochem. 112*, 39 (1981).

101. L. Ögren, I. Csiky, L. Risinger, L. G. Nilsson, and G. Johansson, *Anal. Chim. Acta 117*, 71 (1980).

102. K. Mosbach and B. Danielsson, *Biochim. Biophys. Acta 364*, 140 (1974).

103. A. Johansson, J. Lundberg, B. Mattiasson, and K. Mosbach, *Biochem. Biophys. Acta 304*, 217 (1973).

104. J. C. Wasilewski, P. T.-S. Pei, and J. Jordan, *Anal. Chem. 36*, 2131 (1974).

105. M. A. Marini and C. J. Martin, in *Methods in Enzymology,* Vol. 27 (C. H. W. Hirs· and S. N. Timasheff, eds.). Academic Press, New York, 1973, p. 590.

106. S. N. Pennington, *Enzyme Technol. Digest 3*, 105 (1974).

107. C. L. Cooney, J. C. Weaver, S. R. Tannenbaum, D. Faller, A. Shields, and M. Jahnke, in *Enzyme Engineering,* Vol. 2 (E. K. Pye and L. B. Wingard, eds.), Plenum Press, New York, 1974, p. 411.

108. K. Mosbach, B. Danielsson, A. Borgerud, and M. Scott, *Biochim. Biophys. Acta 403*, 256 (1975).

109. L. D. Bowers, L. Canning, C. Sayers, and P. Carr, *Clin. Chem. 22*, 1314 (1976).

110. S. P. Fulton, C. L. Cooney, and J. C. Weaver, *Anal. Chem. 52*, 505 (1980).

111. B. Danielsson, *Appl. Biochem. Biotechnol. 7*, 127 (1982).

112. B. Danielsson, The Enzyme Thermistor, Ph.D. thesis, University of Lund, 1979.

113. K. Mosbach and B. Danielsson, *Anal. Chem. 53*, 83A (1981).

114. C. J. Martin and M. A. Marini, Microcalorimetry in Biochemical Analysis, *CRC Reviews 8*, 221 (1979).

115. J. K. Grime, *Anal. Chim. Acta 118*, 191 (1980).

116. R. Schifreen, A. Hanna, L. Bowers, and P. Carr, *Anal. Chem. 49*, 1929 (1977).

117. B. Mattiasson, B. Danielsson, and K. Mosbach, *Anal. Lett. 9*, 217 (1976).

118. B. Mattiasson, C. Borrebaeck, B. Sanfridsson, and K. Mosbach, *Biochim. Biophys. Acta 483*, 221 (1977).

119. C. Borrebaeck, J. Borjesson, and B. Mattiasson, *Clin. Chim. Acta 86*, 267 (1978).

120. B. Danielsson, B. Mattiasson, and K. Mosbach, *Pure Appl. Chem. 51*, 1443 (1979).

121. K. Mosbach, B. Danielsson, *Biochim. Biophys. Acta 364*, 140 (1974).

122. M. Aizawa, Y. Watanabe, and S. Suzuki, *J. Solid Phase Biochem. 4*, 131 (1979).

123. A. Johansson, J. Lundberg, B. Mattiasson, and K. Mosbach, *Biochim. Biophys. Acta 304*, 217 (1973).

124. K. Mosbach, B. Danielsson, A. Borgerud, and M. Scott, *Biochim. Biophys. Acta 403*, 256 (1975).

125. B. Danielsson and K. Mosbach, Technical Center, University of Lund, Sweden, unpublished results.

126. B. Mattiasson, K. Mosbach, and A. Svensson, *Biotechnol. Bioeng. 19*, 1556 (1977).

127. G. G. Guilbault, B. Danielsson, C. F. Mandelius, and K. Mosbach, *Anal. Chem.*, in press, 1983.

128. B. Mattiasson, K. Svensson, C. Borrebaeck, S. Jonsson, and G. Kronwall, *Clin. Chem. 24*, 1770 (1978).

129. B. Mattiasson, K. Mosbach, and A. Svensson, *Biotechnol. Bioeng. 19*, 1643 (1977).

130. H. L. Schmidt, G. Krisam, and G. Grenner, *Biochim. Biophys. Acta 429*, 283 (1976).

131. B. Danielsson, K. Gadd, B. Mattiasson, and K. Mosbach, *Clin. Chim. Acta 81*, 163 (1977).

132. W. Marconi, in *Enzyme Engineering*, Vol. 4 (G. Broun, L. Wingard, and G. Maneckeo, eds.). Plenum Press, New York, 1978, p. 179.

133. B. Mattiasson, *FEBS Lett. 77*, 107 (1977).

134. L. D. Bowers, P. W. Carr, and R. S. Schifreen, *Clin. Chem. 22*, 1427 (1976).

135. B. Mattiasson, B. Danielsson, C. Humansson, and K. Mosbach, *FEBS Lett. 85*, 203 (1978).

136. B. Mattiasson, E. Riecke, D. Munnecke, and K. Mosbach, *J. Solid Phase Biochem. 4*, 263 (1979).

137. B. Mattiasson and C. Borrebaeck, in *Enzyme Labelled Immunoassay of Hormones and Drugs* (S. B. Pal, ed.). Walter de Gruzter, Berlin, New York, 1978, p. 91.

138. B. Mattiasson, B. Danielsson, F. Winquist, H. Nilsson, and K. Mosbach, *Appl. Environ. Microbiol. 41*, 903 (1981).

139. L. J. Forrester, D. Yourtree, and H. D. Brown, *Anal. Lett. 7*, 599 (1974).

140. C. F. Mandenius, B. Danielsson, and B. Mattiasson, *Acta Chem. Scand. B34*, 463 (1980).

141. I. Satoh, B. Danielsson, and K. Mosbach, *Anal. Chim. Acta 131,* 255 (1981).

142. L. D. Bowers, L. M. Canning, C. N. Sayers, and P. W. Carr, *Clin. Chem. 22,* 1314 (1976).

143. C. Tran-Minh and D. Vallin, *Anal. Chem. 50,* 1874 (1978).

144. P. Kirch, J. Danzer, G. Krisam, and H. Schmidt, in *Enzyme Engineering,* Vol. 4 (G. Broun, G. Maneck, and L. Wingard, eds.). Plenum Press, New York, 1978, p. 217.

145. B. Danielsson, K. Gadd, B. Mattiasson, and K. Mosbach, *Anal. Lett. 9,* 987 (1976).

146. L. M. Canning and P. W. Carr, *Anal. Lett. 8,* 359 (1975).

147. J. Ford and M. DeLuca, *Anal. Biochem. 110,* 43 (1981).

148. Y. Lee, I. Jablonski, and M. DeLuca, *Anal. Biochem. 80,* 496 (1977).

149. N. N. Ugarova, L. Bovko, and I. Berezin, *Anal. Lett. 13,* 881 (1980).

150. L. Brovko, N. Kost, and N. Ugarova, *Biochem. SSR 45,* 1199 (1980).

151. L. Brovko, N. Ugarova, T. Vasileva, V. Dombrovskii, and I. Berezin, *Biochem. SSR 43,* 633 (1978).

152. L. Brovko and N. Ugarova, *Biochem. SSR 45,* 607 (1980).

153. D. T. Bostick and D. M. Hercules, *Anal. Chem. 47,* 447 (1975).

154. D. C. Williams, G. F. Hiff, and W. R. Seitz, *Clin. Chem. 22,* 372 (1976).

155. J. P. Auses, S. Cook, and J. Maloy, *Anal. Chem. 47,* 244 (1975).

156. R. Toftgård, T. Anfalt, and A. Graneli, *Anal. Chim. Acta 99,* 383 (1978).

157. S. Brolin, A. Ågren, B. Ekman, and I. Sjoholm, *Anal. Biochem. 78,* 577 (1977).

158. E. Jablonski and M. DeLuca, *Clin. Chem. 25,* 1622 (1979).

159. E. Jablonski and M. DeLuca, *Proc. Natl. Acad. Sci. 73,* 3848 (1976).

160. G. G. Guilbault and D. N. Kramer, *Anal. Chem. 37,* 1665 (1967).

161. B. F. Rocks, *Proc. Soc. Anal. Chem. 10,* 164 (1973).

162. J. N. Miller, B. F. Rocks, and D. T. Burns, *Anal. Chim. Acta 86,* 93 (1976).

163. W. E. Hornby, in Summer Symposia on Enzymes in Anal. Chem., Amherst, Ma., June 14, 1977.

164. F. Gorus and E. Schram, *Arch. Int. Physiol. Biochem. 85,* 981 (1977).

165. P. Kamoun and O. Douay, *Clin. Chem. 24,* 2033 (1978).

166. N. Potezny, R. Bais, P. D. O'Loughlin, J. B. Edwards, A. M. Rofe, and R. A. J. Conyers, *Clin. Chem. 29,* 16 (1983).

5

Applications of Enzyme
Layers and Other Techniques.
Commercial Instrumentation

I. APPLICATIONS OF ENZYME LAYERS

In this chapter assays utilizing semisolid surface fluorescence,
enzyme stirrers, dip-stick methodology, mass spectrometry, and other
techniques will be discussed. Table 1 lists these applications.

A. Semisolid Surface Fluorescence

Fluorescent procedures are several orders of magnitude more sensitive
than colorimetric methods and thus have replaced the colorimetric ones
in numerous instances. Further advantages are greater selectivity
(because two wavelengths are used) and an accuracy independent of
region of measurement.

   Previous fluorometric methods, although they have been improve-
ments over other prior art methods of determining enzyme activity,
have not eliminated all the problems associated with enzymatic analyses. Fluorometric analysis depends on the production of a fluorescent compound as a result of enzyme activity between a substrate and
an enzyme. The rate of production of the fluorescent compound is
related to both the enzyme concentration and the substrate concentration. This rate can be quantitatively measured by exciting the fluorescent compound as it is produced and by recording the quantity of
fluorescence emitted per unit of time with a fluorometer.

   Fluorometrically measuring enzyme activities or enzymatic reactions is usually done by wet chemical methods that rely on the reaction of a substrate solution with an enzyme solution. Unfortunately,

TABLE 1 Applications of Enzyme Layers

Substance	Enzyme	Reactor type	Detection system	Ref.
Acid or alkaline phosphatase		SSSF	Fluorescence	1, 2
Cholesterol	Cholesterol esterase/ cholesterol oxidase	Enzyme stirrer	Amperometric	3
		SSSF	Fluorescence	24
Cholinesterase		SSSF	Fluorescence	4
Creatine	Creatine kinase	SSSF	Fluorescence (NADH)	5
Ethanol	Alcohol dehydrogenase	Enzyme stirrer	Fluorescence	6
Galactose	Galactose oxidase/ peroxidase	Layer	Color	7
Glucose	Hexokinase/ glucose-6-phosphate dehydrogenase	SSSF	Fluorescence	8
		Stirrer	Fluorescence	9
	Glucose oxidase/ peroxidase	Enzyme stirrer	Fluorescence	10
		Layer	Color	7
	Glucose dehydrogenase	Stirrer	Fluorescence	11
γ-Glutamyl transpeptidase		SSSF	Fluorescence	12
Glutamate-oxaloacetate transaminase	Malate dehydrogenase	SSSF	Fluorescence	13
Glutamate-pyruvate transaminase	Lactate dehydrogenase	SSSF	Fluorescence	13
Hexokinase/gluco-phosphoisomerase	Glucose-6-phosphate dehydrogenase	Membrane	Fluorescence	14
α-Hydroxybutyrate dehydrogenase		SSSF	Fluorescence	13
Lactate dehydrogenase		SSSF	Fluorescence	15
NADH	Alcohol dehydrogenase (EC 1.1.1.1)	Membrane	Single-ion monitoring (mass spectrometric)	16
Nitrate	NADH-nitrate reductase	Enzyme stirrer	Potentiometric	17
Urea	Urease	Enzyme stirrer	Potentiometric	18
	Urease	Membrane	Single-ion monitoring (mass spectrometric)	16
		Layer	Color	21
	Urease/glutamate dehydrogenase	SSSF	Fluorescence	19
Uric acid	Uricase	Membrane	Fluorescence	20
		SSSF	Fluorescence	22, 23

wet chemical methods involve the preparation of costly substrates, cofactors or coenzymes, and enzyme solutions.

For example, when determining the presence and concentration of an enzyme, a substrate must be accurately measured and dissolved in a large amount of buffer solution, usually about 100 ml, to prepare a stock solution. The enzyme reaction is then usually carried out by measuring a certain volume of stock substrate solution into an optical cuvette, adding a measured amount of enzyme solution to the substrate solution, and recording the change in absorbance emitted from the resultant solutions.

When determining the concentration of a specific substrate kinetically in an enzymatic catalyzed reaction, the procedure is more cumbersome and costly. A relatively large amount of expensive enzyme, usually 0.1 ml of stock solution, is needed to make the reaction proceed at a conveniently measurable rate. These enzyme solutions must be prepared fresh daily. This standard wet chemical method requires considerable technician time and relatively large quantities of expensive substrates or enzymes.

Guilbault and co-workers [1] developed solid-surface fluorometric methods, using a "reagentless" system, for the assay of enzymes, substrates, activators, and inhibitors. An attachment to an Aminco filter fluorometer has been adapted to accept, instead of a glass cuvette, a metal slide (a cell), painted black to reduce the background. A silicone rubber pad is placed on the slide. All the reagents for a quantitative assay are placed in a form of solid reactant film on the surface of the pad. The sample of the fluid containing the substance to be assayed is then added to the pad. The change in fluorescence with time is measured and equated to the concentration of the substance determined.

These reagent pads are simple to prepare, and hundreds could be conveniently manufactured at one time. They are stable for months or longer when stored under specified conditions. There is no need for the cumbersome, time-consuming preparation of reagents when performing an analysis because essentially all the reagents for a quantitative assay are already present on the pad. If samples are hard to obtain, the pad method could be another great advantage since only 3-25 µl of sample is required.

1.   *EXPERIMENTAL: PREPARATION OF SILICONE*
     RUBBER PADS

Silicone rubber (Dow-Corning Glass and Ceramic Adhesive, Dow-Corning, Midland, Michigan) pads were prepared by pressing uncured silicone rubber between a glass plate and a stainless steel mold (Fig. 1), both of which were lined by a piece of glassine paper (Eli Lilly and Co.). The surfaces contacted with the silicone rubber were prelubricated with a thin layer of Dow-Corning silicone stopcock grease (Dow-Corning Co., Midland, Mich.). The silicone rubber was kept in the mold at room temperature for 2 days to cure. The cured strips were then removed, wiped, and washed briefly with concentrated KOH solution to remove the grease. Next the pads were washed with H_2O and dried at 80°C for 1 hr. The strips were cut to individual pads 6 mm in width. About 20 individual pads can be obtained from one strip.

The reactant film on the pad may be formed by dissolving the reagents and buffer in a suitable solvent, depositing the reactant

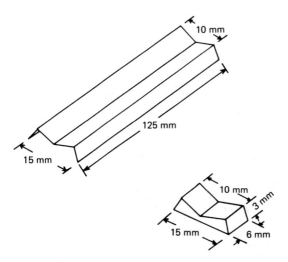

FIG. 1 Silicone rubber pads of 6-mm width (lower left) are cut from a 125-mm strip (upper right).

solution on the silicone pad (so that the solution spreads evenly over the pad), and evaporating off the solution by vacuum or lyophilization. The reagent may also be applied to the pad in a polymeric film such as polyacrylamide or some stabilizer can be added if an enzyme is present in the reagent. Either substrate or enzyme and/or coenzyme can be deposited on the pad in film form, depending on whether the substance to be assayed is an enzyme or a substrate. The pads can be stored in a dark, cold place or in a refrigerator or desiccator before use.

Two methods were used to apply the reactant film to the pad. The first method, a batch method, involved putting a solution of the reagent onto the strip of silicone rubber, allowing the solution to evaporate, and then cutting the strip into the desired pad dimensions. The prepared pads were then stored. The second method involved cutting the individual pad from the strip and then applying the reagent to the pad. The latter method was found to be preferable.

The color of the silicone rubber affects both the background and rate of change of fluorescence. With any of the filter systems, the background and rate of change of fluorescence increase in the order: black < gray < clear < white. Each possible combination of pad color and filter was examined, and it was found that the most accurate results could be obtained if a gray silicone rubber pad was used. The silicone materials can retain a reactant film on their surfaces for an indefinite time and permit the direct measurement of fluorescence from their surface when an appropriate second reagent solution is applied onto the first reactant film. Background interference due to light scattering and nonspecific fluorescence are minimal compared with other materials.

An Aminco filter fluorometer set on its end (Fig. 2) was used for all the fluorometric measurements. The fluorometer was supported by two wooden blocks placed parallel to the primary filter holder to prevent electric noise and to make it convenient to change the primary filter when needed. A 5-in. linear recorder was used to display the result obtained.

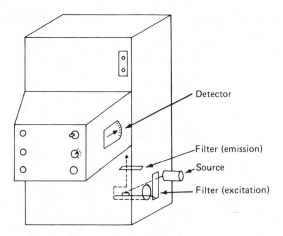

FIG. 2   An Aminco filter fluorometer equipped with a cell and cell
holder.

Two kinds of cells and cell holders were specially constructed
to hold the pad in the optical path of the fluorometer. The first
kind, which was easier to construct, was designed by Guilbault and
Vaughn [2] to study reactions primarily at room temperature. The
second kind, which was a modified version of the first kind but had
temperature control, was designed to study reactions above 30°C.
Only the second model is discussed here because the same device can
be used for room temperature or high-temperature studies.
     The cell holder consisted of an Aminco cuvette adapter (Catalog
No. J4-7330) with water circulating around it to maintain constant
temperature. Temperature of the circulating water was regulated by
a constant-temperature water bath capable of maintaining ±0.1°C.
Black binders were placed on both sides of the two entrance and exit
slits so that smaller slits, about two-thirds the length of the pad
used, would restrict the radiation that entered and left the cell
cavity. The cell was constructed of a cylindrical aluminum rod with
a slot, approximately twice the length of the pad, located toward
the end of the rod. The depth of the slot was such that the pad,
with its contents, received the full beam of incident radiation.
The cell was painted a dull black to avoid scattered light. A

drawing of the cell and cell holder appears in Fig. 3. A drawing of
the pad inside the cell holder and its relationship with the incident
beam is given in Fig. 4.

2. METHODS

The concentration of substrate participating in an enzyme reaction can be calculated in one of two general ways. The first method
measures, by chemical, physical, or enzymatic analysis, either the
total change that occurs in the end product or in the unreacted
starting material. In this method, large amounts of enzyme and small
amounts of substrate are used to ensure a complete reaction. In the
second method, which is a kinetic method, the initial rate of reaction is measured, in one of many conventional ways, by following the
production of product or the disappearance of the substrate. In this

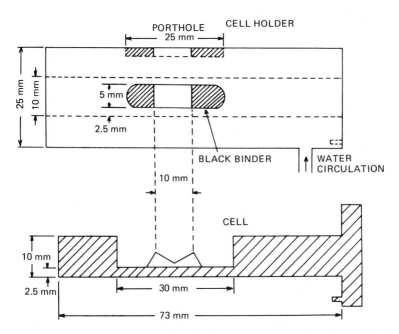

FIG. 3 The cell (lower drawing) is constructed of a cylindrical
aluminum rod with a slit, approximately twice the length of the pad,
located toward the end of the rod. The cell holder (top drawing)
consists of an Aminco cuvette adapter with water circulating around
it to maintain constant temperature.

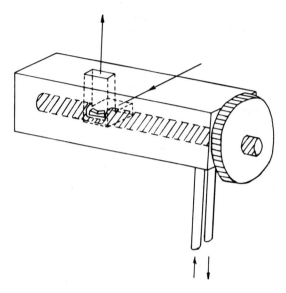

FIG. 4 This shows the pad inside the cell holder and the relationship of the pad to the incident beam.

method, the rate of reaction is a function of the concentration of
substrate, enzyme, inhibitor, and activator.

 However, because enzymes are catalysts, and as such affect the
rate and not the equilibrium of reactions, their concentration and
activity must be measured by the rate or kinetic method. Similarly,
activators and inhibitors that affect the enzyme's catalytic effect
can be measured only by the rate method. While, as pointed out
above, the substrate can be measured either by a total change or by
a rate method, the latter method is faster because the initial reaction can be measured without waiting for equilibirum to be established. The accuracy and precision of both methods are comparable.
The following is a general description of the rate method using
fluorometric silicone rubber pad procedures.

 The refrigerator-stored pads containing all reagents, including
the "immobilized" enzymes, are allowed to reach room temperature inside a desiccator. Then 25-90 μl of water or buffer solution maintained at a certain temperature is added to each pad placed in the

cell holder, followed by 3-35 μl of the sample solutions to be
assayed, also at the same temperature. After mixing with the tip
of the syringe used to deliver the sample, the cell is placed into
the cell holder and into the light path of the fluorometer.

The recorder is turned on immediately and the change in fluorescence (Δf) is recorded. A calibration plot of Δf/min versus concentration is prepared and used for subsequent analyses of enzymes or
substrates. The silicone rubber pads can be reused after they have
been cleaned with detergent and water.

The number of fluorometric systems that can be monitored by
this type of pad method is as many as the number of fluorometric
systems known. Some of the more important systems that have been
studied and some for which methods have been developed will be presented later.

3. ADVANTAGES OF THE SILICONE RUBBER PAD METHOD

The pads with solid reactant film on them are not yet commercially available. In order to use this pad method at the present
time, one has to prepare one's own pads and reactant films. This
is bothersome for a small clinical laboratory. However, the silicone
rubber pad method has many advantages and should find wide acceptance
for many applications.

a. Time

Since essentially all the reagents that are necessary for a
quantitative assay are present on the pads, there is no need for
further reagent preparation. If a sample for glucose assay is presented at 2 A.M., the technician would only have to open a bottle
marked "glucose test," pick up a pad, add 10 μl of buffer solution
and a 20-μl sample of serum, and read out the results directly.
Total time is about 4 min using an initial rate method. With the
pad method, this would require about 10 bottles of pads, one for
each of 10 common tests.

b. Sample Size and Cost

The pad method is a microprocedure and as such microamounts of
reagents and sample are required. On the creatine pad method, for

example, the cost of the reagents is only 1/27 that of a regular
spectrophotometric method, and only as little as 3 μl of serum
sample are needed.

   c. Sensitivity and Linearity

   Fluorescent procedures are several orders of magnitude more
sensitive than colorimetric methods. The accuracy of colorimetric
methods is limited to an error of about 2-3% over a narrow range
(0.15-0.85 absorbance unit). Because fluorescence methods are based
on the production of a signal over zero signal, the accuracy of these
methods is independent of scale reading and remains the same over a
three- to fourfold linear range of concentration. The linear range
of concentration is further improved in the pad method. Self-quench-
ing interferences in fluorometric assays are minimized because of the
small distance the fluorescence beam travels in the sample drop of
the pad (less than 2 mm).

   d. Temperature

   In conventional methods, temperature must be critically controlled in all enzyme assays based on a kinetic approach. This can
be inconvenient and bothersome. The fluorescent pad method proposed
here is temperature-independent for those reactions run at 25°C
because the silicone rubber pad used to support the sample is not
heat-conductive. Thus, provided that the sample of serum, blood,
or urine is at the same temperature (25°C), the temperature of the
environment does little to affect the results.

   e. Stability

   Many substrates and enzyme solutions used in the present clinical procedures are unstable and new solutions must be prepared fresh
daily. o-Toluidine and peroxidase in glucose assay, NADH in LDH assay,
and hexokinase and G-6-phosphate dehydrogenase (PDH) in creatine phosphokinase (CPK) assay are examples of this. Yet, when the reagents
are placed in a solid form on the silicone rubber matrix, they can be
kept for weeks or months with no deterioration. The enzymes are held
in a freeze-dried state in the presence of stabilizers and thus are
in an "immobilized" state prior to analysis.

4. *ASSAY OF IMPORTANT SUBSTRATES WITH*
   IMMOBILIZED ENZYMES

a. *Urea Nitrogen in Serum*

The determination of serum urea nitrogen is presently the most popular screening test for the evaluation of kidney function. Elevated urea nitrogen values are found in patients with acute glomerulonephritis, chronic nephritis, polycystic kidney, nephrosclerosis, and tubular necrosis.

A fluorometric method for the determination of urea in serum was developed using a coupled enzyme system [19]:

$$\text{Urea} \xrightarrow{\text{urease}} 2NH_4^+ \tag{1}$$

$$NADH + 2NH_4^+ + \alpha\text{-ketoglutarate} \xrightarrow[\text{dehydrogenase}]{\text{glutamate}} NAD + \text{glutamic acid} \tag{2}$$

The rate of disappearance of NADH fluorescence (λ_{ex} = 365 nm; λ_{em} = 460 nm) was proportional to the content of urea in serum. Reaction conditions and reagent concentrations were similar to that in the spectrophotometric reagent kit by Calbiochem but less NADH was used. The rate of decrease in the fluorescence of NADH, triggered by the addition of urea, was measured. The system worked fine for assay of serum urea. The calibration plot is linear up to 25 mg urea/dl. The method affords a rapid, simple, and inexpensive means for urea assay, the results of which correlate well with the automatic diacetylmonoxime method (correlation coefficient .998).

b. *Creatine in Urine*

Creatine tolerance tests, which measure the ability of an individual to retain a test dose of creatine, are of great diagnostic value in indicating extensive muscle destruction and disease of the kidney.

The following reaction scheme was investigated to develop a sensitive fluorometric method and later on a fluorometric pad procedure for creatine in biological samples [5]:

$$\text{Creatine} + \text{ATP} \xrightarrow{\text{CPK}} \text{creatine phosphate} + \text{ADP} \qquad (3)$$

$$\text{Phosphoenolpyruvate (PEP)} + \text{ADP} \xrightarrow{\text{PK}} \text{pyruvate} + \text{ATP} \qquad (4)$$

$$\text{Pyruvate} + \text{NADH} \xrightarrow{\text{LDH}} \text{lactate} + \text{NAD} \qquad (5)$$

The rate of fluorescence change of NADH, $\Delta f/\text{min}$, was proportional to the amount of creatine in the sample.

This scheme was successfully applied to urine creatine. The content of creatine in urine is about 10 times that in serum. Fluorogenic substances in urine interfere but can be easily removed by activated carbon. There was little or no loss of creatine as determined by a spectrophotometric method, when 10 ml of urine was treated with 0.5 g of activated carbon followed by filtration. The overall reaction occurred at pH 9 with tris buffer. A linear calibration curve was obtained for 0-20 and 0-100 mg% creatine. The 0- to 20-mg% curve covered the normal range of urine creatine. The amounts of NADH, ATP, PEP, Mg, buffer mixture, and LDH-PK mixture were optimized. Two-fifths of the amount of NADH, as in the spectrophotometric method, gave a reasonable slope change, $\Delta f/\text{min}$, and about 3 min of linearity of the slope was obtained.

The complexity of this solution mixture made its preparation difficult; for many laboratories it would be impractical. A silicone pad method in which all the reagents were lyophilized or immobilized would be an ideal answer to this problem. However, the crystalline lyophilized reagent is stable only for about 48 hr under refrigeration. Attempts are now being made to add some stabilizer to the mixture, such as mannitol, gum, or ammonium sulfate, and/or to lyophilize the reagents separately on a pad.

c. Uric Acid in Serum

Determination of serum uric acid is most helpful in the diagnosis of gout. Elevated levels of uric acid are found in patients with familiar idiopathic hyperuricemia and in decreased renal function.

A direct reaction rate method was investigated for the sensitive determination of serum uric acid, based on a fluorometric Schoenemann reaction. The rate of formation of indigo white, $\Delta f/min$ (λ_{ex} = 395 nm, λ_{em} = 470 nm), is a measure of the uric acid content in the sample [22].

Optimum conditions, such as pH, buffer system, and organic solvent of the hydrogen peroxide-phthalic anhydride, were studied. The optimum pH of the uricase reaction (pH 9.4) coincided with the optimum pH using phosphate buffer for the hydrogen peroxide-phthalic anhydride reaction (pH 9.1). Therefore, an overall pH of 9.4 was used. Phosphate buffer was found to be superior to tris and borate buffers.

Acetone or methyl cellulsolve both were good solvents for phthalic anhydride.

$$\text{Uric acid} \xrightarrow{\text{Uricase}} H_2O_2 \tag{6}$$

The rate of formation of hydrogen peroxide in the uricase-uric acid reaction was a function of uricase enzyme. Enough uricase enzyme was used so that in about 5 min the formation of hydrogen peroxide was completed, before the addition of the coupling fluorogenic reagents. The system works fine and gives a linear response of 0-5 mg% of uric acid. Similar results were observed with the silicone pad method on which the enzyme and the buffer were placed.

Difficulties were encountered in serum samples, however. Protein had to be removed by either TCA, acetone, or tungstate methods. Still inhibition was observed. Amino acids and ammonium hydroxide were found to inhibit the reaction so that a lower value of uric acid was found.

Another fluorometric reaction system that holds great promise
for a specific analysis of serum uric acid on silicone pads was
developed [23]. The method consists of oxidizing uric acid with
uricase to hydrogen peroxide which in turn oxidizes homovanillic
acid (I) to a highly fluorescent material (II). The rate of production of fluorescence is proportional to the amount of uric acid
present.

I. NONFLUORESCENT   II. FLUORESCENT

$(\lambda ex = 315$ nm; $\lambda em = 425$ nm$)$

A calibration curve was constructed from a series of standards
and was linear from 0.1 to 14 mg/dl. The method is simple and has
good precision and accuracy. Comparison of the results obtained by
this method with the standard enzymatic spectrophotometric method
(disappearance of uric acid at 293 nm) gave a coefficient of variation of .983.

d. Cholesterol Assay

A method has been developed based on the following sequential
enzymatic reactions:

$$\text{Cholesterol esters} + H_2O \xrightarrow{\text{cholesterol ester hydrolase}} \text{fatty acid} + \text{cholesterol} \quad (10)$$

$$\text{Cholesterol} + O_2 \xrightarrow{\text{cholesterol oxidase}} \text{cholest-4-en-3-one} + H_2O_2 \quad (11)$$

$$H_2O_2 + \text{homovanillic acid} \xrightarrow{\text{peroxidase}} \text{fluorescent dimer} \quad (12)$$

The initial rate of fluorescent dimer formation is monitored and
is proportional to the concentration of total cholesterol. A linear
relationship was found in the range 0-400 mg/dl. Results had good
precision and accuracy [24].

e. *Assay of Glucose*

Kiang et al. [8] proposed a simplified assay of glucose via a solid-surface fluorescence method. Hexokinase and glucose-6-phosphate dehydrogenase are immobilized onto a silicone rubber pad and the NADH produced is measured. The pads are usable for a month after preparation and a good C.V. (3%) is obtained.

f. *Assay of Nitrate Ion*

In a prototype system that should be applicable to many other oxyanions, Kiang et al. [17] described a semisolid surface fluorescence (SSSF) method for nitrate ion using the NADH-dependent nitrate reductants from *Chlorella vulgaris*:

$$NO_3^- + NADH \xrightarrow{\quad E \quad} NAD + NO_2^- \tag{13}$$

By placing the enzyme and NADH on a pad, the addition of NO_3^- effects conversion of NADH to NAD, which is measured fluorometrically. Good results were obtained. Attempts to extend this methodology to the assay of nitrite was only partially successful, because of contamination of the nitrite reductase, isolated from *Azotobacter chroococcum*, with NADH oxidase. The latter interferes in the assay by effecting a spontaneous oxidation of NADH to NAD.

$$NO_2^- + 3NADH \xrightarrow{\dfrac{nitrite}{reductase}} NH_3 + 3NAD$$

Both enzymes are highly specific, working only with nitrate or nitrite as substrate, respectively. The linear range was about 10^{-3} to 10^{-5} M substrate.

5. *ASSAY OF ENZYMES WITH IMMOBILIZED SUBSTRATES AND/OR ENZYMES*

Guilbault and Zimmerman [4] developed a SSSF method for assay of cholinesterase using immobilized N-methyl indoxyl acetate on a silicone rubber pad. As little as 10^{-5} units of enzymes can be detected.

Nonfluorescent
N - methyl indoxyl acetate

Fluorescent
N - methyl indoxyl

Guilbault and Vaughn [2] in 1971 first developed an SSSF method
for assay of acid and alkaline phosphatases. The substrate naphthol
AS-BI phosphate is used, a nonfluorescent compound, which is cleaved
by acid phosphatase at pH 5 or alkaline phosphatase at pH 8, to the
highly fluorescent naphthol AS-BI. The influence of substrate, drop
volume, and shape of the drop on the background fluorescence was
studied, as well as the effect of potential interferences, such as
bilirubin in blood.

Rietz and Guilbault [1] used 4-methyl umbelliferone phosphate
as substrate, for the assay of acid and alkaline phosphatase. The
highly fluorescent 4-methyl umbelliferone is measured, the $\Delta F/\Delta t$
rate of formation being proportional to the concentration of this
enzyme.

Rietz and Guilbault [12] developed an assay for γ-glutamyl
transpeptidase using SSSF. The reagents, N-γ-L-glutamyl-α-naphthyl-
amide and glycyl glycine are placed on the pad surface, and the fluorescence measured:

$$\text{N-}\gamma\text{-L-glutamyl-}\alpha\text{-naphthylamide + Gly Glyc} \xrightarrow{\text{GT}} \alpha\text{-naphthylamine}$$
$$(\lambda_{ex} = 342, \ \lambda_{em} = 445 \text{ nm}) \quad (16)$$

Only 10-50 μl of serum is required, and a direct assay is effected
in 2-3 min. The range is 26-265 U/liter with a precision of 2-3%.

Later Rietz and Guilbault [13] described procedures for the assay
of glutamate-oxaloacetate transaminase and glutamate-pyruvate transaminase using SSSF. Dade (Div. American Hospital Supply, Miami) tab-
lets, containing all reagents necessary for a spectrophotometric assay,
were dissolved in water, then 30 μl of the solution was added to the

silicone rubber surface. After addition of 10 μl of serum, the
fluorescence change due to NADH is measured, and ΔF/min is plotted
versus activity of GOT or GPT. The pads were stable for 3-4 days,
and a linear range of 2.2-106 U/liter was obtained. In this same
paper a method was described for α-hydroxybutyrate dehydrogenase
using SSSF. A layer of substrates placed on the pad was stable for
up to a month.

   Zimmerman and Guilbault [15] described an SSSF procedure for
assay of lactate dehydrogenase.

$$\text{Lithium lactate} + \text{NAD} \xrightarrow{\text{LDH}} \text{pyruvate} + \text{NADH} \tag{17}$$

Fifty microliters of 1 mM NAD and 20 μl of 1 mM Li lactate were
placed onto a silicone rubber pad. Upon addition of 20 μl of sample
(LDH), the fluorescence of NADH is produced and monitored. The
linear range was 160-820 U/ml with 3% C.V. The pads, prepared with
immobilized lactate and NAD, were stable for 30 days with a day-to-
day variation of about 3%.

B.   The Immobilized Enzyme Stirrer

In all of the applications above the immobilized enzyme has been
stationary, either in a column or in a thin layer, and the solution
has been stirred and passed over the enzyme or brought into contact
with it. However, there are even more potential applications of
immobilized enzymes in the wide variety of assays performed in clinical laboratories with soluble enzymes in diagnostic kits by spectrophotometric and fluorometric methods.

   If the enzyme can be directly attached to the stirrer used to
mix solutions, a significant advantage for enzymatic analysis can be
realized. The enzyme stirrer would simultaneously mix the solution
and promote the enzymatic transformation, the products being measured
in the same reaction chamber by electrochemical, spectrophotometric,
or fluorometric methods. The stirrer could then be moved, manually
or automatically, to the next sample chamber, or the chamber could
be automatically flushed and the next sample introduced. Thus, a
rapid, simple, and inexpensive assay can be achieved.

Guilbault and Stakbro [18] described an immobilized urease
enzyme stirrer for the assay of urea in blood, with an air-gap electrode to monitor the ammonium ion produced. In the electrode construction used (Fig. 5), which is based on the air-gap electrode,
the enzyme stirrer is simply placed in the microchamber which contains sample and buffer, the only reagent required. As constructed,
the stirring bar has a fairly even layer of enzyme over two-thirds
of its surface and still turns easily. After a pH_e value has been
recorded (about 2-3 min), the stirrer is picked out, washed for 10
sec under a distilled water tap, and patted dry; the cell is pulled
back tightly into position and the next assay can then be performed.
Less than 1 min is necessary to wash the stirrer and cell and replen-
ish the electrolyte layer on the electrode surface.

The stirrer is stable and economical, is useful for several
hundred assays, and permits a very fast assay in a highly accurate
and reproducible manner.

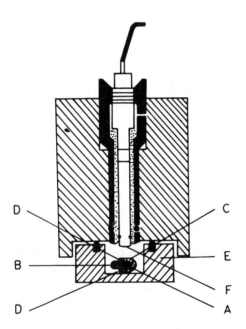

FIG. 5 Air-gap electrode and immobilized enzyme stirrer. (From
Ref. 18.)

Such enzyme stirrers can be made for use in spectrophotometric and fluorometric assays. Although a very useful stirrer results from the use of enzyme placed on a Teflon-coated stirring bar, other such enzyme stirrers can be made in many other ways, such as by direct attachment of the enzyme to a glass stirrer by simple methods. Many assays done with soluble enzymes can probably be performed with an immobilized enzyme stirrer with significant advantages, and the stirrer may eventually replace the soluble enzyme in most diagnostic kits.

Kuan et al. [6] described a fluorometric method for ethanol using immobilized alcohol dehydrogenase. Ethanol is oxidized by alcohol dehydrogenase to aldehyde, with the simultaneous conversion of NAD to NADH; the increase in fluorescence of NADH is measured at 465 nm (λ_{ex} = 365 nm). The calibration curve is linear up to 250 mg/dl of ethanol. Deproteinization is required prior to the assay of samples; recoveries were 97-102%. The stability of the enzyme, reproducibility, and specificity were discussed.

Kiang et al. [9] have described an immobilized hexokinase stirrer, useful for the assay of glucose concentrations in human blood plasma. The procedure used was a fluorometric rate method, measuring the formation of NADPH catalyzed by immobilized glucose-6-phosphate dehydrogenase and hexokinase held in a tiny stirrer. Over 800 assays (2 months) stability was observed with no loss of activity.

In another procedure these same workers [10] proposed an immobilized glucose oxidase/peroxidase stirrer for assay of glucose. Homovanillic acid is used as the fluorometric substrate; the rate of production of fluorescence is measured and related to the glucose concentration:

$$\text{Glucose} \xrightarrow{\text{glucose oxidase}} H_2O_2 \tag{18}$$

$$H_2O_2 + \text{homovanillic acid} \xrightarrow{\text{peroxidase}} \text{Ox homovanillic acid} \tag{19}$$
$$\text{(fluorescent)}$$

Kuan et al. [11] determined plasma glucose with a stirrer containing immobilized glucose dehydrogenase. The device is useful for at least 2 months and 500 assays. The reaction was monitored kinetically and linearity was observed to 4 g glucose/liter.

$$\text{Glucose} + \text{NAD} \xrightarrow[\text{dehydrogenase}]{\text{glucose}} \text{NADH} \tag{20}$$

The technique has good accuracy, specificity, and operational simplicity.

Huang et al. [3] described an enzyme stirrer (Fig. 6) containing immobilized cholesterol esterase and cholesterol oxidase for the assay of cholesterol in serum. The peroxide liberated in the enzymatic reaction is monitored with a Pt electrode placed directly above the enzyme stirrer (Fig. 7). The oxidase is very stable, but the stability

$$\text{Cholesterol esters} \xrightarrow{\text{esterase}} \text{cholesterol} \tag{21}$$

$$\text{Cholesterol} \xrightarrow{\text{oxidase}} \text{H}_2\text{O}_2 \tag{22}$$

of the stirrer was limited by the loss of activity of the less stable esterase. Nevertheless several hundred assays were possible with one stirrer. The entire assay can be performed in 10-12 min at a cost of less than 10 cents per assay. The calibration curve is linear from 0-50 g/liter of cholesterol, and the results correlate well with those obtained by the standard Abel method with a correlation coefficient

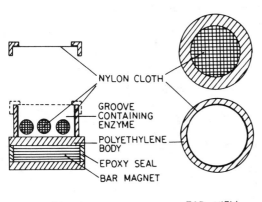

CROSS SECTION   TOP VIEW

FIG. 6 Cross section of the rotating porous stirrer.

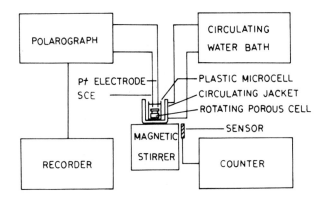

FIG. 7 Instrumental setup for use of the enzyme rotating porous
cell with amperometric monitoring.

of .9992. Ascorbic acid and bilirubin do not interfere at concentrations of up to 100 ng/liter.

C. Visual Color Tests Using Immobilized Enzymes

A test for glucose in urine has been developed by workers at Miles
[7]. The test, called Clinistix, contains two immobilized enzymes
(glucose oxidase and peroxidase) which are adsorbed onto a cellulose
mat. When the dip-and-read test is moistened with a urine containing
glucose, a color change develops within 10 sec according to the equations:

$$\beta\text{-D-glucose} + H_2O + O_2 \xrightarrow[\text{glucose oxidase}]{\text{immobilized}} \text{gluconic acid} + H_2O_2$$

(23)

$$H_2O_2 + \text{o-tolidine} \xrightarrow[\text{peroxidase}]{\text{immobilized}} \text{oxidized o-tolidine}$$
(24)
$$\text{(blue)}$$

A somewhat comparable system which utilizes immobilized galactose
oxidase provides a specific test for galactose in urine:

$$\text{D-galactose} + H_2O + O_2 \xrightarrow[\substack{\text{galactose}\\\text{oxidase}}]{\text{immobilized}} \text{D-galactohexodialose} + H_2O_2$$

(25)

$$H_2O_2 + o\text{-tolidine} \xrightarrow{\text{immobilized}} \text{oxidized } o\text{-tolidine} \qquad (26)$$
$$\text{peroxidase}$$
$$\text{(blue)}$$

This test is called Galactostix.

Immobilized enzymes have also been used in creating rapid test systems using whole blood for the assay of blood sugar and blood urea [21]. The reaction for the Azostix, the blood urea test, is:

$$\text{Urea} + 3H_2O + \xrightarrow{\text{immobilized}} 2NH_4OH + CO_2 \qquad (27)$$
$$\text{urease}$$

$$NH_4OH + \text{bromthymol blue} \longrightarrow \text{bromthymol blue} \qquad (28)$$
$$\text{(yellow)} \qquad\qquad \text{(green to blue)}$$

The acid base indicator, bromthymol blue, changes color from yellow to green-blue because of the NH_4OH liberated.

After the enzymes and chromogens have been adsorbed on the cellulose fibers with each of these systems, the fibers are coated with ethyl cellulose. Glucose, galactose, or urea readily and rapidly diffuse through the ethyl cellulose membrane, but the presence of the coating prevents absorption of the cells and allows them to be washed away. The strips are completely disposable.

Analogous chemistry is incorporated into the Kodak Clinical Analyzer, which uses discrete, disposable slides, one for each test. All the reagents, immobilized enzymes, substrates, and chromogens are incorporated into the slide in layers. When the sample is added, the substrate to be assayed (e.g., glucose) diffuses through and a color develops which is read by the analyzer. Glucose and urea as well as many other chemistries are currently offered.

D. Mass Spectrometry

The interfacing of a mass spectrometer with immobilized enzyme systems has been described by Weaver et al. [16]. By counting the volatile molecules produced by an immobilized enzyme-catalyzed reaction interfaced to a mass spectrometer via a semipermeable membrane, a detector with high sensitivity, specificity, and speed is obtained. Immobilized urease was used to assay urea with CO_2 as the volatile

product, and alcohol dehydrogenase was used to assay NAD with ethanol
as the volatile product.

E. Other Techniques

Guilbault and co-workers [25-27] demonstrated that immobilized enzymes can be used directly for the assay of air pollutants by coating a piezoelectric crystal device with the insolubilized reagent.
The crystal, oscillating at a frequency of 9 or 14 MHz, contains the
immobilized reagent (alcohol oxidase for assay of alcohol [25], carbonic anhydrase for assay of CO_2 [26], or formaldehyde dehydrogenase
for assay of formaldehyde [27]). When the substrate is adsorbed
onto the immobilized enzyme surface from the gas phase, a change in
frequency is observed, according to the Sauerbrey equation:

$$\Delta F = -\text{constant} \ (\Delta \text{ mass}) \tag{29}$$

These studies are the first demonstration that an immobilized enzyme
can be used directly in air, with no solution present, for the assay
of molecules present in the gas phase.

     This will undoubtedly be one of the exciting new areas of re-
search in the future.

II. COMMERCIAL INSTRUMENTS

A. General Discussion

In response to the high interest in the analytical use of immobilized
enzymes, several companies have introduced products to the consumer
(Table 2).

     Technicon and Carla Erba have offered enzyme coils, useful on
any automated analyzer that utilizes a flow system, for the assay of
glucose, urea, uric acid, and creatinine. The products of Technicon
have temporarily been removed from the market, but will possibly be
reoffered in response to the high demand.

     Instruments with electrochemical detectors are offered by Setric,
Leeds and Northrup, Yellow Springs, Midwest Research, and Owens Illi-
nois (Technicon) for carbohydrates (glucose, sucrose, lactose, maltose,
galactose), urea, uric acid, cholesterol, and lactate.

TABLE 2   List of Commercial Instruments Using Immobilized Enzymes

Company	Instrument(s)
1. Carla Erba (Milano, Italy)	Enzyme coils for glucose, urea, uric acid, creatinine
2. Leeds & Northrup (North Wales, PA)	Glucose, lactose, sucrose, maltose electrochemical analyzers
3. Midwest Research Institute (Kansas City, MO)	CAM and IEM analyzers for pesticides in air and water
4. Owens-Illinois (Kimble Division), Technicon (Ardsley, NY)	BUN, glucose electrochemical analyzer
5. Roche (Nutley, NJ)	Model 640 lactate analyzer
6. Setric (Paris, France)	Lactate analyzer (AL7 and ALC7)
7. Solea-Tacussel (Lyon, France)	Glucose electrode
8. Technicon Corp. (Tarrytown, NY)	Glucose, urea, uric acid, enzyme coils
9. Universal Sensors (New Orleans, LA)	Enzyme electrodes for glucose, urea, uric acid, alcohol, amino acids, and others on request
10. Yellow Springs Instrument Co. (Yellow Springs, OH)	Glucose, galactose, uric acid, cholesterol electrochemical analyzers

Single self-contained enzyme electrodes are offered by Solea-Tacussel (Lyon, France), a glucose electrode, and by Universal Sensors (P.O. Box 736, New Orleans 70148), which offers electrodes for almost any metabolite of interest.

B.   Yellow Springs Instrument Company Glucose Analyzer

The first commercial instrument to incorporate an immobilized enzyme electrode was the Yellow Springs Glucose Analyzer.   The probe and membrane system within the analyzer are shown in Fig. 8.   The probe is a Pt anode, poised at +0.7 V, at which H_2O_2 is oxidized, and a silver cathode.   Three membranes cover the probe: a polycarbonate membrane, a layer of immobilized glucose oxidase, and a cellulose acetate membrane.   The outer polycarbonate membrane excludes passage

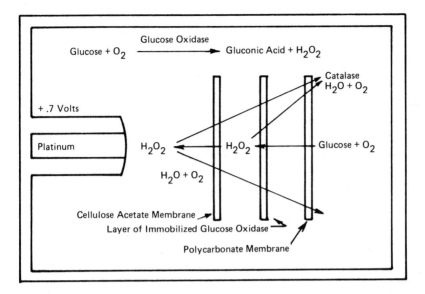

FIG. 8   Membrane sensor for Yellow Springs Instrument Company glucose analyzer.

of glucose oxidase into the solution or passage of catalase present in the buffer into the membrane composite. Proteins and other macromolecules that might poison the probe are also excluded from passing into the membrane layer. The secondary layer, which contains immobilized glucose oxidase, converts the glucose and oxygen that have diffused through the polycarbonate membrane to gluconic acid and hydrogen peroxide. The final layer contains a cellulose acetate membrane of such small pore size that only molecules the size of hydrogen peroxide, oxygen, and water can contact the probe. Substances of larger molecular weight that may be electrochemically active, such as ascorbic acid, are prevented from contact with the probe by this cellulose acetate membrane.

A diagram of the instrument is shown in Fig. 9. A buffer supply is used to clean and fill the sample chamber, which has a total volume of about 350 μl. To allow a more rapid equilibration to 37°C, a thermistor is present in the sample chamber to compensate the probe signal for small changes in temperature.

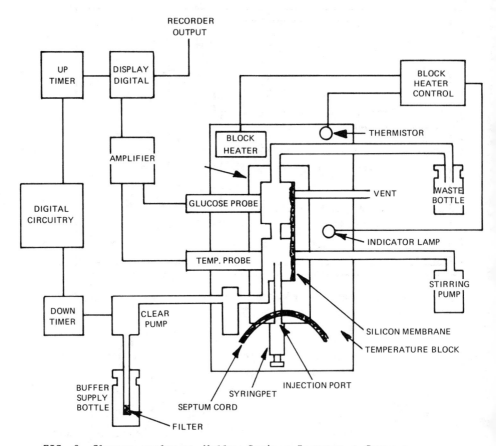

FIG. 9   Glucose analyzer, Yellow Springs Instrument Company.

This instrument operates in a kinetic mode.   When a sample is injected, glucose diffuses to the glucose oxidase layer and the generated hydro peroxide is sensed by the platinum anode.   After a short time, the rate of formation and diffusion of hydrogen peroxide to the probe equals the rate of diffusion away from the probe at which point the answer is displayed digitally.

This instrument is carefully designed to meet the requirements of a clinical environment.   The choice of membranes allows the measurement of glucose without appreciable interferences.   Analysis of each sample requires less than 1 min.   Since the measurement is kinetic, the temperature must be maintained at a constant 37°C with a heater block, and small variations owing to sample addition are corrected for with an additional thermistor in the sample chamber. Finally, catalase present in the buffer solution causes the breakdown of hydrogen peroxide that diffuses into the sample chamber and eliminates back diffusion to the probe.

Instruments are also offered for uric acid (uricase), cholesterol (cholesterol oxidase and esterase), and galactose (galactose oxidase).

C. Leeds and Northrup Analyzer

Leeds and Northrup has introduced a family of Enzymax analyzers (Fig. 10), which consist of single- and dual-channel glucose analyzers, a dual-channel glucose/sucrose and a dual-channel glucose/lactose instrument. The analysis of sugars is of particular impor-

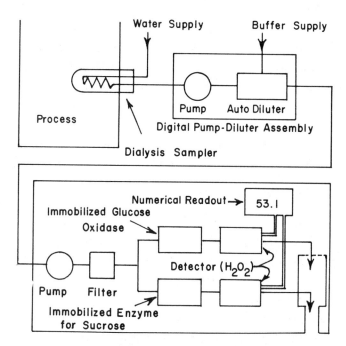

FIG. 10 Diagram of Enzymax sucrose/glucose analyzer with accessories.

tance in the food-processing industry, and the instruments were specifically designed for continuous, on-stream process control.

A dual-stream glucose/sucrose analyzer shown in Fig. 10 illustrates the flow scheme common to all the instruments. A continuous
sample is removed from a process stream through a dialysis thimble.
After dilution with a buffer the stream is split and passes through
packed beds of immobilized enzyme reagents. The enzyme is, of course,
chosen for its specificity to the substrate in question. Hence, for
the analysis of glucose, lactose, and sucrose, the immobilized enzymes would be glucose oxidase, β-galactosidase + glucose oxidase,
and invertase + mutarotase + glucose oxidase, respectively. In each
case, a common product, hydrogen peroxide, is generated and detected.
A potentiostatic three-electrode system is used for this purpose, the
current produced at a carbon anode being directly proportional to
peroxide concentration. The single-channel glucose analyzer lacks
the sucrose stream, whereas in the dual-channel glucose analyzer the
second stream contains a packed bed without any immobilized enzyme.
Thus, in the dual-channel instrument, glucose concentration can be
measured in the presence of any electrochemically active species
that may be present in the sample stream. In this instrument, interferences are circumvented in much the same way as in a double-beam
spectrophotometer.

The future of the Enzymax analyzer is in doubt, but recent
offerings were glucose, lactose, maltose, sucrose, and galactose.
The reader is invited to request more information.

D.   Kimble Division of Owens-Illinois

Under the leadership of Dr. Don Gray and the able assistance of Dr.
Mel Keyes and Dr. Barry Watson, Owens-Illinois introduced several
ago a BUN analyzer, followed by a glucose analyzer. Patent rights
to these two instruments were purchased by Technicon, which sells
the instruments in Europe only.

The blood urea nitrogen analyzer uses a gas-sensing electrode,
specifically for determining ammonia, coupled with an immobilized
urease preparation.

Urease is frequently employed for urea nitrogen determination
because of its high specificity for urea. In this instrument, urease
is immobilized to a porous alumina support that allows many hundreds
of separate analyses to be made before enzyme replacement is necessary. The method relies on the total conversion of urea to ammonia,
and the flow system used is shown schematically in Fig. 11. After
the initial hydrolysis reaction, the products leaving the enzyme
column are mixed with a second flow stream of sodium hydroxide that
raises the pH from 7.5 to more than 11. As a result, ammonium ions
are quantitatively converted to ammonia gas which is then sensed by
the ammonia electrode.

The electrode potential is monitored electronically and is
logarithmically related to ammonia, and thus to urea, concentration

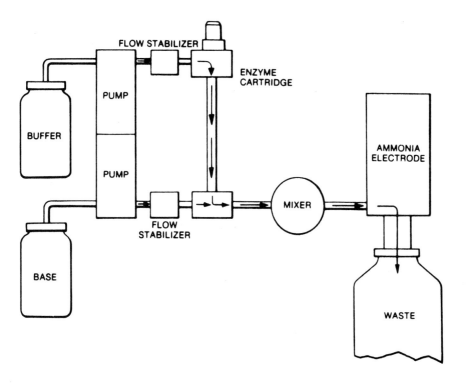

FIG. 11   Flow diagram and reactions of Kimble BUN analyzer. (Reprinted
from B. Watson and M. H. Keyes, *Anal. Lett.* *9*, 713 (1976), courtesy of
Marcel Dekker, Inc., New York.)

according to the Nernst equation. The instrument requires a 20-μl
sample injected directly onto the enzyme column, and linearity extends to 100 mg/dl. Note that the physical separation of enzyme and
sensor functions permits the optimum pH requirements of each to be
maintained. The performance characteristics of the instrument have
been determined in a clinical situation and are excellent. Two
thousand assays can be performed with a C.V. of 1%.

The point has already been made that many sensors, potentiometric, amperometric, and even colorimetric, offer little in the way
of specificity. This is not the case with gas-sensing membrane
probes. The ammonia electrode, shown in cross-section in Fig. 12,
with suitable choice of hydrophobic membrane and electrolyte filling
solution can be made very specific toward ammonia. The only known
interferences are low molecular weight amines that are not present

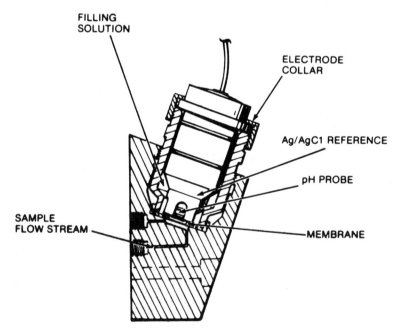

FIG. 12 Ammonia electrode developed for Kimble BUN analyzer.
(Reprinted from B. Watson and M. H. Keyes, *Anal. Lett. 9*, 713
(1976), courtesy of Marcel Dekker, Inc., New York.)

in significant concentrations in physiological fluids. Thus, an
electrochemical sensor can sometimes reinforce the specificity of
the enzyme reactions. Following this success, a glucose analyzer
was developed using immobilized glucose oxidase and a Pt electrode
to detect the H_2O_2 liberated. The flow diagram is similar to that
of Fig. 11.

A comparison of four instruments that assay glucose is given in
Table 3. The O-I analyzer is one of the best developed to date.

E. Midwest Research (CAM and IEM)

Immobilized cholinesterase can be used for the collection and detection of enzyme inhibitors in both air and water. The immobilized
enzyme functions like a dosimeter which, after exposure to the inhibitors in air or water, is then analyzed for its residual enzyme activity. Any loss in enzyme activity can then be related to the quantity
of inhibitor to which the enzyme was exposed. A detection system
based on this principle has several potential advantages. First,
there is great sensitivity due to the "biological amplification" by
the enzyme itself. Enzymes that have large turnover numbers are
quite sensitive detectors of enzyme inhibitors because inhibition
of one active site can produce a large reduction in the amount of
substrate reaction products formed. For example, in the case of
butyryl cholinesterase (acylcholine acyl hydrolase, EC 3.1.1.8) in
which the turnover number is approximately 84,000, inhibition of one
active site by one inhibitor molecule could reduce the formation of
substrate reaction products by 84,000 molecules in tests lasting
1 min. Second, the immobilized enzyme sensor possesses specificity
for its inhibitors; this makes possible the construction of detection systems with relatively few interference problems. Third, the
enzyme system is readily automated so that operation of the system
can proceed while the system is unattended. Fourth, in many cases
the responsiveness of the system to pollutants is related to their
animal toxicities. Fifth, the system can be adapted to monitoring
air through the use of a concentrator which absorbs the inhibitors
from a large volume of air into a small stream of water. With such

TABLE 3   Comparison of Commercial Glucose Analyzers

Feature	Leeds & Northrup glucose analyzer	Yellow Springs Instrument glucose analyzer	Technicon (Technizyme) glucose module	Owens-Illinois (Kimble)
Method	Equilibrium	Kinetic	Kinetic	Equilibrium
Enzyme	Glucose oxidase	Glucose oxidase (soluble catalase)	Hexokinase and G-6-PDH	Glucose oxidase
Detector	Amperometric probe for H_2O_2	Amperometric probe for H_2O_2	Spectrophotometric (340 nm)	Amperometric probe for H_2O_2
Maximum samples/hr	--	40	150 (SMA)	60
Sample method	Continuous	Discrete (25 µl)	Dialysis	Discrete (20 µl)
Precision (s.d.)	1% for 0-50 ppm (5 mg/dl)	1.3 mg/dl up to 150 mg/dl	C.V. 2%	C.V. 1%
Interferences	None	None	None	None

a concentrator it has been possible to detect subnanogram quantities
of sarin per liter of air in 9 min--a sensitivity level not readily
attained on a real-time basis with the usual air-monitoring equipment.

An electrochemical apparatus used to monitor the activity of
the enzyme in urethane pads continuously was first developed by
Bauman et al. [28] as is shown in Fig. 13. This figure shows the
details of the enzyme pad, O ring, and disk electrode assembly. The
disk electrodes were prepared by punching 1/16-in.-diameter holes
into a 1-in. circular piece of 0.003-in.-thick platinum sheet that
has a 3/8 x 1/4 in. handle (available special from J. Bishop and Co.,
Malvern, PA). A pad 3/4 in. in diameter, prepared as described above,
is then placed into a 1 x 3/4 x 1/8 in. O ring, the electrodes are
placed above and below the pad, and the pad and electrodes are placed
into a Millipore microanalysis filter holder.

The filter holder was held together with the clamp provided
with the filter, and the waste was collected in a 250-ml filter
flask. The substrate, 5 x 10^{-4} M butyrylthiocholine iodide in 0.1 M
tris buffer, pH 7.40, is pumped over the pad using a positive dis-
placement liquid pump, with a delivery rate of 1.0 ml/min. Air and

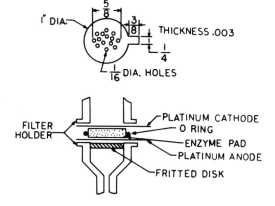

FIG. 13 Details of enzyme pad, O ring, and grid electrode assembly.
(From Ref. 28.)

water (containing possible cholinesterase inhibitors, e.g., pesticides) were sucked through the enzyme pad by means of a Brailsford
blower. A constant current of 2 μA was applied across the electrodes,
and the change in potential that occurred was monitored with a high-
impedance electrometer and was automatically recorded.

As long as the enzyme cholinesterase is active in the pad, the
butyrylthiocholine iodide is hydrolyzed to the easily oxidizable thiocholine. At a constant current of 2 μA, a potential of about 150 mV
(Fig. 14) will be established across the cell assembly pictured in
Fig. 13. Since the electrooxidation of the thiol takes place at the
anode, it is important that the anode be located at the downstream
surface of the pad where the concentration of hydrolysis product is
greatest:

$$\text{Acetylthiocholine iodide} \xrightarrow{\text{ChE}} \text{thiocholine} + \text{acetic acid} \quad (30)$$

If an inhibitor is present in air or water that reduces the activity
of enzyme, less thiol is formed and the potential rises to that of
the iodide/iodine couple, 350-400 mV (Fig. 14, E_f). If the inhibitor
is a reversible one, it can be removed from the pad by substrate flow,

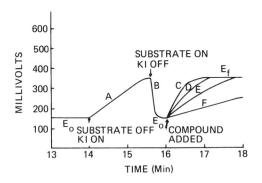

FIG. 14 Typical operation and response curves of experimental apparatus. (A) Flow rate, 1 ml/min; substrate solution, off; KI solution,
on. (B) Flow rate, 1 ml/min; substrate solution, on; KI solution,
off. (C-F) Various concentrations of the pesticide Systox added to
water stream. (From Ref. 28.)

allowing subsequent determinations. If the inhibitor is an irreversible one, a new enzymatic pad must be used. An improvement of this electrochemical system is now being marketed by Midwest Research (Kansas City, MO) for the assay of pesticides in water.

Two instruments have been developed by Midwest Research Institute to measure organophosphates and carbamates. These instruments are unique in that they measure inhibitors of an enzyme reaction rather than the substrate. The immobilized enzyme monitor (IEM) measures air samples while the continuous aqueous monitor (CAM) is designed for water samples. Both instruments operate by measuring the inhibition of catalysis of cholinesterase caused by organophosphates or carbamates. Thus in their absence, butyrylthiocholine iodide is converted to butyric acid and thiocholine iodide. The rate of formation of the easily oxidized thiocholine is monitored by an electrochemical cell in which cholinesterase immobilized on an

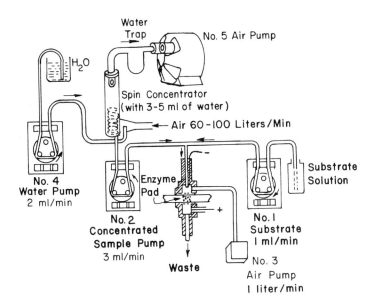

FIG. 15   Immobilized enzyme monitor (IEM).

open-pore disk of polyurethane foam is placed between two electrodes
having an applied current of 2 μA (Fig. 15). The voltage is about
200 mV in the absence of inhibitors and 500 mV in the presence of
inhibitors. Note that for the IEM picture in Fig. 15, the scrubber
allows for the extraction of material from 60-100 liters/min of air
to a water stream (2 ml/min).

Because these instruments require good flow characteristics for
both air and water as well as good contact between the cell electrodes, the method of immobilization is critical. After testing ion-
exchange materials such as cotton and cheesecloth, it was determined
that open-pore polyurethane foam had the best physical properties.
A successful immobilization of cholinesterase was accomplished by
first absorbing starch gel-entrapped cholinesterase on aluminum
hydroxide gel, which is then applied to the polyurethane foam.

Since both instruments are designed for constant or nearly constant monitoring, the response time is not as important as for clinical instruments. For example, CAM (Fig. 16) operates by flowing
water to be tested over the immobilization cholinesterase for 2 min
to collect the enzyme inhibitors. Excess water is blown out of the

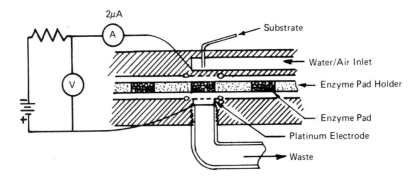

Fig. 16 Electrochemical cell for CAM.

electrochemical cell and the residual enzyme activity measured automatically. Thus a signal might be generated every 3-4 min. When
the concentration of inhibitor becomes dangerously high, an alarm
will go off. These instruments will probably find considerable
application in monitoring levels of hazardous pesticides during
manufacturing, formulations, and packaging operations.

F. Sectric and Roche Analyzers

Sectric (Toulouse, France) offers an AL7 lactate analyzer (sequential
samples) and an ALC7 analyzer (continuous samples).

Both instruments are based on the oxidation of lactic acid by
ferricyanide in the presence of lactate dehydrogenase to produce the
electroactive ferrocyanide ion. The linear range of the instruments
is 0.05-7.0 mM, precision 2%, time of response 20-30 sec, with 20-30
samples/hr assayable. Two hundred assays are possible with one enzyme
cartridge.

An instrument based on the sample principle of assay is offered
by Roche, the Lactate Analyzer 640e. The range is 0-12 mmole/liter
lactate, with a precision of ±0.2 mmole/liter and a reproducibility
of calibration of ±2%. The time of response is 40-60 sec, and 20-30
samples/hr can be assayed. Only 100 μl of sample is required.

G. Tacussel

Solea-Tacussel (Lyon, France) markets a glucose enzyme electrode
that utilizes a membrane of collagen onto which is covalently linked
glucose oxidase. The measurement unit that is solid with the glucose
electrode is shown in Fig. 17 and the electrode is shown in Fig. 18.
The range of assay is 10^{-3} to 10^{-7} M, with reproducibility of 0.5-2%.
The probe keeps a linear response for several months when used at
25-30°C. Its typical sensitivity is 1.5 μA/mmole per liter of glucose. The result for glucose is displayed 45 sec after sample injection, and an interchangeable membrane is capable of at least
1000 assays or 3 months.

FIG. 17   Measurement unit for glucose GLUC 1 electrode.

H.   Enzyme Coils

Technicon originally offered for sale bound enzyme coils (Fig. 19) which are adaptable as modules for the SMA and SMAC systems.   As illustrated in Fig. 19, the enzyme hexokinase and glucose-6-phosphate dehydrogenase are coimmobilized by attachment to the inner wall of

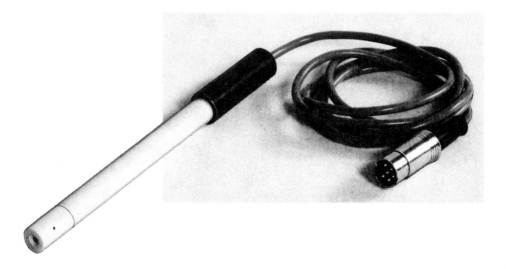

FIG. 18   GLUC 1 electrode with interchangeable enzymatic membrane for the determination of dissolved glucose concentration.

a coiled plastic tube. The design of the enzyme reactor allows for
easy installation and replacement and is compatible with the sample
introduction and segregation systems traditional with the pioneering
company in automated clinical chemistry.

The spectrophotometric sensor, common to most Technicon instruments, is retained and monitors the absorbance due to NADH formation,
the concentration of which is linearly related to the glucose content
in the original sample.

The reusable feature of immobilized enzyme systems allows continuous sampling for a 2-week period before coil replacement is neces-

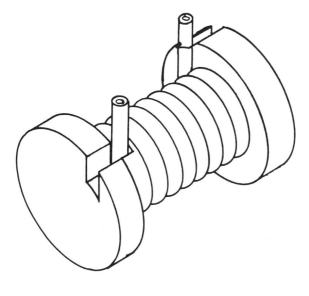

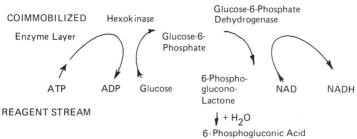

FIG. 19 Technicon bound enzyme coil with illustration of reactions
and reagents.

FIG. 20   Carla Erba enzyme coil for glucose.

sary. The company claims a 50% cost saving compared with similar methods employing soluble enzyme reagents.

These coils have been temporarily removed from the market, but similar tubes are offered by Carla Erba (Milano) as shown in Fig. 20. Such coils are sold for assay of glucose, urea, uric acid, and creatinine, and have a lifetime of 20,000 assays per coil.

I. Universal Sensors

In 1981 a new company was set up to sell the enzyme electrodes developed by Guilbault and co-workers. These are all self-contained electrodes which plug into a pH meter either directly (urea, L-lysine, creatinine, L-methionine, L-tyrosine) or via a cheap $150 adapter (glucose, uric acid, galactose, etc.). The electrodes cost about $450, with new enzyme tips about $50 each.

REFERENCES

1. B. Rietz and G. G. Guilbault, *Clin. Chem.* 21, 1791 (1975).

2. G. G. Guilbault and A. Vaughn, *Anal. Chim. Acta 55*, 107 (1971).

3. N. Huang, S. S. Kuan, and G. G. Guilbault, *Clin. Chem. 23*, 671 (1977).

4. R. L. Zimmerman and G. G. Guilbault, *Anal. Chim. Acta 58*, 75 (1972).

5. G. G. Guilbault and H. Lau, *Clin. Chim. Acta 53*, 209 (1974).

6. J. C. W. Kuan, S. S. Kuan, and G. G. Guilbault, *Anal. Chim. Acta 100*, 229 (1978).

7. J. Campbell, A. S. Chawla, and T. M. S. Chang, *Anal. Biochem.* *83,* 330 (1977).

8. S. W. Kiang, J. W. Kuan, S. S. Kuan, and G. G. Guilbault, *Clin. Chem. 21,* 1799 (1975).

9. S. W. Kiang, J. W. Kuan, S. S. Kuan, and G. G. Guilbault, *Clin. Chim. Acta 78,* 495 (1977).

10. S. W. Kiang, J. W. Kuan, S. S. Kuan, and G. G. Guilbault, *Clin. Chem. 22,* 1378 (1976).

11. J. W. Kuan, S. S. Kuan, and G. G. Guilbault, *Clin. Chem. 23,* 1058 (1977).

12. B. Rietz and G. G. Guilbault, *Clin. Chem. 21,* 715 (1975).

13. B. Rietz and G. G. Guilbault, *Anal. Chim. Acta 77,* 191 (1975).

14. R. A. P. Harrison, *Anal. Biochem. 61,* 500 (1974).

15. G. G. Guilbault and R. Zimmerman, *Anal. Lett. 3,* 133 (1970).

16. J. C. Weaver, M. K. Mason, J. A. Jarrell, and J. W. Peterson, *Biochim. Biophys. Acta 438,* 296 (1976).

17. C. H. Kiang, S. S. Kuan, and G. G. Guilbault, *Anal. Chem. 50,* 1323 (1978).

18. G. G. Guilbault and W. Stakbro, *Anal. Chim. Acta 76,* 237 (1975).

19. J. W. Kuan, H. K. Lau, and G. G. Guilbault, *Clin. Chem. 21,* 67 (1975).

20. P. Kamoun and O. Douay, *Clin. Chem. 24,* 2033 (1978).

21. Ames, Division of Miles Laboratories, Clinistix and Azostix Product Information, 1983.

22. G. G. Guilbault and H. Lau, Presented at NIAMD Contractor's Conference, Bethesda, Maryland, 1973; *Clin. Chem. 19,* 1045 (1973).

23. J. W. Kuan, S. S. Kuan, and G. G. Guilbault, *Clin. Chim. Acta 64,* 19 (1975).

24. N. Huang, J. W. Kuan, and G. G. Guilbault, *Clin. Chem. 21,* 1605 (1975).

25. G. G. Guilbault, C. F. Mandelius, K. Mosbach, and B. Danielsson, *Anal. Chem. 55,* 1582 (1983).

26. G. G. Guilbault and A. Suleiman, unpublished results, University of New Orleans, 1983.

27. G. G. Guilbault, *Anal. Chem. 55,* 1682 (1983).

28. G. Bauman, L. Goodson, G. G. Guilbault, and D. Kramer, *Anal. Chem. 37,* 1378 (1965).

appendix 1
Numbering and Classification of Enzymes

Recommendations (1972) on Enzyme Nomenclature of the International Union of Pure and Applied Chemistry and the International Union of Biochemistry, Commission on Biochemical Nomenclature and Classification of Enzymes, Elsevier, Amsterdam, 1973, pp. 17-22.

1. Oxidoreductases

    1. 1 Acting on the CH-OH group of donors
        1. 1. 1 With NAD^+ or $NADP^+$ as acceptor
        1. 1. 2 With a cytochrome as acceptor
        1. 1. 3 With oxygen as acceptor
        1. 1. 99 With other acceptors

    1. 2 Acting on the aldehyde or keto group of donors
        1. 2. 1 With NAD^+ or $NADP^+$ as acceptor
        1. 2. 2 With a cytochrome as acceptor
        1. 2. 3 With oxygen as acceptor
        1. 2. 4 With a disulfide compound as acceptor
        1. 2. 7 With an iron-sulfur protein as acceptor
        1. 2. 99 With other acceptors

    1. 3 Acting on the CH-CH group of donors
        1. 3. 1 With NAD^+ or $NADP^+$ as acceptor
        1. 3. 2 With a cytochrome as acceptor
        1. 3. 3 With oxygen as acceptor
        1. 3. 7 With an iron-sulfur protein as acceptor
        1. 3. 99 With other acceptors

1. 4   Acting on the CH-NH$_2$ group of donors

    1. 4. 1   With NAD$^+$ or NADP$^+$ as acceptor

    1. 4. 3   With oxygen as acceptor

    1. 4. 4   With a disulfide compound as acceptor

    1. 4. 99   With other acceptors

1. 5   Acting on the CH-NH group of donors

    1. 5. 1   With NAD$^+$ or NADP$^+$ as acceptor

    1. 5. 3   With oxygen as acceptor

    1. 5. 99   With other acceptors

1. 6   Acting on NADH or NADPH

    1. 6. 1   With NAD$^+$ or NADP$^+$ as acceptor

    1. 6. 2   With a cytochrome as acceptor

    1. 6. 4   With a disulfide compound as acceptor

    1. 6. 5   With a quinone or related compound as acceptor

    1. 6. 6   With a nitrogenous group as acceptor

    1. 6. 7   With an iron-sulfur protein as acceptor

    1. 6. 99   With other acceptors

1. 7   Acting on other nitrogenous compounds as donors

    1. 7. 2   With a cytochrome as acceptor

    1. 7. 3   With oxygen as acceptor

    1. 7. 7   With an iron-sulfur protein as acceptor

    1. 7. 99   With other acceptors

1. 8   Acting on a sulfur group of donors

    1. 8. 1   With NAD$^+$ or NADP$^+$ as acceptor

    1. 8. 2   With a cytochrome as acceptor

    1. 8. 3   With oxygen as acceptor

    1. 8. 4   With a disulfide compound as acceptor

    1. 8. 5   With a quinone or related compound as acceptor

    1. 8. 6   With a nitrogenous group as acceptor

    1. 8. 7   With an iron-sulfur protein as acceptor

    1. 8. 99   With other acceptors

1. 9   Acting on a heme group of donors

    1. 9. 3   With oxygen as acceptor

    1. 9. 6   With a nitrogenous group as acceptor

    1. 9. 99   With other acceptors

1. 10   Acting on diphenols and related substances as donors
    1. 10. 2   With a cytochrome as acceptor
    1. 10. 3   With oxygen as acceptor

1. 11   Acting on hydrogen peroxide as acceptor

1. 12   Acting on hydrogen as donor
    1. 12. 1   With NAD^+ or $NADP^+$ as acceptor
    1. 12. 2   With a cytochrome as acceptor
    1. 12. 7   With an iron-sulfur protein as acceptor

1. 13   Acting on single donors with incorporation of molecular oxygen (oxygenases)
    1. 13. 11   With incorporation of two atoms of oxygen
    1. 13. 12   With incorporation of one atom of oxygen (internal monooxygenases or internal mixed function oxidases)
    1. 13. 99   Miscellaneous (requires further characterization)

1. 14   Acting on paired donors with incorporation of molecular oxygen
    1. 14. 11   With 2-oxoglutarate as one donor and incorporation of one atom each of oxygen into both donors
    1. 14. 12   With NADH or NADPH as one donor and incorporation of two atoms of oxygen into one donor
    1. 14. 13   With NADH or NADPH as one donor and incorporation of one atom of oxygen
    1. 14. 14   With reduced flavin or flavoprotein as one donor and incorporation of one atom of oxygen
    1. 14. 15   With a reduced iron-sulfur protein as one donor and incorporation of one atom of oxygen
    1. 14. 16   With reduced pteridine as one donor and incorporation of one atom of oxygen
    1. 14. 17   With ascorbate as one donor and incorporation of one atom of oxygen
    1. 14. 18   With another compound as one donor and incorporation of one atom of oxygen
    1. 14. 99   Miscellaneous (requires further characterization)

1. 15   Acting on superoxide radicals as acceptor

1. 16   Oxidizing metal ions
     1. 16. 3   With oxygen as acceptor

1. 17   Acting on CH_2 groups
     1. 17. 1   With NAD^+ or $NADP^+$ as acceptor
     1. 17. 4   With a disulfide compound as acceptor

2. Transferases

    2. 1   Transferring one-carbon groups
       2. 1. 1   Methyltransferases
       2. 1. 2   Hydroxymethyl-, formyl-, and related transferases
       2. 1. 3   Carboxyl- and carbamoyltransferases
       2. 1. 4   Amidinotransferases

    2. 2   Transferring aldehyde or ketonic residues

    2. 3   Acyltransferases
       2. 3. 1   Acyltransferases
       2. 3. 2   Aminoacyltransferases

    2. 4   Glycosyltransferases
       2. 4. 1   Hexosyltransferases
       2. 4. 2   Pentosyltransferases
       2. 4. 99 Transferring other glycosyl groups

    2. 5   Transferring alkyl or aryl groups, other than methyl groups

    2. 6   Transferring nitrogenous groups
       2. 6. 1   Aminotransferases
       2. 6. 3   Oximinotransferases

    2. 7   Transferring phosphorus-containing groups
       2. 7. 1   Phosphotransferases with an alcohol group as acceptor
       2. 7. 2   Phosphotransferases with a carboxyl group as acceptor
       2. 7. 3   Phosphotransferases with a nitrogenous group as acceptor
       2. 7. 4   Phosphotransferases with a phospho group as acceptor

2. 7. 5   Phosphotransferases with regeneration of donors
           (apparently catalyzing intramolecular transfers)
2. 7. 6   Diphosphotransferases
2. 7. 7   Nucleotidyltransferases
2. 7. 8   Transferases for other substituted phospho groups
2. 7. 9   Phosphotransferases with paired acceptors

2. 8   Transferring sulfur-containing groups
        2. 8. 1   Sulfurtransferases
        2. 8. 2   Sulfotransferases
        2. 8. 3   CoA-transferases

3. Hydrolases

    3. 1   Acting on ester bonds
            3. 1. 1   Carboxylic ester hydrolases
            3. 1. 2   Thiolester hydrolases
            3. 1. 3   Phosphoric monoester hydrolases
            3. 1. 4   Phosphoric diester hydrolases
            3. 1. 5   Triphosphoric monoester hydrolases
            3. 1. 6   Sulfuric ester hydrolases
            3. 1. 7   Diphosphoric monoester hydrolases

    3. 2   Acting on glycosyl compounds
            3. 2. 1   Hydrolyzing O-glycosyl compounds
            3. 2. 2   Hydrolyzing N-glycosyl compounds
            3. 2. 3   Hydrolyzing S-glycosyl compounds

    3. 3   Acting on ether bonds
            3. 3. 1   Thioether hydrolases
            3. 3. 2   Ether hydrolases

    3. 4   Acting on peptide bonds (peptide hydrolases)
            3. 4. 11   α-Aminoacylpeptide hydrolases
            3. 4. 12   Peptidylamino acid or acylamino acid hydrolases
            3. 4. 13   Dipeptide hydrolases
            3. 4. 14   Dipeptidylpeptide hydrolases
            3. 4. 15   Peptidyldipeptide hydrolases
            3. 4. 21   Serine proteinases
            3. 4. 22   SH-proteinases

3. 4. 23 Acid proteinases

3. 4. 24 Metalloproteinases

3. 4. 99 Proteinases of unknown catalytic mechanism

3. 5 Acting on carbon-nitrogen bonds, other than peptide bonds

   3. 5. 1 In linear amides

   3. 5. 2 In cyclic amides

   3. 5. 3 In linear amidines

   3. 5. 4 In cyclic amidines

   3. 5. 5 In nitriles

   3. 5. 99 In other compounds

3. 6 Acting on acid anhydrides

   3. 6. 1 In phosphoryl-containing anhydrides

   3. 6. 2 In sulfonyl-containing anhydrides

3. 7 Acting on carbon-carbon bonds

   3. 7. 1 In ketonic substances

3. 8 Acting on halide bonds

   3. 8. 1 In C-halide compounds

   3. 8. 2 In P-halide compounds

3. 9 Acting on phosphorus-nitrogen bonds

3. 10 Acting on sulfur-nitrogen bonds

3. 11 Acting on carbon-phosphorus bonds

4. Lyases

4. 1 Carbon-carbon lyases

   4. 1. 1 Carboxylyases

   4. 1. 2 Aldehydelyases

   4. 1. 3 Oxoacid lyases

   4. 1. 99 Other carbon-carbon lyases

4. 2 Carbon-oxygen lyases

   4. 2. 1 Hydrolyases

   4. 2. 2 Acting on polysaccharides

   4. 2. 99 Other carbon-oxygen lyases

4. 3   <u>Carbon-nitrogen lyases</u>

    4. 3. 1   Ammonia lyases

    4. 3. 2   Amidine lyases

4. 4   <u>Carbon-sulfur lyases</u>

4. 5   <u>Carbon-halide lyases</u>

4. 6   <u>Phosphorus-oxygen lyases</u>

4. 99   <u>Other lyases</u>

5.   <u>Isomerases</u>

  5. 1   <u>Racemases and epimerases</u>

    5. 1. 1   Acting on amino acids and derivatives

    5. 1. 2   Acting on hydroxy acids and derivatives

    5. 1. 3   Acting on carbohydrates and derivatives

    5. 1. 99   Acting on other compounds

  5. 2   <u>Cis-trans isomerases</u>

  5. 3   <u>Intramolecular oxidoreductases</u>

    5. 3. 1   Interconverting aldoses and ketoses

    5. 3. 2   Interconverting keto and enol groups

    5. 3. 3   Transposing C=C bonds

    5. 3. 4   Transposing S-S bonds

    5. 3. 99   Other intramolecular oxidoreductases

  5. 4   <u>Intramolecular transferases</u>

    5. 4. 1   Transferring acyl groups

    5. 4. 2   Transferring phosphoryl groups

    5. 4. 3   Transferring amino groups

    5. 4. 99   Transferring other groups

  5. 5   <u>Intramolecular lyases</u>

  5. 99   <u>Other isomerases</u>

6.   <u>Ligases (Synthetases)</u>

  6. 1   <u>Forming carbon-oxygen bonds</u>

    6. 1. 1   Ligases forming aminoacyl-tRNA and related compounds

6. 2   Forming carbon-sulfur bonds

    6. 2. 1   Acid-thiol ligases

6. 3   Forming carbon-nitrogen bonds

    6. 3. 1   Acid-ammonia ligases (amide synthetases)

    6. 3. 2   Acid-amino acid ligases (peptide synthetases)

    6. 3. 3   Cycloligases

    6. 3. 4   Other carbon-nitrogen ligases

    6. 3. 5   Carbon-nitrogen ligases with glutamine as
            amido-N donor

6. 4   Forming carbon-carbon bonds

6. 5   Forming phosphate ester bonds

appendix 2
Immobilized (Insoluble) Enzymes

1. Acylase (Aminoacylase, EC 3.5.1.14)
   Enzygel - Boehringer
2. Alcohol dehydrogenase (EC 1.1.1.1)
   Agarose - Sigma
   Polyacrylamide - Sigma
3. Aldolase (EC 4.1.2.13)
   Polyacrylamide - Sigma
4. α-Amylase (EC 3.2.1.2)
   Enzacryl - Aldrich
   Polyacrylamide - Sigma
5. Amyloglucosidase (EC 3.2.1.3)
   Enzacryl - Aldrich
   Enzygel - Boehringer
6. Asparaginase (EC 3.5.1.1)
   Agarose - Sigma
7. Bromelain (EC 3.4.4.24)
   Carboxymethyl cellulose - Merck, Gallard Schlesinger, Miles
8. Carboxypeptidase A (EC 3.4.2.1)
   Agarose - Sigma
9. Catalase (EC 1.11.1.6)
   Enzacryl - Aldrich
10. Cholinesterase, acetyl, electric eel (EC 3.1.1.7)
    Agarose - Sigma

11. α-Chymotrypsin (EC 3.4.21.1)

    Agarose - Miles

    Carboxymethyl cellulose - Merck, Gallard Schlesinger, Sigma

    Dextran gel - Schwartz Mann

    Enzacryl - Aldrich

    Enzygel - Boehringer

12. Creatine phosphokinase/hexokinase/glucose-6-phosphate

    Dehydrogenase (EC 2.7.3.2/2.7.1.1/1.1.1.49)

    Agarose - Sigma

13. Deoxyribonuclease I (EC 3.1.4.5)

    Sepharose - Worthington

14. Dextranase (EC 3.2.1.11)

    Enzacryl - Aldrich

15. Ficin (EC 3.4.22.3)

    Carboxymethyl cellulose - Gallard Schlesinger, Merck, Sigma

16. Glucose oxidase (EC 1.1.3.4)

    Carboxymethyl cellulose - Gallard Schlesinger

    Duolase - Diamond Shamrock

    Enzacryl - Aldrich

    Enzygel - Boehringer Mannheim

    Polyacrylamide - Sigma

17. Glucose-6-phosphate dehydrogenase (EC 1.1.1.49)

    Agarose - Sigma

    Polyacrylamide - Sigma

18. β-Glucosidase (EC 3.2.1.21)

    Enzacryl - Aldrich

19. β-Glucuronidase (EC 3.2.1.31)

    Agarose - Sigma

20. Glyceraldehyde-3-phosphate dehydrogenase (EC 1.2.1.12)

    Polyacrylamide - Sigma

21. Glyceraldehyde-3-phosphate dehydrogenase/3-phosphoglycerate

    Phosphokinase (EC 1.2.1.12/2.7.2.3)

    Agarose - Sigma

    Polyacrylamide - Sigma

22. α-Glycerophosphate dehydrogenase/triosephosphate isomerase
     (EC 1.1.1.8/5.3.1.1)
     Polyacrylamide - Sigma

23. Hexokinase (EC 2.7.1.1)
     Agarose - Sigma
     Polyacrylamide - Sigma

24. Hexokinase/glucose-6-phosphate dehydrogenase (EC 2.7.1.1/1.1.1.49)
     Agarose - Sigma

25. Lactase (EC 3.2.1.23)
     Duolase - Diamond Shamrock

26. L-lactate dehydrogenase (EC 1.1.1.27)
     Polyacrylamide - Sigma

27. Malate dehydrogenase (EC 1.1.1.37)
     Polyacrylamide - Sigma

28. Neuraminidase, types VI-A and X-A (EC 3.2.1.18)
     Agarose - Sigma

29. Papain (EC 3.4.22.2)
     Acid anhydride - Merck
     Agarose - Miles, Sigma
     Carboxymethyl cellulose - Gallard Schlesinger, Merck, Sigma
     Duolase - Diamond Shamrock
     Maleic anhydride and divinyl ether copolymer - Merck

30. Pepsin (EC 3.4.23.1)
     Disks - Worthington

31. Peroxidase (EC 1.11.1.7)
     Cow's milk (lactoperoxidase), Sepharose - P-L, Worthington
     Horseradish, Agarose - Sigma

32. Phosphatase, alkaline (EC 3.1.3.1)
     Agarose - Sigma

33. Phosphoglucose isomerase (EC 5.3.1.9)
     Polyacrylamide - Sigma

34. 3-Phosphoglyceric phosphokinase (EC 2.7.2.3)
     Polyacrylamide - Sigma

35. Phospholipase A$_2$ (EC 3.1.1.4)

   Agarose - Sigma

   Membrane-bound - New England Enzyme

36. Pronase or protease (*S. griseus*) (EC 3.4.4.-)

   Agarose - Miles

   Carboxymethyl cellulose - Merck, Sigma

37. Protease (EC 3.3.3.1)

   B. amyloliquefaciens, Agarose - Sigma

   B. subtilis, Carboxylmethyl cellulose - Merck

38. Proteinase K (EC 3.2.1.41)

   Hexylaminosuccinilate cellulose - Merck

39. Pyruvate kinase (EC 2.7.1.40)

   Agarose - Sigma

   Polyacrylamide - Sigma

40. Ribonuclease A (EC 3.1.4.22)

   Acid anhydride and divinyl ether copolymer - Merck

   Agarose - Miles, Sigma

   Carboxymethyl cellulose - Merck, Miles, Gallard Schlesinger

   Enzygel - Boehringer

   Polyacrylamide - Sigma

   Sepharose - Worthington

41. Trypsin (EC 3.4.4.4)

   Agarose - Miles, Sigma

   Carboxymethyl cellulose - Gallard Schlesinger, Merck

   1-Carboxymethyl histidine 119 - Schwartz-Mann.

   Enzacryl - Aldrich

   Enzygel - Boehringer

   Maleic anhydride and divinyl ether copolymer - Merck

   Polyacrylamide - Sigma

   Sepharose - Calbiochem, Worthington

42. Trypsin inhibitor - Lima Bean

   Sepharose - Worthington

43. Trypsin inhibitor - Ovomucoid

   Sepharose - Worthington

44. Trypsin inhibitor - Soybean
        Carboxymethyl cellulose - Merck
        Succinylaminohexyl cellulose - Merck
45. Urease (EC 3.5.1.5)
        Enzacryl - Aldrich
        Enzygel - Boehringer
        Polyacrylamide - Sigma
46. Uricase (EC 1.7.3.3)
        Enzacryl - Aldrich

Note: A series of prepackaged enzyme coil reactor columns are available from Carla Erba, Milan, and from Technicon, Ardsley, New York. Carla Erba: creatininase, glucose oxidase, urease, uricase; Technicon: hexokinase/glucose-6-phosphate dehydrogenase, uricase.

appendix 3
Purified Enzymes (Soluble)

1. Acetate kinase (ATP; acetate phosphotransferase, EC 2.7.2.1)

   E. coli - Boehringer, P-L, Serva, Sigma

2. S-Acetyl coenzyme A synthetase (acetate: CoA ligase, AMP forming, EC 6.2.1.1)

   Baker's yeast - Boehringer, Serva, Sigma

3. Acetylesterase (acetic ester hydrolase, EC 3.1.1.6)

   Orange peel - Sigma

4. α-N-Acetyl-galactosaminidase (EC 3.2.1.49)

   Charonia lampus - Miles, Sigma

5. N-Acetyl-β-D-glucosaminidase (2-acetamido-2-deoxy-β-D-glucoside acetamidodeoxyglucohydrolase, EC 3.2.1.30)

   A. niger - Sigma

   Bovine kidney - Boehringer, Sigma, Serva

   Charonia lampus - Miles

   Human placenta - Sigma

   Jack beans - Sigma

   Type A - Bovine epididymis - Sigma

   Type A - Porcine placenta - Sigma

   Type B - Bovine epididymis - Sigma

   Type B - Porcine placenta - Sigma

   Type B - Human placenta - Sigma

6. β-N-Acetylhexosaminidase (EC 3.2.1.52)

   Charonia lampus - Miles

   Turbo cornutus - Miles

7. N-Acetylneuraminic acid aldolase (N-acetylneuraminate pyruvate lyase, EC 4.1.3.3)

   C. perfringens - Sigma

8. Acetyltrypsin

   Bovine pancreas - Koch Light; Miles, Schwartz-Mann, Sigma

9. Aconitase [aconitate hydratase; citrate (isocitrate) hydrolase, EC 4.2.1.3]

   Porcine heart - Sigma

10. Actomyosin (myosin ATPase)

    Rabbit muscle - Sigma

11. Acylase I (aminoacylase, N-acylamino acid amidohydrolase, EC 3.5.1.14)

    Porcine kidney - BDH; Biozyme; Boehringer, Calbiochem;
        General Biochemical, Gibco, Koch-Light, Miles, NBC, Paesel,
        Schwartz-Mann, Serva, Sigma, U.S. Biochemical

12. Acylase II (EC 3.5.1.14)

    Porcine kidney - Paesel

13. Acylase III (EC 3.5.1.14)

    Porcine kidney - Paesel

14. Acyl coenzyme A oxidase

    Candida - Sigma

15. Acyl coenzyme A synthetase (acid: coenzyme A ligase, AMP forming, EC 6.2.1.3)

    Pseudomonas sp - Sigma

16. Adenosine deaminase (adenosine aminohydrolase, EC 3.5.4.4)

    Calf - intestinal mucosa - Boehringer, BDH, U.S. Biochemical,
        Calbiochem, P-L, Sigma

    Bovine spleen - Sigma

17. Adenosine-5'-phosphosulfate kinase

    Sigma[*]

18. Adenosine-5'-triphosphatase (ATPase; ATP phosphohydrolase, EC 3.6.1.3)

    Dog kidney - Sigma

    Porcine cerebral cortex - Sigma

---

[*]Available on special request - inquire.

19. Adenosine-5'-triphosphate sulfurylase (ATP: sulfate adenylyl-
    transferase, EC 2.7.7.4)

    Yeast - Sigma

20. 5-Adenosyl-L-homocysteine hydrolase (EC 3.3.1.1)

    Rabbit erythrocytes - Sigma

21. Adenylate kinase (see Myokinase)

22. 5'-Adenylic acid deaminase (AMP aminohydrolase; Schmidt's
    deaminase; 5'-AMP deaminase, EC 3.5.4.6)

    Rabbit muscle - Sigma

    Calf brain - New England Enzyme*

23. Agarase (agarose-3-glycanhydrolase, EC 3.2.1.81)

    Pseudomonas atlantica - Calbiochem, Sigma

24. L-Alanine aminotransferase (L-alanine: 2-oxoglutarate amino-
    transferase, EC 2.6.1.2) - also called

    Glutamate - pyruvate transaminase

    Porcine heart - Biozyme, Boehringer, Calbiochem, Koch-Light,
       Miles, Paesel, P-L, Schwartz-Mann, Serva, Sigma, U.S. Bio-
       chemical

25. L-Alanine dehydrogenase (L-alanine: NAD oxidoreductase,
    deaminating, EC 1.4.1.1)

26. Alcalase (hydrolytic enzyme acting on most proteins)

    Bacillus - Novo

27. Alcohol dehydrogenase (alcohol: NAD oxidoreductase, EC 1.1.1.1)

    Equine liver - Aldrich, Boehringer, Calbiochem, Koch-Light,
       NBC, Schwartz-Mann, Sigma, U.S. Biochemical

    Human liver - New England Enzyme*

    Leuconostoc mesenteroides - Boehringer

    Yeast - Aldrich, Boehringer, BDH, Calbiochem, Koch-Light,
       Merck, Miles, NBC, Paesel, P-L, Schwartz-Mann, Serva, Sigma,
       U.S. Biochemical, Worthington

28. Alcohol oxidase (alcohol: oxygen oxidoreductase, EC 1.1.3.13)

    Candida boidinii - Boehringer, Sigma

29. Aldehyde dehydrogenase (aldehyde: NAD(P) oxidoreductase,
    EC 1.2.1.5)

    Baker's yeast - Boehringer, Sigma

30. Aldolase (D-fructose-1,6-diphosphate-D-glyceraldehyde-3-phosphatelyase, EC 4.1.2.13)

    Rabbit muscle - Aldrich, Boehringer, Calbiochem, Koch-Light, NBC, P-L, Schwartz-Mann, Serva, Sigma, U.S. Biochemical, Worthington

    Trout muscle - Sigma

    Yeast - U.S. Biochemical

31. Amine oxidase (monoamine oxidase, amine: oxygen oxidoreductase, deaminating, EC 1.4.3.4)

    Human plasma - Miles; New England Enzyme[*]

    Bovine plasma - Sigma

32. Amino acid arylamidase (α-aminoacyl peptide hydrolase, EC 3.4.11.2)

    Hog kidney - Boehringer

33. D-Amino acid oxidase (D-amino: oxygen oxidoreductase, deaminating, EC 1.4.3.3)

    Porcine kidney - Aldrich, BDH, Biozyme, Boehringer, General Biochemical, Gibco, Koch-Light, NBC, New England Enzyme,[*] P-L, Schwartz-Mann, Serva, Sigma, U.S. Biochemical, Worthington

34. L-Amino acid oxidase (L-amino acid: oxygen oxidoreductase, deaminating, EC 1.4.3.2)

    Ancistrodom rhodostoma - Serva

    Bothrops atrox - Sigma

    Crotalus adamanteus - Aldrich, BDH, Calbiochem, Koch-Light, P-L, Schwartz-Mann, Sigma, U.S. Biochemical, Worthington

    Crotalus atrox - Serva, Sigma

    Crotalus durissus terrificus - Boehringer

35. Aminoacyl t-RNA synthetase [L-amino acid; t-RNA ligase (AMP), EC 6.1.1.n]

    Baker's yeast - Sigma

    Bovine liver - Gibco, Sigma

    E. coli - Gibco, Miles, P-L, Sigma

    Rabbit liver - Sigma

36. δ-Amino levulinate dehydratase (porpholvilinogen synthetase,
    5-aminolevulinate hydrolase, EC 4.2.1.24)
    Bovine liver - Sigma

37. Aminopeptidase, cytosol [leucine aminopeptidase, α-aminoacyl
    peptide hydrolase (cytosol), EC 3.4.11.1]
    Bovine - Serva
    Porcine kidney - Aldrich, BDH, Biozyme, Boehringer, General
       Biochem, Gibco, K&K, Koch-Light, NBC, P-L, Schwartz-Mann,
       Sigma, U.S. Biochemical, Worthington
    Porcine pancreas - Merck

38. Aminopeptidase, microsomal [α-aminoacyl peptide hydrolase
    (microsomal), EC 3.4.11.2] - also called leucine aminopeptidase
    or arylamidase
    Human placenta - Sigma
    Porcine kidney microsomes - NBC, P-L, Paesel, Sigma
    Tritirachicum album limber - Merck, Serva

39. α-Amylase (1,4-α-D-glucan glucanohydrolase, EC 3.2.1.1)
    A. niger - Calbiochem, Rapides
    A. oryzae (taka-diastase) - Calbiochem, Koch-Light, Miles,
       Novo (Fungamyl), Rapides, Sankyo, Serva, Sigma
    Bacillus licheniformis - Novo (Tetramyl), Searle, Sigma*
    Bacillus subtilus - Boehringer, Calbiochem, Gist-Brocades,
       Koch-Light, K&K, Merck, Miles, NBC, Novo, Rapides, Schwartz-
       Mann, Searle, Serva, Sigma, U.S. Biochemical
    Barley meal - Sigma
    Porcine pancreas - Aldrich, BDH, Boehringer, Calbiochem,
       General Biochemical, Gibco, Koch-Light, Merck, Miles, NBC,
       P-L, Schwartz-Mann, Sigma, U.S. Biochemical, Worthington

40. β-Amylase (1,4-α-D-glucan maltohydrolase, EC 3.2.1.2)
    Barley - BDH, Calbiochem, Gibco, Koch-Light, K&K, Merck, NBC,
       Rapides, Schwartz-Mann, Serva, Sigma, U.S. Biochemical
    Sweet potato - BDH, Boehringer, Calbiochem, Serva, Sigma,
       U.S. Biochemical

41. γ-Amylase (exo-1,4-glucosidase; 1,4-α-D-glucan glucohydrolase,
    EC 3.2.1.3)
    Mucor javanicus - Calbiochem

42. Amyloglucosidase (glucoamylase; 1,4-α-D-glucan glucohydrolase, EC 3.2.1.3)

    A. niger - BDH, Boehringer, Calbiochem, Gist-Brocades, K&K, Koch-Light, Merck, Novo (Spiritamylase 150), Rapides, Sigma

    A. oryzae - Sigma

    Endomyces - Sigma

    Rhizopus mold - Sigma

    Rhizopus nircus - Miles

43. Apotryptophanase

    E. coli - Sigma

44. Apyrase (adenosine-5'-triphosphatase + adenosine-5'-diphosphatase, EC 3.6.1.3 and 3.6.1.5)

    Potato - Sigma

45. Arginase (L-arginine amidinohydrolase, EC 3.5.3.1)

    Bovine liver - Biozyme, Boehringer, K&K, Koch-Light, NBC, P-L, Schwartz-Mann, Serva, Sigma, U.S. Biochemical

46. L-Arginine decarboxylase (L-arginine carboxy lyase, EC 4.1.1.19)

    E. coli - General Biochem, Gibco, K&K, Koch-Light, NBC, Schwartz-Mann, Sigma, U.S. Biochemical

47. Arginine kinase (ATP: L-arginine-N$^{\omega}$-phosphotransferase, EC 2.7.3.3)

    Lobster tail muscle - Sigma

48. Argininosuccinate lyase (L-argininosuccinate arginine lyase, EC 4.3.2.1)

    Bovine liver - Sigma

    Porcine kidney - Sigma

49. Aromatase (arylesterase, EC 3.1.1.2)

    Human placenta - Schwartz-Mann

50. Arylamidase (see aminopeptidase)

51. Arylamine transferase (acetyl-CoA; arylamine N-acetyl transferase, EC 2.3.1.5)

    Pigeon liver - New England Enzyme[*]

52. Aryl sulfatase (aryl sulfate sulfohydrolase, EC 3.1.6.1)

    Aerobacter aerogenes - Sigma

    Abalone entrails - Sigma

Patella vulgata - Sigma

Helix pomatia - Boehringer, K&K, Koch-Light, Sigma

53. Ascorbate oxidase (L-ascorbate: oxygen oxidoreductase, EC 1.10.3.3)

    Cucurbitap - Boehringer, Sigma

54. Asparaginase (L-asparagine amidohydrolase, EC 3.5.1.1)

    E. coli - Aldrich, Boehringer, Calbiochem, Gibco, NBC, P-L, Schwartz-Mann, Sempa-Chimie, Sigma, U.S. Biochemical, Worthington

    Erwinia carotovora - BDH

55. Aspartase (L-aspartate ammonia lyase, EC 4.3.1.1)

    B. cadaveris - Sigma

56. L-Aspartate aminotransferase (L-aspartate: 2-oxoglutarate aminotransferase, EC 2.6.1.1) - also called glutamic-oxalo-acetic transferase

    Chicken heart - New England Enzyme[*]

    Porcine heart - Aldrich, Biozyme, Boehringer, Calbiochem, Koch-Light, Miles, NBC, P-L, Paesel, Schwartz-Mann, Serva, Sigma, U.S. Biochemical, Worthington

57. L-Aspartate carbamoyltransferase (carbamoylphosphate: L-Aspartate carbamoyltransferase, EC 2.1.3.2)

    E. coli - New England Enzyme[*]

    S. faecalis - Sigma

58. Aspartate decarboxylase (L-aspartate 4-carboxylyase, EC 4.1.1.12)

    A. chromobacter sp - U.S. Biochemical

    A. faecalis - New England Enzyme,[*] Sigma[*]

    C. welchii NGTC 6784 - NBC

59. Bromelain (EC 3.4.22.4)

    Pineapple (*Ananas sativus*) - BDH, Boehringer, Calbiochem, Gist-Brocades, K&K, Koch-Light, Merck, NBC, Schwartz-Mann, Serva, Sigma, U.S. Biochemical

60. Calculase

    Adv. Biofractures; NBC, Paesel, Sigma, U.S. Biochemical

61. Carbamate kinase (carbamyl phosphokinase, ATP: carbamate phos-
    photransferase, EC 2.7.2.2)

    S. faecalis - P-L, Sigma

62. Carbonic anhydrase (carbonate hydrolase, EC 4.2.1.1)

    Bovine blood - Worthington

    Human erythrocytes - Sigma, U.S. Biochemical, Worthington

    Bovine erythrocytes - BDH, Biozyme, Boehringer, Calbiochem,
       Koch-Light, NBC, Schwartz-Mann, Serva, Sigma, U.S. Bio-
       chemical, Worthington

    Dog erythrocytes - Sigma

    Rabbit erythrocytes - Sigma

63. Carboxy esterase (carboxylic ester hydrolase, esterase,
    EC 3.1.1.1)

    Beef liver - New England Enzyme[*]

    Porcine liver - Boehringer, Sigma

64. Carboxypeptidase A (peptidyl-L-amino acid hydrolase, EC 3.4.12.2)

    Bovine pancreas - Aldrich, Boehringer Calbiochem, General
       Biochem, Gibco, Merck, NBC, P-L, Schwartz-Mann, Sigma,
       U.S. Biochemical, Worthington

    Bovine pancreas - PMSF treated - P-L, Sigma

    Bovine pancreas - DFP treated - Sigma

65. Carboxypeptidase B (peptidyl-L-lysine (L-arginine) hydrolase,
    EC 3.4.12.3)

    Bovine pancreas - Aldrich, Boehringer, Merck, Schwartz-Mann

    Porcine pancreas - NBC, P-L, Schwartz-Mann, Sigma, U.S. Bio-
       chemical, Worthington

    Porcine pancreas - DFP treated - Sigma

66. Carboxypeptidase C (peptidyl-L-amino acid (L-proline) hydrolase,
    EC 3.4.12.1)

    Bovine pancreas - NBC

67. Carboxypeptidase G (EC 3.4.12.10)

    P. stutzeri - New England Enzyme[*]

68. Carboxypeptidase P (peptidyl-L-tyrosine hydrolase, EC 3.4.12.12)

    Penicillum janthinellum - Sigma

69. Carboxypeptidase Y (peptidyl-L-alanine hydrolase, EC 3.4.12.11)
    Baker's yeast - Boehringer, Calbiochem, Sigma, U.S. Bio-
    chemical, Worthington

70. Carnitine acetyltransferase (acetyl coA: carnitine-0-acetyl-
    transferase, EC 2.3.1.7)
    Pigeon breast muscle - Boehringer, Sigma

71. Catalase (peroxide: peroxide oxidoreductase, EC 1.11.1.6)
    A. niger - Calbiochem, Sigma
    Bovine liver - Aldrich, BDH, Boehringer, Calbiochem, Canada
        Packers, General Biochem, Gibco, Koch-Light, K&K, Merck,
        Miles, NBC, P-L, Schwartz-Mann, Searle, Serva, Sigma,
        U.S. Biochemical, Worthington
    Dog liver - Sigma
    Fungal - U.S. Biochemical
    Micrococcus lysodeikticus - Merck, NBC

72. Catechol-0-methyltransferase (5-adenosyl-L-methionine: catechol-
    0-methyltransferase, EC 2.1.1.6)
    Beef liver - Miles, Paesel
    Porcine liver - Sigma

73. Cathepsin B (EC 3.4.22.1)
    Sigma*

74. Cathepsin C (dipeptidyl peptidase; dipeptidylpeptide hydrolase,
    EC 3.4.14.1)
    Bovine spleen - Boehringer, New England Enzyme,* Sigma

75. Cathepsin D (EC 3.4.23.5)
    Bovine spleen - Sigma

76. Cellulase (1,4-(1,3;1,4)-β-D-glucan-3(4)-glucanohydrolase
    EC 3.2.1.4)
    A. niger - Calbiochem, Koch-Light, Rapides, Schwartz-Mann,
        Sigma, U.S. Biochemical
    Basidiomycetes - Merck
    T. viride - Adv. Biofractures, Aldrich, BDH, Boehringer,
        Gibco, K&K, Merck, NBC, Paesel, Schwartz-Mann, Sigma,
        U.S. Biochemical, Worthington
    T. fusca - Sigma*

77. Cephalosporinase (Cephalosporin amido-β-lactam hydrolase, EC 3.5.2.8)

   Enterobacter cloacae - Miles

78. Ceramide trihexosidase

   Human placenta - New England Enzyme[*]

79. Ceruloplasmin

   Human - Miles, Sigma

80. Chitinase (chitodertrinase; poly(1,4-β-(2-acetamido-2-deoxy-D-glucoside) glycanohydrolase, EC 3.2.1.14)

   Mold - Adv. Biofractures, Calbiochem, Koch-Light, NBC, Paesel, Schwartz-Mann, U.S. Biochemical

   S. griseus - Sigma

81. Chloroperoxidase (chloride peroxidase; chloride:hydrogen peroxide oxidoreductase, EC 1.11.1.10)

   Caldariomycetes fumago - Sigma

82. Chloramphenicol acetyltransferase (EC 2.3.1.99)

   E. coli - P-L

83. Cholesterol esterase (sterol ester acylhydrolase, EC 3.1.1.13)

   Bovine pancreas - Abbott, Miles, P-L

   Microbial - Boehringer, Miles, U.S. Biochemical

   Porcine pancreas - Sigma, Worthington

   Pseudomonas fluorescens - Sigma

84. Cholesterol oxidase (cholesterol: oxygen oxidoreductase, EC 1.1.3.6)

   Nocardia erythropolis - Boehringer, Sigma

   Pseudomonas - Sigma

   Streptomyces sp - Sigma, U.S. Biochemical

85. Choline acetyltransferase (acetyl CoA: choline O-acetyltransferase, EC 2.3.1.6)

   Bovine brain - Sigma

   Human placenta - Sigma

86. Choline kinase (ATP: choline phosphotransferase; choline phosphokinase, EC 2.7.1.32)

   Yeast - Boehringer, Sigma, Worthington

87. Choline oxidase (choline: oxygen 1-oxidoreductase, EC 1.1.3.17)

   Alcaligenes sp - Sigma

88. Cholinesterase, acetyl (true cholinesterase; acetylcholine
    hydrolase, EC 3.1.1.7)
    Bovine erythrocytes - Koch-Light, K&K, NBC, Schwartz-Mann,
    Serva, Sigma, U.S. Biochemical
    Electric eel - Aldrich, Boehringer, Koch-Light, P-L, Schwartz-
    Mann, Serva, Sigma, Worthington

89. Cholinesterase, butyryl (pseudocholinesterase; acylcholine acyl
    hydrolase, EC 3.1.1.8)
    Horse serum - Aldrich, BDH, Calbiochem, Koch-Light, Merck,
    NBC, Paesel, Schwartz-Mann, Sigma, U.S. Biochemical,
    Worthington
    Human serum - Boehringer, Sigma

90. Choloylglycine hydrolase ($3\alpha,7\alpha,12\alpha$-trihydroxy-5-β-cholan-2,4-
    oylglycine amidohydrolase, EC 3.5.1.24)
    Clostridium perfringens (*welchii*) - Sigma

91. Chondroitinase ABC (chondoitin ABC lyase, EC 4.2.2.4)
    Flavobacterium heparinum - Miles
    Proteus vulgaris - Miles, Sigma

92. Chondroitinase AC (chondroitin AC lyase, EC 4.2.2.5)
    Arthrobacter aurescens - Sigma, Miles
    Flavobacterium heparinium - Miles

93. Chondro-4-sulfatase (chondroitin-4-sulfate sulfohydrolase,
    EC 3.1.6.9)
    Proteus vulgaris - Miles, Sigma

94. Chondro-6-sulfatase (chondroitin-6-sulfate sulfohydrolase,
    EC 3.1.6.10)
    Proteus vulgaris - Miles, Sigma

95. Chromopapain (EC 3.4.4.11)
    Papaya latex - NBC, Schwartz-Mann, Sigma, U.S. Biochemical,
    Worthington

96. α-Chymotrypsin (peptide peptidohydrolase, EC 3.4.21.1)
    Bovine pancreas - Aldrich; BDH, Boehringer, Calbiochem,
    Canada Packers, General Biochemicals, Gibco, K&K, Koch-
    Light, Merck, Miles, NBC, P-L, Paesel, Schwartz-Mann,
    Serva, Sigma, U.S. Biochemical, Worthington

DFP treated - Sigma

TLCK treated - Sigma

97. β-Chymotrypsin (peptide peptidohydrolase, EC 3.4.21.1)

Bovine pancreas - Aldrich, General Biochem, Gibco, Koch-Light, K&K, NBC, Schwartz-Mann, Sigma, U.S. Biochemical

98. δ-Chymotrypsin (EC 3.4.21.1)

Bovine pancreas - Aldrich, BDH, K&K, Koch-Light, NBC, Sigma, Schwartz-Mann

99. γ-Chymotrypsin (EC 3.4.21.1)

Bovine pancreas - Aldrich, K&K, Koch-Light, NBC, Schwartz-Mann, Sigma, U.S. Biochemical

100. α-Chymotrypsinogen A

Bovine pancreas - Aldrich, Boehringer, BDH, Calbiochem, K&K, Koch-Light, Merck, Miles, NBC, Schwartz-Mann, Serva, Sigma, U.S. Biochemical, Worthington

101. Citrate lyase (citrate oxaloacetate-lyase, EC 4.1.3.6)

Aerobacter aerogenes - Boehringer

Enterobacter aerogenes - P-L, Sigma

102. Citrate synthase (citrate oxaloacetate lyase (CoA acetylating), EC 4.1.3.7)

Pigeon breast - Sigma

Porcine heart - Boehringer, P-L, Sigma

103. Clostripain (chlostridiopeptidase B, EC 3.4.22.8)

Chlostridium histolyticum - Sigma, Worthington

104. Collagenase (clostridiopeptidase A, EC 3.4.24.3)

C. histolyticum - Adv. Biofractures, Aldrich, BDH, Boehringer, Calbiochem, General Biochem, Gibco, K&K, Koch-Light, Merck, NBC, Paesel, P-L, Schwartz-Mann, Serva, Sigma, U.S. Biochemical, Worthington

105. Collagenase/dispase (EC 3.4.24.3/EC 3.4.24.4)

Achromobacter ionphagus/Bacillus polymyxa - Boehringer

106. Creatinase (creatine amidinohydrolase, EC 3.5.3.3)

Sigma*

107. Creatine kinase (creatine phosphokinase; ATP: creatine N-phosphotransferase, EC 2.7.3.2)

Bovine heart - Biozyme, Serva, Sigma

Chicken muscle - New England Enzyme[*]

Porcine heart - U.S. Biochemical, Calbiochem, Paesel

Rabbit brain - Sigma

Rabbit muscle - Aldrich, BDH, Biozyme, Boehringer, Calbio-
chem, General Biochem, Gibco, Koch-Light, K&K, Miles, NBC,
P-L, Paesel, Serva, Schwartz-Mann, Sigma, U.S. Biochemical,
Worthington

108. Creatininase (creatinine deiminase, EC 3.5.4.21)

Pseudomonas - BDH, Carla-Erba[*]

109. Creatininase (creatinine amidohydrolase, EC 3.5.2.10)

Alcaligenes sp. - Boehringer, Sigma, U.S. Biochemical

Pseudomonas - Sigma

110. Cytochrome C peroxidase (microperoxidase; ferrocytochrome C:
hydrogen peroxide oxidoreductase, EC 1.11.1.5)

Equine heart - Calbiochem, Miles, New England Enzyme,[*] Sigma

111. Cytochrome C reductase (NADH dehydrogenase, NADH: (acceptor)
oxidoreductase, EC 1.6.99.3)

Porcine heart - Koch-Light, NBC, Sigma

112. Cytochrome oxidase (ferrocytochrome C: oxygen oxidoreductase,
EC 1.9.3.1)

Bovine heart - Biozyme, Paesel, Serva, Sigma

113. Deoxyribonuclease I (DNase I, deoxyribonucleate-5'-oligo-
nucleotidohydrolase, EC 3.1.4.5)

Bovine pancreas - Aldrich, BDH, Boehringer, Calbiochem,
Canada Packers, General Biochem, Gibco, K&K, Koch-Light,
Merck, Miles, NBC, P-L, Schwartz-Mann, Serva, Sigma, U.S.
Biochemical, Worthington

114. Deoxyribonuclease II (DNase II, deoxyribonucleate-3'-oligo-
nucleotidohydrolase, EC 3.1.4.6)

Bovine spleen - Biozyme, Calbiochem, Koch-Light, Miles,
Paesel, Schwartz-Mann, Sigma

Porcine spleen - Aldrich, Calbiochem, Gibco, NBC, P-L, Schwartz-
Mann, Serva, Sigma, U.S. Biochemical, Worthington

Rat - Merck

115. Deoxyribonuclease S

   Aspergillus oryzae - Calbiochem

116. Deoxyribonucleic acid ligase (ATP) (polynucleotide synthetase; DNA ligase, EC 6.5.1.1)

   E. coli - Miles, P-L, New England Enzyme,[*] Sigma,[*] Worthington

117. Deoxyribonucleic acid ligase (NAD) (polynucleotide synthetase (NAD) EC 6.5.1.2)

   E. coli - P-L

118. Deoxyribonucleic acid (DNA) polymerase (DNA nucleotidyltransferase; deoxynucleosidetriphosphate: DNA deoxynucleotidyltransferase, EC 2.7.7.7)

   α, Calf thymus - Gibco, Miles, Paesel, P-L, Sigma,[*] U.S. Biochemical, Worthington

   γ, Calf liver - P-L, U.S. Biochemical

   Fetal calf liver - Worthington

   E. coli - Boehringer, Gibco, Miles, P-L, New England Enzyme,[*] Worthington

   T-4 infected *E. coli* - P-L, Paesel, Worthington

   Micrococcus lysodeikticus - Miles, P-L, Sigma, U.S. Biochemical

119. Dextranase (α-1,6-D-glucan-6-glucanohydrolase, EC 3.2.1.1)

   Bacillus sp - Beckman, Calbiochem, Dextran Products

   Penicillum sp - Canada Packers, Koch-Light, NBC, Serva, Sigma, U.S. Biochemical, Worthington

120. Diamine oxidase [amine: oxygen oxidoreductase (deaminating) (pyridoxyl containing), EC 1.4.3.6]

121. Diaphorase (lipoamide dehydrogenase; NADH: lipoamide oxidoreductase, EC 1.6.4.3)

   Bovine heart - Biozyme, Miles, Paesel

   C. kluyveri - Aldrich, BBC, Calbiochem, General Biochem, Gibco, NBC, P-L, Schwartz-Mann, Sigma, Whatman, Worthington, U.S. Biochemical

   Microbial - Boehringer

   Porcine heart - BDH, Biozyme, Boehringer, Calbiochem, K&K, Miles, Schwartz-Mann, Sempa-Chimie, Serva, Sigma

   Tortila yeast - Sigma

122. Diastase (taka-diastase; α- and β-amylase)

    Aspergillus oryzae - Adv. Biofractures, Koch-Light, Rohm &
Haas, Schwartz-Mann, Serva, Sigma, U.S. Biochemical

    Malt - BDH, K&K, NBC, Rapides, Schwartz-Mann, Sigma

123. Dihydrofolate reductase (tetrahydrofolate dehydrogenase;
5,6,7,8-tetrafolate: NADP oxidoreductase, EC 1.5.1.3)

    Bovine liver - Sigma

    Chicken liver - Sigma

    E. coli - New England Enzyme[*]

    Pigeon liver - Sigma

    Rat liver - Sigma

    S. faecalis - New England Enzyme[*]

124. Dihydro-orotate dehydrogenase (L-4,5-dihydro-orotate: oxygen
oxidoreductase, EC 1.3.3.1)

    Zymobacterium croticum - Sigma

125. Dihydropteridine reductase (reduced NADP: 6,7-dihydropteridine
oxidoreductase, EC 1.6.99.7)

    Sheep liver - Sigma

126. Dihydropyrimidinase (5,6-dihydropyrimidine amido hydrolase,
EC 3.5.2.2)

    Sigma[*]

127. DPNase (see NAD nucleosidase)

128. DNA nucleotidylexotransferase (terminal transferase) (nucleoside-
triphosphate: DNA deoxynucleotidyhexotransferase, EC 2.7.7.31)

    Calf thymus - Miles

129. Elastase (pancreatopeptidase, EC 3.4.21.11)

    Fermentation - Adv. Biofractures, Calbiochem, General Biochem,
Gibco, Schwartz-Mann

    Porcine pancreas - Aldrich, BDH, Biozyme, Boehringer, Calbio-
chem, General Biochem, Gibco, Koch-Light, Merck, Miles,
NBC, P-L, Paesel, Schwartz-Mann, Serva, Sigma, U.S. Biochemical, Worthington

130. Emulsin (see β-Glucosidase)

131. Endoglycosidase

    P. pneumariae - Miles, Sigma[*]

    S. griseus - Miles, Sigma[*]

132. Endonuclease (single-stranded nucleate 5'-oligonucleotido-
    hydrolase, EC 3.1.4.21)

    Neurospora crassa - Boehringer, Miles

133. Endo restriction nucleases (see Restriction endonucleases)

134. Endopeptidase

    Bacterial - Sigma

135. Enolase (phosphopyruvate hydratase, 2-phospho-D-glycerate
    hydrolase, EC 4.2.1.11)

    Rabbit muscle - Boehringer, Calbiochem, K&K, Koch-Light,
        NBC, Schwartz-Mann, Sigma, U.S. Biochemical
    Yeast - Calbiochem, Sigma

136. Enteropeptidase (enterokinase, EC 3.4.21.9)

    Porcine intestine - Miles, Sigma
    Porcine duodenum, NBC, U.S. Biochemical

137. Erepsin

    Porcine duodenum - K&K, U.S. Biochemical

138. Esperase (detergent hydrolase)

    Novo

139. Estradiol dehydrogenase - Sigma

140. Exonuclease III (deoxyribonucleate, double-stranded 5'-
    nucleotidohydrolase, EC 3.1.4.27)

    E. coli - Boehringer, Miles, New England Enzyme,[*] P-L,
        Sigma,[*] Worthington

141. Fatty acid thioesterase

    E. coli - New England Enzyme[*]

142. Fibrinolysin (plasmin, fibrinase, EC 3.4.21.7)

    Bacteria - Adv. Biofractures, Calbiochem, General Biochem,
        Gibco, Paesel, Schwartz-Mann, Sigma
    Bovine blood - Koch-Light, Sigma

143. Ficin (EC 3.4.22.3)

    Fig tree latex - Boehringer, Calbiochem, General Biochemical,
Gibco, Gist-Brocardes, K&K, Koch-Light, Merck, NBC,
Schwartz-Mann, Sigma, U.S. Biochemical

144. Figlu enzymes (formimino L-glutamic acid transferase and
formimino tetrahydrofolate cyclodeaminase, EC 2.1.2.5 and
4.3.1.4)

    Pig liver - BDH, Sigma

145. Formaldehyde dehydrogenase (formaldehyde: NAD oxidoreductase,
EC 1.2.1.1)

    Pseudomonas putida - Sigma

    Yeast - U.S. Biochemical

146. Formate Dehydrogenase (Formate: NAD oxidoreductase, EC 1.2.1.2)

    Pseudomonas oxalaticus - Sigma

    Yeast - U.S. Biochemical

147. Formyltetrahydrofolate synthetase (formate: tetrahydrofolate
ligase, ADP forming, EC 6.3.4.3)

    C. cylindrosporum - Koch-Light

148. Fructose-1,6-diphosphatase (D-fructose-1,6-diphosphate-1-
phosphohydrolase, EC 3.1.3.11)

    Bovine liver - New England Enzyme[*]

    Rabbit muscle - Boehringer, Sigma

    Rabbit liver - Aldrich, P-L, Sigma

    Tortula yeast - Sigma

149. Fructose-6-phosphate kinase (phosphofructokinase; ATP: D-
fructose-6-phosphate-1-phosphotransferase, EC 2.7.1.11)

    Rabbit muscle - Boehringer, P-L, Sigma

150. Fructose-6-phosphate kinase, pyrophosphate-dependent
(6-phosphofructokinase, pyrophosphate); pyrophosphate: D-
fructose-6-phosphate-1-phosphotransferase, EC 2.7.1.90)

    Propionibacterium freudenreichii - Sigma

151. β-Fructofuranosidase (β-fructosidase; invertase; saccharase;
sucrase; β-D-fructofuranoside fructohydrolase, EC 3.2.1.26)

    Candida utilis - Sigma

Yeast - BDH, Boehringer, Calbiochem, Gist-Brocandes, K&K,
Koch-Light, Merck, Miles, NBC, Rapides, Schwartz-Mann,
Searle, Serva, Sigma, U.S. Biochemical

152. L-Fucose dehydrogenase (6-deoxy-L-galactose: NAD-1-oxidoreductase, EC 1.1.1.122)

Pig liver - Miles

153. α-L-Fucosidase (α-L-fucoside fucohydrolase, EC 2.3.1.51)

Beef kidney - Boehringer, Sigma

Bovine epididymis - Sigma

C. lampus - Miles

T. cornutus - Miles

154. Fumurase (L-malate hydrolase; fumarate hydratase, EC 4.2.1.2)

Porcine heart - Boehringer, Calbiochem, Koch-Light, NBC,
P-L, Schwartz-Mann, Sigma

Chicken heart - Sigma

Rabbit liver - Sigma

Yeast - Sigma,* U.S. Biochemical

155. Gabase (γ-amino butyric acid transaminase/succinic semialdehyde dehydrogenase, EC 2.6.1.19/1.2.1.16)

Pseudomonas fluorescens - Aldrich, Boehringer, NBC, Sigma,
U.S. Biochemical, Worthington

156. Galactokinase (ATP: D-galactose 1-phosphotransferase,
EC 2.7.1.6)

Yeast - Sigma

157. β-Galactose dehydrogenase (D-galactose: NAD 1-oxidoreductase,
EC 1.1.1.48)

Pseudomonas fluorescens - Boehringer, P-L, Sigma

158. Galactose oxidase (D-galactose: oxygen 6-oxidoreductase,
EC 1.1.3.9)

A. niger - Rapides

Dactylium dendroides - Aldrich, General Biochem, Gibco, K&K,
NBC, Schwartz-Mann, Sigma, Worthington

Polyporous circinatus - P-L, U.S. Biochemical

159. Galactose 1-phosphate uridyltransferase (UDP glucose: α-D-galactose 1-phosphate uridyltransferase, EC 2.7.7.12)

Calf liver - Boehringer, Sigma[*]

Yeast - Sigma

160. α-Galactosidase (melibiase; α-D-galactoside galactohydrolase,
     EC 3.2.1.22)

     A. niger - Sigma

     Green coffee beans - Boehringer, P-L, Sigma

     T. cornutus - Miles

161. β-Galactosidase (lactase; β-D-galactoside galactohydrolase,
     EC 3.2.1.23)

     Aspergillus niger - Calbiochem, Sigma

     Bovine liver - Sigma

     C. lampus - Miles

     E. coli - Aldrich, Boehringer, K&K, NBC, New England Enzyme,[*]
        P-L, Sigma, Worthington

     Jack beans - Sigma

     Saccharomyces fragilis - NBC, Sigma

     Yeast - BDH, Gist-Brocades

162. Galactosyltransferase (lactose synthase; UDP-galactose: D-
     glucose 4-β-galactosyltransferase, EC 2.4.1.22)

     Bovine milk - Sigma

163. Gelatinase

     Fungal - Schwartz-Mann, U.S. Biochemical

164. Gentamicin acetyltransferase (Acetyl CoA: gentamicin C N^3-
     acetyltransferase, EC 2.3.1.60)

     E. coli - P-L, Sigma, U.S. Biochemical

165. β-Glucanase (β-1,3-glucan hydrolase, EC 3.2.1.39)

     BDH, Novo, Searle

166. Gluconate kinase (laminarinase, ATP: D-gluconate 6-phospho-
     transferase, EC 2.7.1.12)

     E. coli - Boehringer, Sigma

167. Glucosaminidase

     Bovine spleen - Calbiochem

168. Glucose dehydrogenase (β-D-glucose: NAD oxidoreductase,
     EC 1.1.1.47)

Bacillus cereus - Merck, Sigma[*]

Calf liver - Sigma

169. Glucose oxidase (β-D-glucose: oxygen-1-oxidoreductase, EC 1.1.3.4)

A. niger - Aldrich, BDH, Beckman, Biozyme, Boehringer, Calbiochem, General Biochem, Gibco, K&K, Koch-Light, Merck, Miles, NBC, P-L, Paesel, Schwartz-Mann, Serva, Sigma, U.S. Biochemical, Worthington

Penicillum notatum - Koch-Light

170. Glucose oxidase/catalase (EC 1.1.3.4/1.11.1.6)

K&K, Searle (Ovazyme, Fermcoenzyme)

171. Glucose-6-phosphatase (D-glucose-6-phosphate phosphohydrolase, EC 3.1.3.9)

Rabbit liver - Sigma

172. Glucose-6-phosphate dehydrogenase (D-glucose-6-phosphate, NADP-1-oxidoreductase, EC 1.1.1.49)

Baker's yeast - Beckman, BDH, Boehringer, Calbiochem, General Biochemical, K&K, Koch-Light, Merck, NBC, Paesel, Schwartz-Mann, Serva, Sigma, U.S. Biochemical, Whatman

Bovine adrenals - Sigma

Brewer's yeast - P-L

Human erythrocytes - New England Enzyme[*]

L. mesenteroides - Aldrich, BBC, Beckman, Calbiochem, Boehringer, P-L, Paesel, Sigma, U.S. Biochemical, Whatman, Worthington

Tortula yeast - Calbiochem, NBC, P-L, Sigma

173. Glucose phosphate isomerase (see phosphoglucose isomerase)

174. α-Glucosidase (maltase: α-D-glucoside-glucohydrolase, EC 3.2.1.20)

A. niger - Rapides, Sigma

Yeast - Boehringer, Calbiochem, P-L, Serva, Sigma

Baker's yeast - Sigma

Brewer's yeast - Sigma

Rice - Sigma[*]

175. β-Glucosidase (emulsin, β-D-glucoside glucohydrolase,
      EC 3.2.1.21)

      Almonds - Aldrich, BDH, Boehringer, Calbiochem, General
      Biochemical, Gibco, K&K, Koch-Light, Miles, NBC, P-L,
      Paesel, Schwartz-Mann, Serva, Sigma, U.S. Biochemical,
      Worthington

176. β-Glucuronidase (β-D-glucuronide glucuronohydrolase,
      EC 3.2.1.31)

      Abalone entrails - Sigma

      Bovine liver - Aldrich, BDH, Calbiochem, Gibco, NBC, P-L,
      Serva, Schwartz-Mann, Sigma, U.S. Biochemical, Worthington

      E. coli - Boehringer, Sigma

      Helix aspersa - Sigma

      Helix pomatia - Boehringer, Sigma

      Patella vulgata - BDH, Calbiochem, Gibco, Koch-Light, Miles,
      P-L, Schwartz-Mann, Sigma, U.S. Biochemical

      Scallop - Sigma[*]

177. β-Glucuronidase/aryl sulfatase (EC 3.2.1.31/3.1.6.1)

      Helix pomatia - Boehringer, Calbiochem, Gibco, Merck, Serva,
      Sigma (Glusulase)

178. L-Glutamate decarboxylase (L-glutamate-1-carboxylase,
      EC 4.1.1.15)

      C. welchii - Sigma

      E. coli - Aldrich, BDH, K&K, Koch-Light, NBC, Serva, Sigma,
      U.S. Biochemical, Worthington

179. L-Glutamate dehydrogenase (L-glutamate: NAD oxidoreductase,
      deaminating, EC 1.4.1.3)

      Bovine liver - BDH, Beckman, Biozyme, Boehringer, Calbiochem,
      K&K, Merck, Miles, NBC, P-L, Paesel, Schwartz-Mann, Serva,
      Sigma, U.S. Biochemical, Whatman

      Chicken liver - New England Enzyme[*]

      Porcine liver - Biozyme

      Rat liver - Sigma

      Yeast - Sigma,[*] U.S. Biochemical

180. Glutamate oxaloacetate transaminase (GOT, see Aspartate
     aminotransferase)

181. Glutamate pyruvic transaminase (GPT, see Alanine glutamate
     transaminase)

182. Glutaminase (L-glutamine amidohydrolase, EC 3.5.1.2)

     E. coli - Aldrich, K&K, NBC, New England Enzyme,* P-L,
     Schwartz-Mann, Sigma, U.S. Biochemical, Worthington
     Porcine kidney - Sigma

183. L-Glutamine decarboxylase

     E. coli - Sigma

184. L-Glutamine synthetase (L-glutamate ammonia ligase (ADP-forming),
     EC 6.3.1.2)

     Sheep brain - NBC, New England Enzyme,* P-L, Schwartz-Mann,
     Sigma

185. γ-Glutamyltransferase [γ-glutamyl peptidase; (γ-glutamyl)
     peptide: amino acid-glutamyltransferase, EC 2.3.2.2]

     Beef liver - Miles, Sigma
     Porcine kidney - Biozyme, Calbiochem, Paesel, Sigma

186. Glutathione reductase (NAD(P)H: oxidized-glutathione oxido-
     reductase, EC 1.6.4.2)

     Bovine liver - Sigma*
     Spinach - Sigma
     Yeast - Boehringer, Calbiochem, Koch-Light, NBC, P-L,
        Schwartz-Mann, Sigma, U.S. Biochemical
     Wheat germ - Sigma

187. Glutathione S-transferase (glutathione S-alkyltransferase;
     glutathione S-aryltransferase; glutathione S-aralkyltrans-
     ferase; glutathione S-alkene transferase; glutathione S-epoxide
     transferase; RX: glutathione R-transferase, EC 2.5.1.18)

     Bovine liver - Sigma
     Equine liver - Sigma
     Porcine liver - Sigma
     Rabbit liver - Sigma
     Rat liver - Sigma

188. Glyceraldehyde-3-phosphate dehydrogenase (GAPDH; D-glycer-
     aldehyde-3-phosphate: NAD oxidoreductase, phosphorylating,
     EC 1.2.1.12)

     Human red blood cells - Sigma[*]

     Rabbit muscle - Aldrich, Boehringer, Calbiochem, K&K, Koch-
          Light, NBC, P-L, Schwartz-Mann, Serva, Sigma, U.S. Bio-
          chemical, Worthington

     Yeast - Aldrich, Boehringer, NBC, Sigma, U.S. Biochemical

189. Glyceraldehyde-3-phosphate dehydrogenase/3-phosphoglycerate
     phosphokinase (EC 1.2.1.12/2.7.2.3)

     Yeast - Boehringer, Sigma

190. Glycerate dehydrogenase (D-glycerate: NAD oxidoreductase,
     EC 1.1.1.29)

     Spinach leaves - Sigma

191. Glycerol dehydrogenase (glycerol: NAD oxidoreductase,
     EC 1.1.1.6)

     Aerobacter aerogenes - New England Enzyme,[*] P-L, U.S. Bio-
          chemical

     Enterobacter aerogenes - Aldrich, Boehringer, Schwartz-Mann,
          Sigma, Worthington

192. Glycerol kinase (ATP: glycerol 3-phosphotransferase,
     EC 2.7.1.30)

     Candida mycoderma - BDH, Biozyme, Boehringer, P-L, Sigma

     E. coli - Beckman, Biozyme, Calbiochem, Paesel, Sigma,
          U.S. Biochemical, Worthington

     Yeast - U.S. Biochemical

193. Glycerol-1-phosphate dehydrogenase (L-glycerol-1-phosphate:
     FAD oxidoreductase, EC 1.1.99.5)

     Rabbit muscle - Biozyme

194. Glycerol-3-phosphate dehydrogenase (L-glycerol-3-phosphate:
     NAD 2-oxidoreductase, EC 1.1.1.8)

     Chicken liver - New England Enzyme

     Rabbit muscle - Beckman, Boehringer, Calbiochem, General
          Biochemical, Gibco, Koch-Light, P-L, Paesel, Schwartz-
          Mann, Serva, Sigma, Whatman

195. Glycerol-3-phosphate dehydrogenase/triosephosphate isomerase
(glycerol-3-phosphate: NAD-2-oxidoreductase/D-glyceraldehyde-
3-phosphate ketol isomerase, EC 1.1.1.8/5.3.1.1)
Rabbit muscle - Boehringer, Calbiochem, Sigma

196. Glycosidases (mixture)
Charonia lampus - Miles
Turbo cornutus - Miles

197. Glycogen synthase (UDP glucose: glycogen 4α-glucosyltransferase,
EC 2.4.1.11)
Rabbit muscle - Sigma

198. Glycollate oxidase (glycollate: oxygen oxidoreductase,
EC 1.1.3.1)
Spinach - Sigma
Sugarbeet - Sigma

199. Glyoxalase I (S-lactoyl-glutathione methylglyoxal-lyase,
isomerizing, EC 4.4.1.5)
Yeast - Boehringer, Sigma

200. Glyoxyase II (S-2-Hydroxyacylglutathione hydrolase, EC 3.1.2.6)
Beef liver - Sigma

201. Glyoxylate reductase (glycolate dehydrogenase, glycollate: NAD
oxidoreductase, EC 1.1.1.26)
Spinach leaves - Boehringer, Calbiochem, Sigma

202. Guanase (guanine deaminase; guanine aminohydrolase, EC 3.5.4.3)
Rabbit liver - Boehringer, P-L, Schwartz-Mann, Sigma

203. Guanylate kinase (guanosine-5'-monophosphate kinase; ATP: GMP
phosphotransferase, EC 2.7.4.8)
Bovine brain - Sigma
Hog brain - Boehringer, Sigma
Human erythrocytes - New England Enzyme[*]

204. Hemicellulase
A. niger - Sigma
Rhizopus mold - General Biochem, Koch-Light, Schwartz-Mann,
U.S. Biochemical

205. Hesperidinase
A. niger - Miles, NBC, Rapides (Rapides C, Clarzyme LS)

206. Hexokinase (ATP: D-hexose-6-phosphotransferase, EC 2.7.1.1)
      Yeast - BBC, BDH, Beckman, Biozyme, Boehringer, Calbiochem,
      General Biochem, Gibco, Koch-Light, Merck, Miles, NBC,
      Paesel, P-L, Schwartz-Mann, Serva, Sigma, U.S. Biochemical,
      Whatman, Worthington
      Baker's yeast - Sigma

207. Hexokinase/glucose-6-phosphate dehydrogenase (EC 2.7.1.1/
      1.1.1.49)
      Yeast - Boehringer, Calbiochem, Sigma

208. Histidase (L-histidine ammonia lyase, EC 4.3.1.3)
      Pseudomonas fluorescens - Aldrich, K&K, NBC, P-L, Schwartz-
      Mann, Sigma

209. Histidine decarboxylase (L-histidine carboxy-lyase, EC 4.1.1.22)
      Cl. welchii - Koch-Light, NBC, Sigma
      E. coli - Sigma
      Lactobacillus sp - Sigma

210. Hyaluronidase (hyaluronate-4-glucanohydrolase, EC 3.2.1.35)
      Bovine testes - Aldrich, BDH, Biozyme, Calbiochem, General
         Biochem, K&K, Koch-Light, Merck, Miles, NBC, P-L, Paesel,
         Schwartz-Mann, Serva, Sigma, U.S. Biochemical, Worthington
      Cl. perfringens - Sigma[*]
      Fungal - Calbiochem
      Ovine testes - BDH, Boehringer, Calbiochem, Gibco, Koch-
         Light, Paesel, Sigma

211. 3-Hydroxy CoA-dehydrogenase (L-3-hydroxy acyl-CoA NAD oxido-
      reductase; β-hydroxy acyl-CoA dehydrogenase, EC 1.1.1.35)
      Bovine liver - Sigma
      Porcine heart - Boehringer, Koch-Light, P-L, Sigma

212. α-Hydroxylbutyrate dehydrogenase (EC 1.1.1.61)
      Porcine heart - Calbiochem

213. 3-Hydroxybutyrate dehydrogenase (β-hydroxybutyrate dehydro-
      genase, D-3-hydroxybutyrate: NAD oxidoreductase, EC 1.1.1.30)
      Pseudomonas lemoignei - Sigma
      Rhodoseudomonas spheroides - Boehringer, Calbiochem, P-L,
         Sigma

214. Hydroxydecanoyl thioester dehydrase

    E. coli - New England Enzyme*

215. Hydroxy madelonitrile lyase (hydroxynitrile lyase; 4-hydroxy-mandelonitrile hydroxy benzaldehyde lyase, EC 4.1.2.11)

    Sorghum seedlings - Sigma

216. 15-Hydroxyprostaglandin dehydrogenase (11α,15-dihydroxy-9-oxoprost-13-enoate-NAD-15-oxidoreductase, EC 1.1.1.141)

    Beef lung - Calbiochem, Miles

217. 3-α-Hydroxysteroid dehydrogenase (3-α-hydroxysteroid: NAD(P) oxicoreductase, EC 1.1.1.50)

    Pseudomonas testosteroni - NBC, U.S. Biochemical, Worthington

218. α,β-Hydroxysteroid dehydrogenase (mixture of α and β enzymes, EC 1.1.1.50,51)

    Pseudomonas testosteroni - Aldrich, Koch-Light, NBC, Schwartz-Mann, Sigma, Worthington

219. β-Hydroxysteroid dehydrogenase (β-hydroxysteroid: NAD(P) oxidoreductase, EC 1.1.1.51)

    Pseudomonas testosteroni - Aldrich, P-L, Sigma

220. 7-α-Hydroxysteroid dehydrogenase (EC 1.1.1. -)

    E. coli - Sigma

221. (3α),20β-Hydroxysteroid dehydrogenase (cortisone reductase; 17,20β,21-trihydroxysteroid: NAD oxidoreductase, EC 1.1.1.53)

    Streptomyces hydrogenans - Boehringer, Calbiochem, Sigma

222. 3β-Hydroxysteroid oxidase

    Sigma*

223. Imidodipeptidase (see Prolidase)

224. Inorganic pyrophosphatase (pyrophosphate phosphohydrolase, EC 3.6.1.1)

    Yeast - Aldrich, Boehringer, Gibco, NBC, P-L, Sigma, U.S. Biochemical

225. Inositol dehydrogenase (myoinositol:NAD-2-oxidoreductase, EC 1.1.1.18)

    Enterobacter aerogenes - Sigma

226. Invertase (see β-Fructofurosidase)

227. Isocitrate dehydrogenase (threo-D_s-isocitrate NADP oxido-
      reductase, decarboxylating, EC 1.1.1.42)
          Porcine heart - BDH, Boehringer, Calbiochem, General Bio-
              chemical, Gibco, K&K, Koch-Light, NBC, P-L, Schwartz-Mann,
              Sigma, U.S. Biochemical
          Yeast - Sigma,[*] U.S. Biochemical
228. Isoleucine RNA synthetase (EC 6.1.1.5)
          E. coli - New England Enzyme[*]
229. Kallikrein (callicrein; padutin; kininogenin; kininogenase,
      EC 3.4.21.8)
          Human plasma - Sigma[*]
          Porcine pancreas - NBC, Sigma
230. Kanamycin-6-acetyltransferase (EC 2.3.1.55)
          E. coli - P-L, U.S. Biochemical
231. α-Ketoglutarate dehydrogenase (EC 1.2.4.2)
          Sigma[*]
232. Lactase (see β-Galactosidase)
233. D-Lactate dehydrogenase (D-lactate: NAD oxidoreductase,
      EC 1.1.1.28)
          Lactobacillus leichmannii - Boehringer, Sigma
234. L-Lactate dehydrogenase (L-lactate: NAD oxidoreductase,
      EC 1.1.1.27)
          Bovine adrenal - Sigma
          Bovine heart - Aldrich, Biozyme, Boehringer, Calbiochem,
              General Biochemical, Gibco, Koch-Light, Miles, NBC, Schwartz-
              Mann, Serva, Sigma, U.S. Biochemical, Worthington
          Bovine muscle - Boehringer, Sigma
          Chicken breast - New England Enzyme[*]
          Chicken heart - New England Enzyme,[*] P-L, Schwartz-Mann, Sigma
          Chicken liver - Sigma
          Dogfish muscle - New England Enzyme[*]
          Dog muscle - Sigma
          E. coli - New England Enzyme[*]
          Horse heart - New England Enzyme[*]
          Horse muscle - New England Enzyme[*]

Human red blood cells - Sigma

Lobster tail - Sigma

Porcine heart - BBC, Biozyme, Boehringer, Calbiochem, Merck,
P-L, Serva, Sigma, U.S. Biochemical, Worthington

Porcine muscle - Boehringer, Merck, Miles, Paesel, Serva,
Sigma

Rabbit heart - Sigma

Rabbit muscle - Aldrich, Biozyme, Boehringer, BDH, Calbiochem,
General Biochem, K&K, Koch-Light, Merck, Miles, NBC, Paesel,
P-L, Schwartz-Mann, Serva, Sigma, U.S. Biochemical, Worth-
ington

Yeast [cytochrome b_2, EC 1.1.2.3] - Aldrich, NBC, Schwartz-
Mann, Sigma, U.S. Biochemical, Worthington

Isoenzymes

Band 1 (H_4) bovine heart - Aldrich, NBC, Sigma

    Human red blood cells - Sigma

    Porcine heart - BDH, Boehringer, Calbiochem, Serva, Sigma

    Rabbit heart - Boehringer, Sigma

Band 2 (H_3M_1) - bovine heart - Sigma

    Human red blood cells - Sigma

Band 3 (H_2M_2) human red blood cells - Sigma

    Bovine muscle - Sigma

Band 5 (M_4) porcine muscle - Boehringer, Sigma

    Human placenta - Sigma

    Rabbit muscle - Aldrich, Boehringer, Calbiochem, Serva, Sigma

235. Lactoperoxidase (see Peroxidase)

236. Laminarinase [1,3-(1,3;1,4)-β-D-Glucan-3-(4)-glucanohydrolase,
EC 3.2.1.6]

    Mollusk - Sigma

237. Leucine aminopeptidase (see Aminopeptidase)

238. Lipase (triacylglycerol acyl hydrolase, EC 3.1.1.3)

    A. niger - Miles

    A. oryzae - Rapides

    Calf gland - Schwartz-Mann

Candida cylindraccae - BDH, General Biochemical, Schwartz-
   Mann, Sempa-Chimie, Sigma, U.S. Biochemical, Worthington
Chromobacterium viscosium - U.S. Biochemical
Geotrichum candidum - Sigma,* U.S. Biochemical
Mucor - U.S. Biochemical
Porcine pancreas - Worthington
Pseudomonas - Sigma
Rhizopus arrhizus - Boehringer, P-L
Rhizopus delamar - Miles, Serva, Sigma*
Wheat germ - Aldrich, BDH, Calbiochem, K&K, Koch-Light, NBC,
   P-L, Schwartz-Mann, Serva, Sigma, U.S. Biochemical
TG - Calbiochem

239. Lipoamide dehydrogenase (see Diaphorase)
240. Lipoprotein lipase
   Microorganisms - Miles
241. Lipoxidase (lipoxygenase; linoleate: oxygen oxidoreductase,
   EC 1.13.1.13)
   Soybean - Aldrich, Biozyme, K&K, Koch-Light, Miles, NBC,
      Paesel, P-L, Schwartz-Mann, Serva, Sigma, U.S. Biochemical
242. Luciferase (EC 1.2.-.-)
   Photobacterium fischeri - Aldrich, Boehringer, K&K, NBC,
      Schwartz-Mann, Sigma, Worthington
   Photonius pyralis - Sigma
243. Lysine decarboxylase (L-lysine carboxylyase, EC 4.1.1.18)
   B. cadaveris - Aldrich, BDH, Calbiochem, General Biochem,
      Gibco, Koch-Light, K&K, NBC, Schwartz-Mann, Sigma, U.S.
      Biochemical
   E. coli - Sigma
244. Lysozyme (muramidase, mucopeptide N-acetylmuranoyl-hydrolase,
   EC 3.2.1.17)
   Chalaropsis sp - Miles, Sigma
   Egg white - Aldrich, BDH, Biozyme, Boehringer, Calbiochem,
      General Biochem, Gibco, K&K, Koch-Light, Merck, Miles, NBC,
      Paesel, P-L, Schwartz-Mann, Serva, Sigma, U.S. Biochemical,
      Worthington
   Human urine - Worthington

245. Macerase

   Rhizopus sp - Calbiochem

246. Malate dehydrogenase (L-malate: NAD oxidoreductase, EC 1.1.1.37)

   B. subtilis - New England Enzyme[*]

   Bovine heart - Sigma, Paesel

   Chicken heart - New England Enzyme,[*] P-L, Sigma

   Human erythrocytes - Sigma

   Human placenta - Sigma

   Pigeon breast muscle - Sigma

   Mitochondria - BDH, Biozyme, Worthington

   Porcine heart - Aldrich, BDH, Biozyme, Boehringer, Calbiochem,
      General Biochem, Gibco, K&K, Koch-Light, Merck, Miles, NBC,
      Paesel, P-L, Schwartz-Mann, Serva, Sigma, U.S. Biochemical,
      Whatman, Worthington

   Yeast - U.S. Biochemical

247. Maltase (see α-Glucosidase)

248. Mandelonitrile lyase (oxynitrilase; mandelonitrile benzyaldehyde
   lyase, EC 4.1.2.10)

   Almonds - Boehringer, Sigma

249. α-Mannosidase (α-D-mannoside mannohydrolase, EC 3.2.1.24)

   Canavalia ensiformis - Boehringer, Sigma

   Turbo cornutus - Miles

250. 5,10-Methylene tetrahydrofolate dehydrogenase (5,10-methylene
   tetrahydrofolate: NADP oxidoreductase, EC 1.5.1.5)

   Yeast - Sigma

251. Microperoxidase (see Cytochrome C peroxidase)

252. Monoamine oxidase (see Amine oxidase)

253. Muramidase (see Lysozyme)

254. Mutarotase (aldolase-1-epimerase, EC 5.1.3.3)

   Hog kidney - Boehringer

255. Mycozyme

   Aspergillus oryzae - Koch-Light

256. Mylase - P

   Aspergillus oryzae - Koch-Light, NBC, Rapides, Schwartz-Mann,
      U.S. Biochemical

257. Myokinase (adenylate kinase; ATP: AMP phosphotransferase,
     EC 2.7.4.3)
     Porcine muscle - Boehringer, Calbiochem, P-L, Sigma
     Rabbit liver - Schwartz-Mann
     Rabbit muscle - BDH, Boehringer, Calbiochem, General Biochem,
     Gibco, K&K, Koch-Light, NBC, P-L, Schwartz-Mann, Sigma,
     U.S. Biochemical
258. Myosin
     Rabbit muscle - Sigma
259. NADase (see NAD nucleosidase)
260. NAD/diaphorase (DPN/diaphorase, EC 1.6.4.3)
     Porcine heart - BDH, Boehringer, Sigma, U.S. Biochemical,
     Worthington
261. NADH peroxidase (NADH: hydrogen peroxide oxidoreductase,
     EC 1.11.1.1)
     S. faecalis - Boehringer
262. NAD kinase (ATP: NAD 2'-phosphotransferase, EC 2.7.1.23)
     Chicken liver - Sigma
     Pigeon liver - Boehringer
263. NAD nucleotidase (NADase; DPN nucleotidase; NAD glycohydrolase,
     EC 3.2.2.5)
     B. subtilis - New England Enzyme[*]
     Bovine spleen - Sigma
     Horse brain and spleen - New England Enzyme[*]
     N. crassa - NBC; Schwartz-Mann, Sigma, U.S. Biochemical,
     Worthington
     Porcine brain - New England Enzyme,[*] Sigma
264. NAD pyrophosphorylase (ATP: NMN adenylyltransferase, EC 2.7.7.1)
     Hog liver - Boehringer
265. Nagarinase (see Subtilisin)
266. Naringinase
     Aspergillus niger - Sigma
267. Neuraminidase (sialidase, acyl neuraminyl hydrolase, EC 3.2.1.18)
     Arthrobacter ureafaciens - Boehringer, Calbiochem
     Cl. perfringens - Aldrich, Boehringer, Gibco, Merck, NBC, P-L,
     Schwartz-Mann, Sigma, U.S. Biochemical, Worthington

Influenza virus - Calbiochem, General Biochem, Schwartz-Mann

Vibrio cholerae - BDH, Calbiochem, General Biochem, Gibco,
Koch-Light, Merck, Schwartz-Mann

268. Nitrate reductase (ferrocytochrome: nitrate oxidoreductase,
EC 1.9.6.1)

E. coli - Sigma, U.S. Biochemical, Worthington

269. Nuclease (nucleate 3'-oliganucleotidylhydrolase, EC 3.1.4.7)

Staphylococcus aureus - Aldrich, Boehringer, Gibco, Merck,
NBC, P-L, Schwartz-Mann, Sigma, U.S. Biochemical, Worthington

N. crassa - P-L

Mung bean - P-L

P_1 - *Penicillum citrinum* - Boehringer, Calbiochem, P-L, Sigma

S_1 - *Aspergillus oryzae* - Calbiochem, Miles, P-L, Sigma,
Worthington

270. Nucleoside-5'-diphosphate kinase (ATP: nucleoside-diphosphate
phosphotransferase, EC 2.7.4.6)

Baker's yeast - Sigma

Bovine liver - Boehringer, Sigma

Human erythrocytes - New England Enzyme*

271. Nucleoside monophosphate kinase (ATP: nucleoside monophosphate
phosphotransferase, EC 2.7.4.4)

Bovine liver - Boehringer, Sigma

272. Nucleoside phosphorylase (purine nucleoside: orthophosphate
ribosyltransferase, EC 2.4.2.1)

Bovine spleen - Boehringer, NBC, Sigma

273. 3'-Nucleotidase (3'-ribonucleotide phosphohydrolase, EC 3.1.3.6)

Rye grass - Sigma

274. 5'-Nucleotidase (5'-ribonucleotide phosphohydrolase, EC 3.1.3.5)

Crotalus adamanteus venom - Sigma

Crotalus atrox venom - Sigma

275. Nucleotide pyrophosphatase (dinucleotide nucleotidylhydrolase,
EC 3.6.1.9)

Crotalus adamanteus venom - Sigma

Crotalus atrox venom - Sigma

276. Octopine dehydrogenase (N^2-1(1-carboxymethyl)-L-arginine: NAD
oxidoreductase (L-arginine forming, EC 1.5.1.11)

    Scallop - Sigma
277. Ornithine carbamyltransferase (ornithine transcarbamylase;
     carbamoylphosphate: L-ornithine carbamoyltransferase,
     EC 2.1.3.3)
     Streptococcus faecalis - New England Enzyme, P-L, Sigma
278. L-Ornithine decarboxylase (L-ornithine carboxy lyase,
     EC 4.1.1.17)
     E. coli - Sigma
279. Orotidine-5'-monophosphate decarboxylase (orotidine 5'-
     phosphate carboxyl lyase, EC 4.1.1.23)
        Baker's yeast - Sigma
        Brewer's yeast - Sigma
280. Orotidine-5'-monophosphate pyrophosphorylase (orotidine-5'-
     phosphate: pyrophosphate phosphoribosyltransferase,
     EC 2.4.2.10)
        Yeast - Boehringer, P-L, Sigma
281. Orotidine-5-phosphate pyrophosphorylase/orotidine decarboxylase
     (mixed enzymes, EC 4.1.1.23/2.4.2.10)
        Yeast Sigma
282. Oxalate decarboxylase (oxalate carboxyl lyase, EC 4.1.1.2)
        Cod muscle - New England Enzyme[*]
        Collybia velutipes - Aldrich, NBC, Schwartz-Mann, Sigma,
           U.S. Biochemical, Worthington
283. Oxalate oxidase (EC 1.2.3.4)
        Sigma[*]
284. Oxynitrilase (see Mandelonitrile lyase)
285. Pancreatin (amylase, lipase, protease, RNase, trypsin, etc.)
        Porcine pancreas - BDH, General Biochemical, Gibco, Koch-
           Light, Merck, NBC, Schwartz-Mann, Serva, Sigma, U.S.
           Biochemical
286. Papain (EC 3.4.22.2)
        Papaya latex - Aldrich, BDH, Boehringer, Calbiochem, General
           Biochem, Gist-Brocardes, K&K, Koch-Light, Merck, NBC, P-L,
           Schwartz-Mann, Serva, Sigma, U.S. Biochemical, Worthington
287. Pectinase [poly-(α-1,4-D-galacturonide)glycanohydrolase,
     EC 3.2.1.15]

Aspergillus niger - General Biochem, Gibco, Koch-Light, K&K,
    NBC, Rapides, Rohm & Haas, Schwartz-Mann, Searle, Serva,
    Sigma, U.S. Biochemical

288. Pectinesterase (pectin methyl esterase; pectin pectyl hydrolase,
    EC 3.1.1.11)

    Orange peel - Sigma

    Tomato - Aldrich, NBC, Schwartz-Mann, Sigma

289. Penicillinase (β-lactamase, penicillin amido-β-lactam hydrolase,
    EC 3.5.3.6)

    Bacillus cereus (lichiformis) - BDH, Calbiochem, Gibco, Koch-
        Light, NBC, Schwartz-Mann, Serva, Sigma, Whatman

    Enterobacter cloacae - Sigma

290. Pentosanase

    Rohm & Haas, Searle

291. Pepsin (EC 3.4.23.1)

    Porcine stomach mucosa - Aldrich, BDH, Biozyme, Boehringer,
        Calbiochem, General Biochem, Gibco, K&K, Koch-Light, Merck,
        Miles, NBC, Paesel, P-L, Schwartz-Mann, Serva, Sigma,
        U.S. Biochemical, Worthington

292. Pepsinogen

    Porcine stomach mucosa - Aldrich, Boehringer, General Bio-
        chem, Gibco, Merck, K&K, NBC, Schwartz-Mann, Sigma, U.S.
        Biochemical, Worthington

293. Peptidase

    Porcine intestinal mucosa - K&K, NBC, Schwartz-Mann, Sigma

294. Peroxidase (donor: H_2O_2 oxidoreductase, EC 1.11.1.7)

    Cow's milk (lactoperoxidase) - Boehringer, Calbiochem, Miles,
        P-L, Sigma, Worthington

    Horseradish - Aldrich, BDH, Biozyme, Boehringer, Calbiochem,
        K&K, Koch-Light, Merck, Miles, NBC, Paesel, P-L, Schwartz-
        Mann, Sempa-Chemie, Serva, Sigma, U.S. Biochemical, Worth-
        ington

295. Phenylalanine ammonia lyase (L-phenylalanine ammonia lyase,
    EC 4.3.1.5)

Potato - Sigma*

Rhototorula glutinis - NBC, P-L, Schwartz-Mann, Sigma, U.S.
   Biochemical

296. Phenylalanine decarboxylase (L-phenylalanine carboxy lyase,
     EC 4.1.1.53)

     Streptococcus faecalis - Koch-Light, Sigma

297. Phenylalanine hydroxylase (phenylalanine-4-monooxygenase;
     L-phenylalanine tetrahydropteridine: oxygen oxidoreductase,
     4-hydroxylating, EC 1.14.16.1)

     Pseudomonas sp - Sigma*

     Rat liver - Sigma

298. Phenylethanolamine N-methyltransferase (5-adenosyl-L-methionine:
     phenylethanolamine N-methyltransferase, EC 2.1.1.28)

     Bovine adrenal medulla - Biozyme, Miles, Paesel, Sigma

     Rabbit adrenals - Paesel

299. Phosphatase, acid (orthophosphoric monoester phosphohydrolase
     (acid optimum), EC 3.1.3.2)

     Potatoes - Boehringer, Calbiochem, Gibco, Miles, P-L, Schwartz-
        Mann, Sigma

     Human prostrate - Calbiochem

     Wheat germ - Aldrich, BDH, K&K, Koch-Light, P-L, NBC, Schwartz-
        Mann, Serva, Sigma, U.S. Biochemical, Worthington

300. Phosphatase, alkaline (orthophosphoric monoester phosphohydrolase
     (alkaline optimum), EC 3.1.3.1)

     Bovine placenta - Sigma

     Bovine (calf) intestine - BDH, Biozyme, Boehringer, Calbiochem,
        K&K, Koch-Light, Miles, P-L, Paesel, Serva, Sigma, U.S. Bio-
        chemical

     Bovine intestine mucosa - Sigma

     Bovine liver - Sigma

     Bovine milk - Sigma

     Bovine kidney - Biozyme, Miles, Worthington

     Chicken intestine - Aldrich, General Biochemical, Gibco, NBC,
        Schwartz-Mann, Sigma, U.S. Biochemical, Worthington

     Dog intestine - Sigma

E. *coli* - Aldrich, Boehringer, Calbiochem, Gibco, Miles, NBC,
New England Enzyme,[*] P-L, Schwartz-Mann, Sigma, U.S. Biochemical, Worthington

Eel intestine - Sigma

Horse intestine - Sigma

Human placenta - Calbiochem, Miles, New England Enzyme,[*]
Sigma, Worthington

Pigeon intestine - Sigma

Porcine placenta - Sigma

Porcine intestine - Sigma

Porcine intestine - Sigma

Porcine kidney - Miles

Porcine placenta - Sigma

Trout intestine - Sigma

301. Phosphodiesterase I (5'-exonuclease; oligonucleate 5'-nucleo-
tidohydrolase, EC 3.1.4.1)

Bothrops atrox - Sigma

Crotalus atrox - Sigma

Crotalus adamanteus venom - Aldrich, BDH, Boehringer, Calbiochem, General Biochem, Gibco, K&K, Koch-Light, Merck,
NBC, P-L, Schwartz-Mann, Sigma, U.S. Biochemical, Worthington

Crotalus durissus terrificus - Sigma

302. Phosphodiesterase (3':5'-cyclic AMP-5'-nucleotidohydrolase,
EC 3.1.4.17)

Beef heart - Boehringer, New England Enzyme,[*] Sigma

303. Phosphodiesterase II (spleen phosphodiesterase; 3'-exonuclease,
oligonucleate 3'-nucleotidohydrolase, EC 3.1.4.18)

Bovine spleen - Aldrich, Koch-Light, Merck, NBC, P-L, Schwartz-
Mann, Sigma, U.S. Biochemical, Worthington

Calf spleen - Boehringer, Gibco

Rat spleen - Merck

304. 3'-Phosphodiesterase 2':3'-cyclic nucleotide (nucleoside 2':3'-
cyclic phosphate-2'-nucleotide hydrolase, EC 3.1.4.37)

Bovine brain - Sigma

305. Phospho(enol)pyruvate carboxylase (orthophosphate: oxalacetate
      carboxylase, phosphorylating, EC 4.1.1.31)

      E. coli - Worthington

      Maize - Sigma, Vickers Limited

      Sorghum - Vickers Limited

      Wheat - Boehringer

306. 6-Phosphofructokinase (see Fructose-6-Phosphate kinase)

307. Phosphoglucomutase (α-D-glucose, 1,6 biphosphate: α-D-glucose-
      1-phosphate phosphotransferase, EC 2.7.5.1)

      Rabbit muscle - Boehringer, Calbiochem, Koch-Light, NBC,
      P-L, Sigma

308. 6-Phosphogluconate dehydrogenase (6-phospho-D-gluconate: NADP
      2-oxidoreductase, decarboxylating, EC 1.1.1.44)

      Yeast - Boehringer, Calbiochem, NBC, P-L, Sigma

309. Phosphoglucose isomerase (D-glucose-6-phosphate-ketol isomerase;
      glucose phosphate isomerase, EC 5.3.1.9)

      Rabbit muscle - NBC, Sigma

      Yeast - Boehringer, Calbiochem, Koch-Light, P-L, Schwartz-
      Mann, Sigma

310. D-3-Phosphoglycerate dehydrogenase (D-3-phosphoglycerate: NAD
      oxidoreductase, EC 1.1.1.95)

      Chicken liver - Sigma

311. 3-Phosphoglycerate kinase (ATP: 3-phospho-D-glycerate-1-
      phosphotransferase, EC 2.7.2.3)

      Rabbit muscle - Sigma

      Yeast - Boehringer, Calbiochem, Koch-Light, NBC, P-L,
      Schwartz-Mann, Sigma

312. 3-Phosphoglycerate kinase/glyceraldehyde-3-phosphate dehydro-
      genase (ATP: 3-phospho-D-glycerate-1-phosphotransferase/D-
      glyceraldehyde-3-phosphate: NAD oxidoreductase (phosphorylating),
      EC 2.7.2.3/1.2.1.12)

      Yeast/rabbit muscle - Boehringer

313. Phosphoglycerate mutase (2,3-diphospho-D-glycerate: 2-phospho-
      D-glycerate, phosphotransferase, EC 2.7.5.3)

      Rabbit muscle - Boehringer, Calbiochem, Koch-Light, NBC,
      P-L, Sigma

314. Phospholipase A (lecithinase A; phosphatide-2-acyl hydrolase, EC 3.1.1.4)

    Bee venom - Boehringer, Calbiochem, Sigma

    Crotalus adamanteus venom - Calbiochem, Koch-Light, Miles, P-L, Sigma, U.S. Biochemical, Worthington

    Crotalus atrox - P-L

    Crotalus durissus terrificus - Boehringer

    Naja Naja venom - Koch-Light, Calbiochem, Sigma

    Porcine pancreas - Boehringer, Sigma

    Vipera russelli - Koch-Light, Sigma

315. Phospholipase C (lecithinase C; phosphatidylcholine choline-phosphohydrolase, EC 3.1.4.3

    B. cereus - Boehringer, Calbiochem, New England Enzyme,[*] Sigma

    Cl. perfringens - Aldrich, Calbiochem, General Biochem, Gibco, K&K, Koch-Light, Miles, NBC, P-L, Schwartz-Mann, Sigma, U.S. Biochemical, Worthington

    Cl. welchii - Calbiochem, Sigma

316. Phospholipase D (phosphatidylcholine phosphatidohydrolase, EC 3.1.4.4)

    Cabbage - BDH, Boehringer, Calbiochem, General Biochem, Gibco, Koch-Light, Miles, NBC, P-L, Schwartz-Mann, Sigma

    Peanuts - Calbiochem, P-L, Serva, Sigma, U.S. Biochemical

317. Phosphomannose isomerase (D-mannose-6-phosphate ketol isomerase, EC 5.3.1.8)

    Yeast - Boehringer, P-L, Sigma

318. Phosphoriboseisomerase (5-ribose-5-phosphate ketol isomerase, EC 5.3.1.6)

    Spinach - Calbiochem, Sigma

    Yeast - Sigma

319. Phosphoribosyl pyrophosphate amidotransferase (EC 2.4.2.14)

    Chicken liver - New England Enzyme[*]

320. Phosphoribulokinase (ATP D-ribulose-5-phosphate-1-phospho-transferase, EC 2.7.1.19)

    Spinach - Sigma

321. Phosphorylase a (α-1,4-glucan: orthophosphate glucosyltrans-
     ferase, EC 2.4.1.1)

     Rabbit muscle - Aldrich, Boehringer, General Biochem, Gibco,
         K&K, NBC, P-L, Schwartz-Mann, Sigma, U.S. Biochemical,
         Worthington

322. Phosphorylase b (α-1,4-glucan: orthophosphate glucosyltrans-
     ferase, EC 2.4.1.1)

     Rabbit muscle - Aldrich, Boehringer, NBC, Schwartz-Mann,
         Sigma, U.S. Biochemical

323. Phosphorylase kinase (ATP: phosphorylase phosphotransferase,
     EC 2.7.1.38)

     Rabbit muscle - General Biochem, Sigma

324. Phosphotransacetylase (acetyl-CoA: orthophosphate acetyltrans-
     ferase, EC 2.3.1.8)

     Clostridium kluyveri - Boehringer, NBC, P-L, Sigma

325. Phytase (phytate-6-phosphatase; myoinositol hexakisphosphate-
     6-phosphohydrolase, EC 3.1.3.26)

     Wheat - Sigma

326. Plasmin (see Fibrinolysin)

327. Plasminogen (profibrinolysin)

     Bovine plasma - Sigma
     Human plasma - Sigma
     Human serum - Worthington
     Porcine blood - Sigma

328. Polynucleotide kinase (ATP: 5'-hydroxypolynucleotide-5'-
     phosphotransferase, EC 2.7.1.78)

     E. coli - Boehringer, Gibco, Miles, New England Enzyme,[*]
         P-L, Paesel, Serva, U.S. Biochemical, Worthington

329. Polynucleotide phosphorylase (polyribonucleotide: orthophosphate
     nucleotidyltransferase, EC 2.7.7.8)

     E. coli - Gibco, P-L
     Micrococcus lysodeikticus (*luteus*) - Aldrich, Boehringer,
         Calbiochem, Koch-Light, Miles, NBC, P-L, Schwartz-Mann,
         Serva, Sigma, U.S. Biochemical, Worthington

330. Polynucleotide synthetase (see DNA ligase)

331. Polyol dehydrogenase (see Sorbitol dehydrogenase)

332. Polyphenol oxidase (tyrosinase, EC 1.10.3.1)

    Mushroom - Aldrich, BDH, Biozyme, K&K, Koch-Light, Miles, NBC, Paesel, P-L, Schwartz-Mann, Serva, Sigma, U.S. Biochemical, Worthington

333. Prolidase (imidodipeptidase; proline dipeptidase; amino acyl-L-proline hydrolase, EC 3.4.13.9)

    Porcine kidney - Biozyme, Miles, Paesel, Serva, Sigma, U.S. Biochemical

334. Prolinase (prolyl dipeptidase; iminodipeptidase; L-prolyl-amino acid hydrolase, EC 3.4.13.8)

    K&K, NBC, U.S. Biochemical

335. Prostaglandin dehydrogenase

    Sigma*

336. Prostaglandin-Δ^9-ketoreductase

    Sigma*

337. Prostaglandin-Δ^{13}-reductase

    Dog lung - Sigma*

338. Prostaglandin synthase (8,11,14-eicosatrienote hydrogen donor: oxygen oxidoreductase, EC 1.14.99.1)

    Bovine testicles - Miles

339. Protease

    A. oryzae - Rohm & Haas, Schwartz-Mann, Sigma*

    Bovine pancreas - Gibco, NBC, Schwartz-Mann, Sigma, U.S. Biochemical

    B. amyloliquenfaciens (*subtilis BPN*) - Sigma

    B. cereus (*microprotease*) - Sigma,* Worthington

    B. polymyxa - Boehringer, Sigma

    B. subtilis (*subtilisin Carlsberg*; subtilopeptidase A; alkaline protease, EC 3.4.21.14) - Gist Brocades, NBC, Rapides, Schwartz-Mann, Serva, Sigma

    B. thermoproteolyticus rokko (thermolysin, EC 3.4.24.4) - Calbiochem, Merck, Serva, Sigma

    Crotalus atrox - Merck, Sigma*

    Papaya - Rapides, Sigma

S. aureus - Miles, Worthington

S. griseus (pronase) - Boehringer, Calbiochem, Koch-Light,
Merck, Serva, Sigma

S. caesputosus - Calbiochem, Sigma

Submaxillaris - Boehringer

340. Proteinase

Bacillus subtilis - Koch-Light, Novo

C - Baker's yeast - New England Enzyme*

K - *Tritirachium album limber* - Boehringer, BDH, K&K, Merck,
NBC, Serva, Sigma, U.S. Biochemical

341. Proteinase A (endopeptidase, EC 3.4.22.9)

Baker's yeast - Sigma

342. Protein kinase (3',5'-cyclic AMP-dependent, EC 2.7.1.37)

Beef heart - Sigma

Rabbit muscle - Sigma

343. Proteolytic enzyme (detergents)

NBC, Novo

344. Pullulanase (pullan-6-glucanhydrolase, EC 3.2.1.41)

Aerobacter aerogenes - Boehringer, Calbiochem, Serva

Enterbacter aerogenes - Sigma

345. Purine nucleoside phosphorylase (purine nucleoside: orthophosphate ribosyltransferase, EC 2.4.2.1)

Calf spleen - P-L

Human erythrocytes - New England Enzyme*

346. Pyroglutamate aminopeptidase (L-pyroglutamyl peptide hydrolase,
EC 3.4.11.8)

Bovine liver - Sigma

347. Pyrophosphatase, nucleotide (dinucleotide nucleotidohydrolase,
EC 3.6.1.9)

Crotalus adamanteus - Sigma

Crotalus atrox - Sigma

348. Pyruvate decarboxylase (2-oxoacid carboxylase, EC 4.1.1.1)

Brewer's yeast - Sigma

349. Pyruvate kinase (ATP: pyruvate 2-O-phosphotransferase,
EC 2.7.1.40)

Human erythrocytes - New England Enzyme[*]

Dog muscle - Sigma

Porcine heart - U.S. Biochemical

Rabbit liver - Sigma

Rabbit muscle - Aldrich, Beckman, BDH, Biozyme, Calbiochem,
Boehringer, General Biochem, Gibco, K&K, Koch-Light, Merck,
Miles, NBC, Paesel, P-L, Schwartz-Mann, Serva, Sigma, U.S.
Biochemical, Whatman, Worthington

350. Pyruvate kinase/lactate dehydrogenase (EC 2.7.1.40/1.1.1.27)

Rabbit muscle - Boehringer, Miles, Sigma

351. Receptor-destroying enzyme

Vibrio cholerae - General Biochem

352. Renin (EC 3.4.99.19)

Porcine kidney - Miles, NBC, Novo (Rennidase), Sigma, U.S.
Biochemical

353. Rennin (chymosin, EC 3.4.23.4)

Calf stomach - Koch-Light, K&K, NBC, Schwartz-Mann, Sigma,
U.S. Biochemical

354. Restriction endonucleases (EC 3.1.23.X)

Acc 1 - *Acinetobacter calcoaceticus* - P-L

Alu 1 - (EC 3.1.23.1) - *Arthrobacter luteus* - Boehringer,
Calbiochem, Miles, P-L, Paesel, Sigma, U.S. Biochemical,
Worthington

Ava 1 - *Anabaena variabilis* - P-L

Ava 11 - *Anabaena variabilis* - P-L

Bal 1 - *Brevibacterium albidium* - P-L, Sigma

Bam H-1 (EC 3.1.23.6) - *Bacillus amyloliquefaciens H* -
Boehringer, Calbiochem, Miles, P-L, Paesel, Sigma, U.S.
Biochemical

Bcl 1 - *Bacillus caldolyticus* - P-L, Sigma

Bgl 1 - (EC 3.1.23.9) - *Bacillus globigii* - Boehringer,
Calbiochem, Miles, P-L, Paesel, Sigma

Bgl 11 - (EC 3.1.23.10) - *Bacillus globigii* - Boehringer,
Miles, Paesel, P-L, Sigma, Worthington

Bst E 11 - *Bacillus stearothermophillus* ET - Boehringer,
P-L, Sigma

Cau 1 - *Chloroflexus aurantiacus* - Sigma[*]

Cau 11 - *Chloroflexus aurantiacus* - Sigma[*]

Cla 1 - *Caryophanon latum* L - Boehringer, Sigma[*]

Eco R 1 (EC 3.1.23.13) -   Ry 13 - Boehringer, Calbiochem,
   Miles, P-L, Paesel, Sigma, U.S. Biochemical, Worthington;
   E. coli Ry 22 - Miles; *E. coli* 1100 (K-12) - Sigma

Eco R 11 - *E. coli* R 245 - Calbiochem, Paesel, U.S. Biochemical

Hae II - *Haemophilus aegypticus* - Calbiochem, Miles, P-L,
   Paesel, Sigma, U.S. Biochemical, Worthington

Hae III - *Haemophilus aegypticus* - Calbiochem, Paesel, P-L
   Sigma, U.S. Biochemical, Worthington

Hga I - *Haemophilus gallinarium* - P-L

Hha I - *Haemophilus haemolyticus* - Calbiochem, Miles, Paesel,
   P-L, Sigma, U.S. Biochemical

Hinc II - *Haemophilus influenza* R_c - Miles, Paesel, P-L,
   Sigma, U.S. Biochemical

Hinc II/Hind III - *Haemophilus influenza* R_d, R_c, Com^{-10} -
   P-L, U.S. Biochemical

Hind II (EC 3.1.23.20) - *Haemophilus influenza* Rd - Boehringer,
   Calbiochem, Sigma,[*] Paesel

Hind III (EC 3.1.23.21) - *Haemophilus influenza* Rd - Boehringer, Calbiochem, Miles, P-L, Paesel, Sigma, U.S. Biochemical, Worthington

Hinf I (EC 3.1.23.22) - *Haemophilus influenza* Rf - P-L, Sigma

Hpa I (EC 3.1.23.23) - *Haemophilus parainfluenza* - Boehringer,
   Calbiochem, Miles, Paesel, P-L, Sigma, U.S. Biochemical

Hpa II (EC 3.1.23.24) - *Haemophilus parainfluenza* - Boehringer,
   Calbiochem, Miles, P-L, Paesel, Sigma, U.S. Biochemical

Hph I (EC 3.1.23.25) - *Haemophilus parahaemolyticus* - P-L

Kpn I - (EC 3.1.23.26) - *Klebsiella pneumoniae* - Miles, P-L,
   Sigma

Mbo I - (EC 3.1.23.27) - *Moraxella bovis* - Paesel, P-L, Worthington

Mbo II (EC 3.1.23.28) - *Moraxella bovis* - Miles, Paesel, P-L,
   Worthington

Pst I (EC 3.1.23.31) - *Providencia stuartii* - Boehringer, Miles, P-L, Sigma, U.S. Biochemical

Pvu I - *Proteus vulgaris* - P-L, Paesel

Pvu II - *Proteus vulgaris* - P-L, Paesel

Sac I - *Streptomyces achromogenes* - P-L, Sigma

Sal I - (EC 3.1.23.37) - *Streptomyces albus* G - Boehringer, Miles, P-L, Sigma, U.S. Biochemical

Sau 3A - *Staphylococcus aureus* 3A - Miles, Sigma

Sin I, *Salmonella infantis* - Sigma

Sma I - *Serratia marcescens* S_b - Boehringer, P-L, Sigma,

Taq I (EC 3.1.23.39) - *Thermus aquaticus* YT1 - Boehringer, P-L, Sigma, Worthington

Tha I - *Thermoplasma acidophilium* - Sigma[*]

Xba I - *Xanthomonas badrii* - Miles, P-L, Sigma, Worthington

Xho I - *Xanthomonas holcicola* - P-L

Xma I - *Xanthomonas malvacearum* - Worthington

355. Reverse transcriptase (RNA-depending DNA polymerase)

Myeloblastosis virus - Boehringer

356. Rhodanese (thiosulfate: cyanide sulfurtransferase, EC 2.8.1.1)

Beef liver - Sigma

357. Ribonuclease A (ribonucleate-3'-oligonucleotidohydrolase, EC 3.1.4.22)

A. clavatus - Sigma[*]

Bovine pancreas - Aldrich, BDH, Biozyme, Boehringer, Calbiochem, Canada Packers, General Biochem, Gibco, K&K, Koch-Light, Merck, Miles, NBC, Paesel, P-L, Schwartz-Mann, Sigma, Serva, U.S. Biochemical, Worthington

Porcine pancreas - Merck

358. Ribonuclease B (ribonucleate-3'-oligonucleotidohydrolase, EC 3.1.4.22)

Bovine pancreas - Aldrich, Koch-Light, Miles, NBC, Schwartz-Mann, Sigma, U.S. Biochemical, Worthington

359. Ribonuclease C

Human plasma - Miles, Paesel, Sigma

360. Ribonuclease III (EC 3.1.4.X)

      E. coli - P-L

361. Ribonuclease D_1

      Porcine pancreas - Calbiochem, NBC

362. Ribonuclease H (hybrid ribonuclease; RNA: DNA hybrid ribonucleo-
tidohydrolase, EC 3.1.4.34)

      E. coli - P-L, Paesel

363. Ribonuclease oxidized (oxidized RNase A)

      Bovine pancreas - Biozyme, Miles, Schwartz-Mann, Sigma

364. Ribonuclease N (guanyloribonuclease; ribonucleate 3'-guanylo-
oligonucleotidohydrolase, EC 3.1.4.8)

      Neurospora crassa - Sigma

365. Ribonuclease Phy 1 (EC 3.1.4.X)

      Physarium polycephalium - P-L, Paesel

366. Ribonuclease S (protein-modified RNase A)

      Bovine pancreas - Schwartz-Mann, Sigma

367. Ribonuclease S protein

      Bovine pancreas - Sigma

368. Ribonuclease S peptide (protease modified RNAase)

      Bovine pancreas - Sigma

369. Ribonuclease T (guanyloribonuclease; ribonucleate 3'-guanylo-
oligonucleotidohydrolase, EC 3.1.4.8)

      A. oryzae - Boehringer, Calbiochem, Gibco, Koch-Light, Merck,
Miles, NBC, P-L, Sankyo, Schwartz-Mann, Serva, Sigma, U.S.
Biochemical, Worthington

370. Ribonuclease T_2 (ribonuclease II; ribonucleate 3'-oligonucleo-
tidohydrolase, EC 3.1.4.23)

      A. oryzae - Calbiochem, Koch-Light, Sankyo, Sigma

371. Ribonuclease U_2 (puryloribonuclease, EC 3.1.27.4)

      Ustilago syhaerogena - Boehringer, Calbiochem, Koch-Light,
Sankyo, P-L, Sigma

372. Ribonucleic acid ligase

      E. coli - P-L, U.S. Biochemical

      T_4-infected *E. coli* - Miles, Paesel

373. Ribonucleic acid polymerase (nucleosidase triphosphate: RNA nucleotidyltransferase; nucleosidetriphosphate: RNA nucleotidyltransferase, EC 2.7.7.6)

    DNA-dependent - Paesel

    E. coli - Boehringer, Calbiochem, Gibco, Miles, P-L, Schwartz-Mann, Sigma, U.S. Biochemical, Worthington

    M. luteus - Miles

    Wheat germ - P-L, Miles, U.S. Biochemical, Worthington

374. tRNA synthetase

    E. coli - New England Enzyme[*]

375. Ribonucleic acid transfer enzyme

    Baker's yeast - Sigma

    Bovine liver - Sigma

    E. coli - Sigma

    Rabbit liver - Sigma

    Wheat germ - Sigma

376. Ribonucleoside diphosphate reductase

    Sigma[*]

377. Ribonucleoside triphosphate reductase

    Bacterial - Sigma[*]

378. Ribonucleotide reductase

    L. leichmannii - New England Enzyme[*]

379. D-Ribulose-1,5-diphosphate carboxylase (3-phospho-D-glycerate carboxylase, dimerizing, EC 4.1.1.39)

    Spinach - Calbiochem, Sigma

380. D-Ribulose-5-phosphate-3-epimerase (EC 5.1.3.1)

    Yeast - Sigma

381. Saccharopine dehydrogenase, NAD, lysine-forming [N^6-(1,3-dicarboxyl propyl)-L-lysine: NAD oxidoreductase (L-lysine forming), EC 1.5.1.7]

    Brewer's yeast - Sigma

382. Secretin

    Porcine - Calbiochem, Sigma

383. Sorbitol dehydrogenase (L-iditol: NAD 5'-oxidoreductase; polyol dehydrogenase, EC 1.1.1.14)

Sheep liver - Boehringer, P-L, Sigma

Candida utilis - Sigma

384. Sphingomyelinase (spingomyelin phosphodiesterase; sphingomyelin choline phosphohydrolase, EC 3.1.4.12)

B. *cereus* - Boehringer

Human placenta - Sigma

385. Streptodornase (see Deoxyribonuclease I)

386. Streptokinase

Calbiochem, K&K, NBC, Sigma, U.S. Biochemical

387. Subtilisin (see Protease)

388. Succinic thiokinase (succinate: CoA ligase (GDP) EC 6.2.1.4)

Pig heart - Boehringer, New England Enzyme,[*] Sigma

389. Sucrase (see β-Fructofuranosidase)

390. Sulfatase (see Arylsulfatase)

391. Superoxide dismutase (superoxide: superoxide oxidoreductase, EC 1.15.1.1)

Bovine blood - Miles, P-L, Sigma

Bovine erythrocytes - Worthington

Dog blood - Sigma

Human blood - Sigma

392. Taka diastase (see α-Amylase from A. *oryzae*)

393. Tautomerase (phenylpyruvate ketoenol isomerase, EC 5.3.2.1)

Bovine kidney - Sigma

Lamb kidney - New England Enzyme[*]

Porcine kidney - Sigma

394. Terminal deoxynucleotidyltransferase (DNA nucleotidylexotransferase, nucleoside-triphosphate DNA deoxynucleotidyltransferase, EC 2.7.7.31)

Calf thymus gland - Gibco, P-L, Paesel, Serva, Sigma,[*] Worthington

395. Thermolysin (see Protease from B. *thermoproteolyticus rokko*)

396. Thiophorase

Pig heart - New England Enzyme[*]

397. Thrombin (EC 3.4.4.13)

Bovine plasma - K&K, Merck, Sigma

Human plasma - Serva

398. Thrombokinase

   Rabbit brain - Koch-Light

399. Thymidylate synthetase (EC 2.1.1.b)

   E. coli and *L. caseri* - New England Enzyme[*]

400. Thyroid peroxidase

   Pig thyroid - New England Enzyme[*]

401. Transaldolase (D-sedoheptulose-7-phosphate: D-glyceraldehyde-
   3-phosphate dihydroxyacetonetransferase, EC 2.2.1.2)

   C. utilis - New England Enzyme[*]

   Yeast - Boehringer, Sigma

402. Transformylase

   Hog liver - New England Enzyme[*]

403. Transfructosylase

   A. niger - Koch-Light

404. Transglucosylase

   A. niger - Koch-Light

405. Transhydrogenase (EC 1.6.1.1)

   Beef heart mitochondria - New England Enzyme[*]

406. Transketolase (D-sedulose-7-phosphate: D-glyceraldehyde-3-
   phosphate glycoaldehyde transferase, EC 2.2.1.1)

   Yeast - Sigma

407. Triosephosphate isomerase (D-glyceraldehyde-3-phosphate-ketol
   isomerase, EC 5.3.1.1)

   Chicken muscle - P-L

   Dog muscle - Sigma

   Rabbit muscle - Boehringer, Calbiochem, K&K, Koch-Light, NBC,
      P-L, Sigma

   Yeast - Boehringer, Sigma

408. Tripeptide synthetase

   Baker's yeast - New England Enzyme[*]

409. Trypsin (peptidyl peptide hydrolase, EC 3.4.21.4)

   Bovine pancreas - Aldrich, BDH, Boehringer, Calbiochem,
      Canada Packers, General Biochem, Gibco, K&K, Koch-Light,
      Merck, Miles, NBC, Novo, P-L, Paesel, Schwartz-Mann,
      Serva, Sigma, U.S. Biochemical, Worthington

Hog pancreas - Biozyme, Koch-Light, Merck, Miles, NBC, Sigma,
U.S. Biochemical

410. Trypsin inhibitor

α-1-Antitrypsin - Worthington

Bovine lung - Serva

Bovine pancreas - Aldrich, Koch-Light, Merck, NBC, P-L, Sigma,
Worthington

Lima beans - Aldrich, Koch-Light, NBC, P-L, Sigma, U.S. Biochemical, Worthington

Ovomucoid - Aldrich, Boehringer, General Biochem, Gibco, Koch-
Light, Merck, Miles, NBC, P-L, Serva, Sigma, U.S. Biochemical,
Worthington

Soybean - Aldrich, BDH, Biozyme, Boehringer, Calbiochem, General Biochem, Gibco, Koch-Light, K&K, Miles, NBC, P-L, Serva,
Sigma, U.S. Biochemical, Worthington

411. Trypsinogen

Bovine pancreas - Aldrich, BDH, Boehringer, Koch-Light, Merck,
Miles, NBC, P-L, Schwartz-Mann, Sigma, U.S. Biochemical,
Worthington

412. Tryptophanase (L-tryptophan indole lyase, deaminating,
EC 4.1.99.1)

E. coli - Sigma

Apo, *E. coli* - Sigma

413. Tyrosinase (see Polyphenol oxidase)

414. Tyrosine decarboxylase (L-tyrosine carboxy lyase, EC 4.1.1.25)

S. faecalis - Aldrich, General Biochem, Gibco, K&K, Koch-
Light, NBC, Schwartz-Mann, Sigma, U.S. Biochemical, Worthington

Apo, *S. faecalis* - Aldrich, Koch-Light, NBC, Sigma, U.S.
Biochemical, Worthington

415. Urease (urea amidohydrolase, EC 3.5.1.5)

Bacillis pasteurii - Aldrich, Worthington, U.S. Biochemical,
Sigma

Canavalia ensiformis (jack beans) - Aldrich, BDH, Beckman,
Boehringer, Calbiochem, General Biochem, Gibco, K&K,

Koch-Light, Merck, Miles, NBC, P-L, Paesel, Schwartz-Mann, Serva, Sigma, U.S. Biochemical, Worthington

Watermelon seeds - BDH

416. Uricase (urate: oxygen oxidoreductase, EC 1.7.3.3)

A. flavis - Boehringer

Bovine kidney - Aldrich, NBC, Schwartz-Mann, Sigma, U.S. Biochemical, Worthington

Candida utilis - BBC, Beckman, Calbiochem, Miles, P-L, Sempa Chemie, Sigma, U.S. Biochemical, Worthington

Porcine kidney - Aldrich, Calbiochem, K&K, Koch-Light, Miles, NBC, Schwartz-Mann, Sigma, U.S. Biochemical

Porcine liver - BDH, Biozyme, Boehringer, Miles, Paesel, Serva, U.S. Biochemical, Worthington,

417. Uridine-5'-diphosphogalactose-4-epimerase (UDP glucose-4-epimerase, EC 5.1.3.2)

C. pseudotropicalis - New England Enzyme[*]

E. coli - New England Enzyme[*]

Galactose-adapted yeast - Sigma

418. Uridine-5'-diphosphoglucose dehydrogenase (UDPG dehydrogenase; UDP glucose: NAD-6-oxidoreductase, EC 1.1.1.22)

Bovine liver - Boehringer, Sigma

419. Uridine-5'-diphosphoglucose pyrophosphorylase (UDPG pyrophosphorylase; UTP: α-D-glucose-1-phosphate uridyltransferase, EC 2.7.7.9)

Baker's yeast - Sigma

Bovine liver - Boehringer, Sigma[*]

420. Uridine-5'-diphosphoglucoronyltransferase (UDP glucuronyltransferase, EC 2.4.1.17)

Bovine liver - Sigma

Rabbit liver - Sigma

421. Uridyltransferase (see Galactose-1-phosphate uridyltransferase)

422. Urokinase (EC 3.4.99.26)

Human urine - Calbiochem, Koch-Light, NBC, Sigma, U.S. Biochemical

423. Uroporphypinogen-I-synthetase (porphobilinogen ammonia lyase,
      polymerizing, EC 4.3.1.8)
         Spinach - New England Enzyme[*]
424. Wheat germ lysate
         Miles
425. Xanthine oxidase (xanthine-O_2-oxidoreductase, EC 1.2.3.2)
         Milk - Aldrich, Biozyme, Boehringer, Calbiochem, General
            Biochem, Gibco, K&K, Koch-Light, Miles, NBC, P-L, Paesel,
            Schwartz-Mann, Serva, Sigma, U.S. Biochemical
         Unpasteurized cream - BDH
426. Xymolean (mixture of trypsin and chymotrypsin)
         Canada Packers
427. Xyulose reductase (xylitol: NADP-4-oxidoreductase, L-xyulose
      forming, EC 1.1.1.10)
         Pigeon liver - Sigma

appendix 4
Suppliers of Enzymes

1. *Aldrich Biochemicals*
   United States

   Europe

   940 W. Saint Paul Avenue, P.O. Box 355, Milwaukee, Wisconsin 53201

   Aldrich-Europa, c/o Janssen Pharmaceutica, B-2340 Beerse, Belgium

2. *Advance Biofractures Corp.*

   35 Wilbur Street, Lynbrook, New York 11563

3. *Beckman*
   United States

   Europe

   6200 El Camino Real, Carlsbad, California 92008

   Beckman International, 41 rue Mariziano, 1227 Geneva, Switzerland

4. *Biologicals Business Center (BBC)*

   480 Democrat Road, Gibbstown, New Jersey 08027

5. *Biozyme Labs*

   Gilchrist-Thomas Estate, Blaenavon, Guent NP4 9RL, South Wales, Great Britain

6. *Boehringer-Mannheim Corp.*
   United States

   Europe

   7941 Castleway Dr., P.O. Box 50816, Indianapolis, Indiana 46250

   Sanhofer Strasse, D-6800 Mannheim 31, Federal Republic of Germany

7. *British Drug Houses, Ltd.*
   United States

   Europe

   Gallard Schlesinger Chem., 584 Mineola Avenue, Carle Place, New York 11514

   BDH Chemicals, Ltd., Poole, Dorset BH12 4NN, England

8. *Calbiochem-Behring*
   United States

   Europe

   P.O. Box 12087, San Diego, California 92112

   Calbiochem GmbH, Postfach 2360, 6300 Lahn 11, Federal Republic of Germany; *also*

		Calbiochem AG, Loewengraben 14, 6000 Lucerne 5, Switzerland
	Asia	Hoechst Singapore, Units 501-514, 5th floor, Soon Wing Industrial Bldg. 2, Soon Wing Road, Circuit Road, P.O. Box 89, Singapore 13/ Singapur
9.	*Canada Packers*	55 Glen Scarlett Road, Toronto, Ontario, Canada
10.	*Dextran Products, Ltd.*	421 Comstock Road, Scarborough, Ontario, Canada
11.	*Diamond Shamrock Chem.*	P.O. Box 829, Redwood City, California 94064
12.	*General Biochemicals*	950 Laboratory Park, Chagrin Falls, Ohio 44022
13.	*Gibco Diagnostics (Grand Island Biological Co.)* United States	P.O. Box 4385, 2801 Industrial Drive, Madison, Wisconsin 53713; *or* 3175 Stanley Road, Grand Island, New York 14072
	Europe	Bio Cult Labs, 3 Washington Road, Paisley PA3 4EP, Scotland
14.	*Gist Brocades*	P.O. Box 1, Delft, Holland
15.	*Koch-Light Labs*	Colnbrook, Buckinghamshire SL3 0B2, England
16.	*K&K*	ICN Pharmaceuticals, 121 Express Street, Plainview, New York 11803
17.	*Merck* United States	EM Laboratories, 500 Executive Boulevard, Elmsford, New York 10523
	Europe	E. Merck AG, 6100 Darmstadt, Federal Republic of Germany
18.	*Miles Laboratories* United States	P.O. Box 2000, Elkhart, Indiana 46515
	Europe	Laboratoire Miles SA, Tour Maine Montparnesse, 33 Av du Maine, 75755 Paris, France; *also* Miles GmbH, Lyoner Strasse 32, 6 Frankfurt/Main 71, Federal Republic of Germany; *also* Research Products Division, Miles Ltd., P.O. Box 37, Stoke Poges, Slough, England SL2 4L4
	Israel	Miles Yeda, Ltd., Kiryat Weizman, P.O. Box 1122, Rehovot, Israel

	Japan	Seikagaku Kogyo Co., Toyo Bldg., No. 3-7 Nihombashi-Hojcho, Chuo-ku, Tokyo 103, Japan
19.	*New England Enzyme Center*	Tufts University, School of Medicine, 136 Harrison Avenue, Boston, Massachusetts 02111
20.	*Novo Industrial* United States	P.O. Box 189, Mamaroneck, New York 10543
	Europe	Novo Alle, DK 2880 Bagsvaerd, Denmark
21.	*Nutritional Biochemicals*	ICN Pharmaceuticals, 26201 Miles Road, Cleveland, Ohio 44128
22.	*Paesel*	Borsigallee 6, D-6000 Frankfurt, Federal Republic of Germany
23.	*P-L Biochemicals, Ltd.* United States	1037 W. McKinley Avenue, Milwaukee, Wisconsin 53205
	Europe	P-L Biochemicals GmbH, Ulmenhof 28, Postfach 51, 5401 St. Goar, Federal Republic of Germany; *also* International Enzymes, York Road, Wimbledon, London SW19 8UB, England
24.	*Rapides Societé*	15 Rue des Comtesses, 59115 Seclin, France
25.	*Rohm & Haas*	Independence Mall West, Philadelphia, Pennsylvania 19105
26.	*Schwartz-Mann* United States	Division of Becton, Dickinson and Co., Orangebury, New York 10962
	Europe	Mann Research Labs, York House, Empire Way, Wembley, Middlesex, England
27.	*Searle Biochemical*	2634 So. Clearbrook Drive, Arlington, Illinois 60005
28.	*Sempa Chimie, Inc.*	58 rue du Dessous des Berges, Paris XIII, France
29.	*Serva*	P.O. Box 105260, Karl Benz Strasse 7, D-6900 Heidelberg, Federal Republic of Germany
30.	*Sigma Chemical* United States	P.O. Box 14508, St. Louis, Missouri 63178
	Europe	Sigma London Chem. Co., Ltd., Fancy Road, Poole, Dorset BH17 7NH, England; *also* Sigma Kemie GmbH Munchen, AM Bahnsteig 7, D-8028 Taufkirchen, Federal Republic of Germany

31. *United States Biochemical*   P.O. Box 22400, Cleveland, Ohio
    Corp.   44122

32. *Vickers Limited*   Vickers America, Greenwich,
                                     Connecticut 06830

33. *Whatman Biochemicals*   Springfield Mill, Maidstone, Kent
                                     ME14 2LE, Great Britain

34. *Worthington*   Worthington-Millipore Corp.,
    United States   Laboratory Products Division,
                                     Freehold, New Jersey 07728

    Europe   Millipore House, Abbey Road,
                                     London NW10 7SP; *also*
                                     10 Ave des Heliotropes, 1030
                                     Brussels, Belgium; *also*
                                     6078 Neu-Isenburg, Siemensstrasse
                                     20, Federal Republic of Germany

Author Index

Adams, R. E., 251[46], 261[46], *305*

Adler, E., 28[18], *73*

Adu-Amankwa, B., 91[143], *110*

Aebi, V., 44[33,36], *74*

Agren, A., 300[157], 302[157], *310*

Ahn, B. K., 127[106], 184[106], *234*

Aikawa, K., 251[19], 262[19], *304*

Aimo, G., 252[56], 278[56], 279 [56], 280[56], *306*

Aizawa, A., 130[160], *237*

Aizawa, M., 129[180], 177[180], 215[298], 224[298], 229[322, 324,325], 230[328,329], *237, 242, 243,* 292[122], *309*

Akerlund, A., 123[46], 124[46], 128[46], 130[46,62], 131[46], 136[62], 162[46], 163[46], 169 [62], 172[46], 199[46,62], *232, 233*

Albisser, A. M., 250[9], *304*

Albu-Weissenberg, M., 83[58], 89[113], 95[58], *106*

Alexander, P. W., 115[11,166], 131[82], 169[82], *231, 233, 237*

Alifano, A., 66[75], *76*

Anderson, B. S., 83[43], *106*

Ando, M., 131[172], 192[172], *237*

Andrade, J., 104[159], *110*

Anfalt, T., 127[66], 131[66], 137 [66], 150[66], 152[66], 154[66], 161[66], 168[66], *233,* 300[156], *310*

Apoteker, A., 128[193], 178[193], *238*

Araki, Y., 126[244], 131[244], 203[244], *240*

Arima, A., 131[172], 192[172], *237*

Arnold, M., 129[229,230], 203[229, 230], 212[230], 214[230,291], 215[230,305], 217[230,291], 219 [230], 220[305], *239, 242*

Arnold, M. A., 214[313,315], 217 [313], 218[315], 220[313], *243*

Arwin, H., 210[274], *241*

Atallah, M., 258[96], *308*

Atkinson, E., 53[53], *75*

Atlas, D., 88[92], *108*

Atrat, P., 214[288], 215[306], 222[288,306], *242*

Aurich, H., 208[249], *240*

Auses, J. P., 300[155], 301[155], *310*

Aussresses, H., 129[275], *241*
Axen, R., 85[79,80], 88[97,109], 89[80], *107, 108*

Baba, S., 114[7], 128[119], 197 [119], *231, 235*
Bais, R., 252[166], *311*
Bancroft, J., 57[55], *75*
Barabino, R. C., 208[273], 210 [273], *241*, 250[50], 257 [50], *306*
Bar-Eli, A., 88[104], 95[121], *109*
Barker, A. S., 104[158], *110*
Barker, S. A., 85[70], *107*
Barnett, L. B., 81[28], *105*
Bartoli, F., 251[20,23,92,95], 252[92,95], 253[92,95], 259 [23], 264[23], 273[95], 276 [95], *305, 307, 308*
Bassi, D., 252[56], 278[56], 279 [56], 280[56], *306*
Baudras, A., 129[145], 193[145], 195[149-151], *236*
Baum, G., 126[83,84], 202[83, 84], 208[83], *233, 234*
Baum, P., 28[17], *73*
Bauman, E. K., 79[21], 97[21], *105*, 128[114], *235*, 215[21], 263[21], *305*
Bauman, G., 343[28], 344[28], *352*
Bennet, T., 1[1], *73*
Berezin, I., 300[149,151], 303 [149,151], *310*
Berezin, I. V., 77[8], 104[8], *104*
Berg, G., 250[27], 256[27], *305*
Bergmeyer, H. U., 77[5], *104*, 250[47], 255[47], 256[47], *306*
Berjonneau, A., 127[102], *234*
Berjonneau, A.-M., 127[182,183], 184[182], 185[183], *237*
Bernath, F., 103[126], *109*
Bernfeld, P., 81[31], *105*
Bertermann, K., 126[223], 130 [223], 201[223], *239*
Bertrand, C., 128[194,195], 129 [194], 130[194], 131[194], 178

[Bertrand, C.], [194], 179[195], 197[194,195], *238*
Bessmann, S., 129[133], *235*
Bhatti, K., 128[130], 173[130], *235*
Bhatti, K. M., 250[36], 255[36], *305*
Bissie, E., 252[64], 267[64], *306*
Blaedel, W., 128[125], 129[125, 142], 174[142], 191[125], *235, 263*, 264[42,43], *305*
Blaedel, W. J., 28[19], 64[67], 66[77,78,80], *73, 75, 76*, 77[6], 104[6], *104*, 145[70], 204[235, 233, 240*
Blumberg, S., 88[92], *108*
Blumenberg, B., 252[65], *306*
Bochner, J., 177[179], *237*
Bogulaski, R. C., 77[16], 97[16], *105*, 145[70], *233*
Boitieux, J., 228[320,321], *243*
Booker, H., 207[251], 208[251], *240*
Borgerud, A., 288[108], 292[124], 293[124], 296[124], *308, 309*
Borjesson, J., 292[119], *309*
Borrebaeck, C., 115[10], 129[155], 193[155], 203[232], *231, 236, 240*, 292[118,119,128], 293[118, 137], 297[118], *309, 310*
Bostick, D. T., 300[153], 301[153], *310*
Bostick, W. D., 251[53], *306*
Bougeois, J.-P., 174[128], *235*
Bouin, J. C., 258[96], *308*
Bourdillon, C., 174[128], *235*
Bovara, R., 252[100], 253[77], 281[100], 282[77], *307, 308*
Bowers, L., 104[141], *110*, 288 [116], *309*
Bowers, L. D., 77[7,17], 104[7], *104, 105*, 245[44], 288[109], 292[134], 293[142], 294[134, 142], *305, 308-310*
Boyd, J., 208[253], *240*
Branson, H. R., 5[2], *73*
Breaux, J., 126[227], 128[174, 175], 167[174,175], 202[227], 204[174,175], *237, 239*

Bretz, N. S., 251[31], 260[31], 305

Bright, H. J., 249[7], 250[7], 254[7], 255[7], 304

Brignac, P., 49[48,49], 75

Brillouet, J. M., 247[40,41], 248[40], 251[40,41], 253[40, 41], 257[40,41], 274[40], 275 [41], 305

Brolin, S., 300[157], 302[157], 310

Broun, G., 116[15], 126[199], 127[182,199], 129[199], 131 [199], 170[199], 174[199], 184 [182,199], 231, 237, 238

Brovko, L., 300[149-152], 303 [149-152], 310

Brower, M., 45[39], 74

Brown, H., 81[32], 105

Brown, H. D., 85[73], 88[101], 107, 108, 251[51], 274[51], 306, 310

Brown, I. W., 90[119], 91[119], 109

Brown, S. R., 85[71], 107

Buck, R., 127[185], 185[185], 215[294], 220[185,294], 238, 242

Buenning, K., 81[26], 105

Bull, H. B., 81[28], 105

Burbaum, S. N., 49[46], 75

Burk, D. J., 21[10], 73

Burke, M., 49[50], 75

Burns, D. T., 300[162], 311

Burris, R. H., 57[57], 75

Bushby, M., 129[276], 241

Caenaro, G., 252[62], 253[62], 267[62], 269[62], 270[62], 271 [62], 277[62], 280[62], 306

Callanan, W. A., 205[241], 240

Calvot, C., 127[102,183], 185 [183], 234, 237

Cambachi, S., 252[61], 253[61], 269[61], 277[61], 278[61], 306

Cambiaghi, S., 252[56], 278[46], 279[56], 280[56], 306

Campbell, J., 77[15], 105, 250 [8], 252[66,69,88], 253[88], 256[8], 257[8], 266[8], 267 [88], 280[88], 304, 306, 307, 331[7], 351

Campbell, M., 128[219], 196[219], 239

Canning, L., 288[109], 308

Canning, L. M., 293[142,146], 294 [142], 310

Cannon, P. L., 31[20], 63[62], 64 [63-65], 73, 75

Caplan, N. O., 104[155], 110

Caplan, S. R., 83[59], 106

Carey, R. N., 252[67], 267[67], 306

Caropresso, A., 252[56], 278[56], 279[56], 280[56], 306

Carr, P., 104[141], 110, 205[239], 240, 288[109,116], 308, 309

Carr, P. W., 77[7,17], 104[7], 104, 105, 245[44], 251[46], 252 [72], 261[46], 283[72], 284[72], 292[134,142], 305-307, 309, 310

Carraway, K. L., 88[99,100], 108

Carrea, G., 252[100], 281[100], 308

Carter, R., 205[240], 240

Cebra, I., 89[110], 90[110], 108

Chang, T. M., 81[30], 105

Chang, T. M. S., 82[35-40], 105, 106, 252[88], 253[88], 267[88], 280[88], 307, 331[7], 351

Chapot, D., 129[153], 130[153], 131[153], 180[153], 236

Chattapadhyaz, S., 81[32], 88[101], 105, 108

Charlton, G., [60], 75

Chawla, A., 252[88], 253[88], 267 [88], 280[88], 307

Chawla, A. S., 331[7], 351

Chen, A., 129[189], 194[189], 239

Chen, A. K., 130[190], 191[190], 238

Chen, B., 128[124], 200[124], 235

Cheng, C., 54[54], 75

Cheng, F., 113[3], 115[9], 126 [186], 128[126], 129[154,334], 130[154], 179[126], 181[154], 188[186], 194[334], 200[334], 230, 231, 235, 236, 238, 244

Chibata, I., 83[49], 106

Chien, P., 209[271], 241

Chirillo, R., 252[62], 253[62], 267[62], 269[62], 270[62], 271 [62], 277[62], 280[62], 306

Chotani, G., 206[242], *240*
Christian, G., 113[3], 115[9],
    127[112], 128[120,126], 129
    [143,147,154,334], 130[154],
    179[126], 181[154], 186[112],
    194[334], 197[120], 200[334],
    202[143], 208[246,247,250],
    209[270], *230, 231, 234, 235,*
    236, 240, 241, 244
Christian, G. D., 65[70,71], 66
    [79], *76*, 126[186], 188[186],
    238, 250[34,35], 258[34], 264
    [35], *305*
Clark, L., 119[31], 121[42], 124
    [42], 129[42], 170[42], *231,*
    232
Clark, L. C., 128[75,219,336],
    161[75], 171[75], 196[219],
    197[336], *233, 239, 244*
Clemens, A., 174[176], *237*
Coliss, J. S., 276[84], *307*
Colombo, J., 44[33,36], *74*
Colowick, S. P., 25[12], *73*
Comtat, M., 129[145], 193[145],
    195[149-151], *236*
Comtat, T. M., 129[211], 193
    [211], *239*
Constantinides, A., 91[143], *110*
Conyers, R. A. J., 252[166], *311*
Cook, S., 300[155], 301[155],
    310
Cooney, C. L., 288[107,110], 293
    [110], 294[110], *308*
Cooper, A. G., 49[46], *75*
Cordonier, M., 129[153], 130
    [153], 131[153], 180[153], *236*
Corey, P. B., 5[2], *73*
Coughlin, R., 120[34,35], *232*
Coulet, P. R., 91[144,145], 92
    [144,149-152], 93[154], *110,*
    128[135,136,194,195], 129[194],
    130[194], 131[194], 177[135-
    140], 178[138,140,194], 179
    [139,195], 197[194,195], 206
    [139,140], *235, 236, 238,* 247
    [38-41], 248[39,40], 251[38-
    41], 253[40,41], 257[38-41],
    274[40], 275[41], *305*
Craven, G. R., 85[78], *107*
Cremonesi, P., 252[100], 253[77],
    281[100], 282[77], *307, 308*

Crochet, K., 207[256], 208[256],
    240
Crook, E. M., 84[65], 88[65], *107*
Crouch, S. R., 252[63], 272[63],
    306
Crutchfield, G., 85[82], 89[82],
    107
Cserfalvi, T., 130[173], 131[191],
    165[173,191], *237, 238*
Csiky, I., 251[101], 282[101], *308*
Cullen, L. F., 130[158], 199
    [158], *236*, 251[24], 259[24],
    305
Czok, R., 28[17], *73*

Dahodwala, S. K., 250[15], 257
    [15], *304*
Danielsson, B., 285[102], 288[102,
    108,111,112], 290[112,117], 291
    [117,131], 292[117,120,121,124-
    127], 293[117,120,124,125,135,
    138,140,141,145], 294[117,120,
    125], 295[120,121,125,135,141],
    296[120,124-126,140], 297[120],
    298[120], *308-310*, 333[25], *352*
Danzer, J., 293[144], *310*
David, A., 126[228], *239*
Davies, P., 127[192], 129[192],
    130[192], 184[192], 194[192],
    238
Davis, A., 251[26], 263[26], *305*
Davis, P., 127[105], 184[105],
    234
Davis, V., 115[12], *231*
Dawson, D., 17[7], *73*
Day, R. A., 84[63], *106*
Delenk, J., 128[118], 197[118],
    235
Delente, J., 114[6], *230*
Dellenbach, R., 252[94], 267[94],
    307
DeLuca, M., 300[147,148,158,159],
    302[158,159], 303[147,148], *310*
Demos, J., 44[32,37], *74*
Deng, I., 126[224], 201[224], *239*
Denton, M. S., 251[53], *306*
Desmet, G., 228[320,321], *243*
Detar, C., 104[164], *111*
Determan, H., 81[26], *105*
Diamandis, E., 200[222], *239*

Diamandis, E. P., 118[28], 231
Diegelman, M., 253[74], 280[74], 307
Dietschy, J., 114[6], 128[118], 197[118], 230, 235
Dinsimore, S. R., 251[53], 306
Dintzis, H. M., 85[69], 98[69], 107
DiPaolantonio, C. L., 215[305], 220[305], 242
Divies, C., 211[307,308], 212 [308], 214[307,308], 219[307, 308], 243
Dixon, M., 57[58], 75
Dodd, L., 174[176], 237
Doig, A. R., 128[129], 129[129], 172[129], 193[129], 209[129], 235
Dombrovskii, V., 300[151], 303 [151], 310
D'Orazio, P., 209[269], 215[295], 226[295], 241, 242
Douay, O., 312[20], 300[165], 302[165], 310, 351
Drash, A. L., 129[203], 177[203], 238
Dreyfus, J., 44[32,37], 74
Dritschilo, W., 249[7,49], 250 [7,49], 251[33], 254[7,33], 255[7], 261[33], 304-306
Duggan, C., 119[31], 128[336], 197[336], 231, 244
Dumontier, M., 209[265], 241
Dunhill, P., 103[129], 109
Dupret, D., 247[39], 248[39], 251[39], 257[39], 305
Durand, P., 126[228], 239
Durliat, H., 129[145,211], 193 [145,211], 195[149-151], 236, 239

Ebashi, S., 44[34,35], 74
Ebersbach, W., 253[99], 281[99], 308
Edstrom, K., 126[96], 234, 249 [3], 250[3], 304
Edwards, B. A., 85[77], 107
Edwards, J. B., 252[166], 311
Ekman, B., 300[157], 302[157], 310
Emond, C., 250[37], 257[37], 305

Emory, C., 128[219], 196[219], 239
Endo, J., 252[93], 253[93], 270 [93], 280[93], 307
Enfors, S., 130[220], 199[220], 239
Enfors, S.-O., 104[156], 110, 129 [177], 175[177], 237
Engasser, J.-M., 92[151], 110
Engstrom, R., 128[125], 129[125], 191[125], 235
Enke, C., 126[224], 201[224], 239
Eppenberger, H., 17[7], 73
Epps, H. M. R., 83[47], 106
Epton, R., 85[70], 107
Erlanger, B., 208[258], 240
Erlanger, B. F., 49[46], 75, 85 [72], 107
Ernbach, S., 88[109], 108
Everse, J., 104[135], 109

Fallen, R., 253[74], 280[74], 307
Faller, D., 288[107], 308
Feldbruegge, D., 252[67], 267[67], 306
Fergerson, D., 208[253], 240
Fillipusson, H., 85[74,75], 107, 245[2], 246[2], 252[59], 253 [82], 266[59], 276[82], 304, 306, 307
Finley, J. W., 253[76], 307
Fischerova-Bergerova, V., 66[72], 76
Fleischmann, W. D., 252[68], 273 [68], 306
Florkin, M., 11[5], 15[5], 33[5], 50[5], 73
Flournoy, D. S., 215[301], 223 [301], 242
Fogt, E., 174[176], 237
Fojie, Y., 44[35], 74
Ford, J., 300[147], 303[147], 310
Forrester, L. J., 251[51], 264 [51], 293[139], 306, 310
Fox, S., 53[53], 75
Fraenkel-Conrat, H., 85[88], 107
Fraser, D., 88[90,94], 107, 108
Free, A. H., 104[132], 109
Freiden, E., 1[1], 73
Froment, B., 129[204], 177[204], 238

Fukui, S., 251[52], 253[79], 275 [79], 276[52], 283[52], *306, 307*

Fulton, S. P., 288[110], 293 [110], 294[110], *308*

Fung, K. W., 127[110], 186[110], *234*

Fuse, N., 83[49], *106*

Gabel, D., 85[79,83], 89[83], *107*

Gadaleta, M. N., 66[75], *76*

Gadd, K., 291[131], 293[145], *309, 310*

Gale, E. F., 83[47], *106*

Gander, R., 250[9], *304*

Gantner, G., 46[41], *74*

Garber, C. C., 252[67], 267[67], *306*

Gautheron, D. C., 91[144,145], 92[144,149,152,153], 93[154], *110*, 128[135,195], 177[135-140], 178[138,140], 179[139, 195], 197[195], 206[139,140], *235, 236, 238,* 247[38-41], 248 [39,40], 251[38-41], 253[40, 41], 257[38-41], 274[40], 275 [41], *305*

Gebauer, C., 207[255], 208[255], 209[255], 230[329], *240, 243*

Gebauer, C. R., 212[309], 214 [309], *243*

Gellf, C., 127[102], *234*

Gellf, G., 127[183], 185[183], *237*

Gibson, K., 127[99], 207[99], 208[99], *234*

Ginsburg, C., 104[135], *109*

Giovenco, S., 251[92], 252[92], 253[92], *307*

Givol, D., 89[110], 90[110], *108*

Glueck, C., 128[219], 196[219], *239*

Godinot, C., 92[152,153], *110*

Goldfeld, M. G., 83[46], *106*

Goldman, R., 83[59], *106*

Goldstein, L., 83[51], 85[81], 88[92], 89[111,112], 95[81, 125], 96[111], *106-109*

Gondo, S., 129[200], 176[200], *238*

Goodson, L., 251[26], 263[26], *305,* 343[28], *352*

Goodson, L. H., 79[21], 97[21], *105,* 128[114], *235,* 251[21,22], 263[21,22], *305*

Gorton, L., 128[130], 173[130], *235,* 250[36,60], 252[60], 255 [36], 272[60], *305, 306*

Gorus, F., 300[164], 302[164], *311*

Goto, K., 68[86], *76*

Goto, M., 209[267,268], *241*

Gough, D., 104[159], *110*

Granelli, A., 127[66], 131[66], 137[66], 150[66], 152[66], 154 [66], 161[66], 168[66], *233,* 300[156], *310*

Gray, D., 250[50], 257[50], *306*

Gray, D. N., 77[9,10], 104[9,10], *104,* 128[141], 208[273], 210 [273], *236, 241*

Gray, P. P., 130[221], 199[221], *239*

Green, M. L., 85[82], *107*

Gregory, F., 16[6], 50[6], *73*

Grenner, G., 292[130], *309*

Griffin, E. G., 77[1], *104*

Grigorov, I., 209[260], *241*

Grimaud, C., 129[146], 193[146], 195[152], *236*

Grime, J. K., 288[115], *308*

Grobler, S. R., 215[316], *243*

Grooms, T., 119[31], 128[336], 197[336], *231, 244*

Gryszkiewicz, J., 103[127], *109*

Guarnaccia, R., 129[201], 176 [201], *238*

Guilbault, G., 63[61], *75,* 127 [101,108], 128[117], 131[191], 165[191], 181[101], 184[108], 185[101], 196[117], 197[117], 206[257], 208[252,257], 209 [252,263], 212[108], 226[337], 227[338], *234, 235, 238, 240, 241, 244*

Guilbault, G. G., 31[20], 39[25], 45[25], 49[47,48,49], 53[52], 63[62], 64[63-66], 66[81], 67 [82,83], 68[84], *73-76,* 77[3,4, 11,12], 79[18,19,21], 80[22], 97[21], 104[4,11,12,18,19,138,

[Guilbault, G. G.], 139,140,163], *104, 105, 109, 110, 111,* 116 [18,19], 117[19,22-24], 118 [23], 122[44,45], 123[45], 124 [45], 125[48-52], 126[50,52,55, 56,93,94,134], 127[51,52,97-99, 104,110,111], 128[54,69,114, 124,134], 129[56,69,76,93,104], 130[77-99,156,173], 131[49,53, 55,58,63,65,73,74,80,165,169, 171], 133[50,53-56], 136[50,53, 54,58,63,65], 139[44,53], 143 [54], 144[52], 145[53,63], 148 [54], 150[54,72], 151[50,52, 54], 152[53], 153[53], 154[53], 155[53], 156[73], 157[50,54-56, 59,69,73,74], 160[53,63,65], 161[54,55,69,76], 162[53-55,63, 69,76], 163[49,50,52-55,59,76], 164[76], 165[173], 166[78,156], 167[63,79,80], 168[53,58,65, 74], 169[49,73], 170[53,63,65, 169], 171[54,69,76,134], 174 [56], 181[49,51], 182[52,104], 183[50,56], 184[104], 186[110, 111], 187[93,94], 192[50,54,55, 69,94], 200[124], 204[165], 205 [53], 207[99], 208[24,99], *231-237,* 245[1], 251[21], 263[21], 292[127], 300[160], *304, 305, 311,* 313[1], 316[2], 321[5,19], 323[22], 324[23,24], 325[4,8, 17], 326[1,2,12,13], 327[15], 328[18], 329[6,9-11], 330[3], 333[25-27], 343[28], 344[28], *351, 352*
Guillot, C., 129[146], 193[146], 195[152], *236*
Gulberg, E., 129[144], 203[144], *236*
Gulberg, E. L., 250[34,35], 258 [34], 264[35], *305*
Gülich, M., 131[216], 168[216], *239*
Gulinelli, S., 251[20,23,95], 252[95], 253[95], 259[23], 264 [23], 273[95], 276[95], *305, 308*
Gunther, G., 28[18], *73*
Gunzel, G., 84[61], *106*
Gupta, V., 85[78], *107*

Haab, W., 46[42], *74*
Habeeb, A. F. S. A., [56], 88 [108], *106, 108*
Hadjiioannou, T. P., 118[25], 129 [143], 200[222], 202[143], *231, 236, 239*
Haga, M., 228[319], *243*
Hagen, A., 250[47], 255[47], 256 [47], *306*
Hahn, Y., 120[32], *231*
Hall, D. A., 113[1], *230*
Halling, P. J., 103[129], *109*
Hamm, R., 46[40], *74*
Hamoir, G., 83[50], *106*
Haneka, H. F., 204[237], *240*
Hann, D. A., 77[17], *105*
Hanna, A., 288[116], *309*
Hansen, E., 208[252], 209[252], *240*
Hansen, E. H., 169[81], *233*
Hanson, D. J., 251[31], 260[31], *305*
Hanss, M., 121[38,39], *232*
Hara, K., 250[17], 262[17], *304*
Hara, T., 119[29], *231*
Harada, K., 128[218], 172[218], 207[218], 208[218], *239*
Harrington, K. J., 88[95], *108*
Harrison, R. A. P., 312[14], *351*
Hart, L., 119[31], 128[336], 197 [336], *231, 244*
Haslem, J., 207[251], 208[251], *240*
Hass, M., 209[265], *241*
Hassan, S., [13], *231*
Hatakeyama, N., 131[172], 192 [172], *237*
Havas, J., 127[111], 186[111], *234*
Havelvala, N., 104[164], *111*
Haynes, R., 83[55], *106*
Henry, J., 42[29,30], 45[38], *74*
Hercules, D. M., 300[153], 301 [153], *310*
Hersh, L. S., 90[118,119], 91 [118,119], *109*
Herzig, D. J., 84[63], *106*
Hewetson, J. W., 130[221], 199 [221], *239*
Hicks, G. P., 28[19], *73,* 77[6], 80[23-25], 104[6], *104, 105,*

[Hicks, G. P.], 121[43], 124 [43], 129[43], 136[61], 170 [43], 232, 233, 249[6], 250 [6], 252[57], 259[6], 265[57], 304, 306
Hiff, G. F., 300[154], 301[154], 310
Higgins, H. G., 88[90,94,95], 107, 108
Hikuma, M., 214[282,283,290, 292], 215[290,296], 219[292], 221[290], 222[282], 223[283], 225[296], 241, 242
Hinsch, W., 252[65,70], 253[70, 78,87,99], 277[85], 278[87], 281[78,99], 283[70], 306-308
Hiramoto, R., 88[108], 108
Hoffman La Roche & Co., 128[243], 129[243], 173[243], 196[243], 240
Hofsten, B., 85[83], 89[83], 107
Holuk, J., [30], 231
Hornby, W., 245[2], 246[2], 250 [8], 256[8], 257[8], 266[8], 304
Hornby, W. E., 77[14,15], 85[74-76], 105, 106, 252[58,59,66, 69,71], 253[71,81,82], 266[58, 59], 276[58,82], 283[71], 300 [163], 301[163], 306, 307, 311
Horvath, C., 251[54], 252[54,94], 253[54,86], 267[54,94], 280 [86], 281[54], 306, 307
Hrabankova, E., 125[51,52], 126 [52], 127[51,52,97], 131[64], 136[64], 144[52], 151[52], 160 [64], 163[52], 168[64], 170 [64], 181[51,52], 182[52], 232-234
Hsiung, C., 127[98], 234
Hsiung, E., 117[22], 231
Huang, N., 128[117], 196[117], 197[117], 235, 324[24], 330[3], 351, 352
Huang, Y., 209[266], 241
Hultin, H. O., 258[96], 308
Humansson, C., 293[135], 295 [135], 309
Hummel, J. P., 83[43], 106
Humphrey, A., 249[7], 250[7], 254[7], 255[7], 304

Humphrey, A. E., 250[15], 257[15], 304
Hussein, W. R., 117[23], 118[23], 231

Ianniello, R., 126[206], 128[127], 174[127], 182[206], 183[206], 235, 238
Igloi, M. P., 251[55], 252[55], 253[87], 277[55,85], 278[87], 306, 307
Ignatov, J., 209[260], 241
Ikeda, S., 251[52], 253[79], 275 [79], 276[52], 283[52], 306, 307
Imai, H., 130[231], 131[231], 203 [231], 240
Imaki, M., 119[29], 231
Inman, D. J., 85[76], 107, 252 [58,69,71], 253[71], 266[58], 276[58], 283[71], 306, 307
Inman, J. K., 85[69], 98[69], 107
Inokuchi, H., 209[267,268], 241
Isambert, M. F., 85[72], 107
Ishiguro, I., 229[325], 243
Itagaki, H., 228[319], 243
Iwase, A., 206[257], 208[257], 240

Jablonski, E., 300[158,159], 302 [158,159], 310
Jablonski, I., 300[148], 303[148], 310
Jacob, H., 66[74], 76
Jacobs, W., 251[26], 263[26], 305
Jacobs, W. B., 251[22], 263[22], 305
Jacobsen, M. A., 252[73], 253[73], 283[73], 285[73], 307
Jaegfeldt, H., 126[335], 188[335], 244, 250[25], 259[25], 305
Jagner, D., 127[66], 131[66], 137 [66], 150[66], 152[66], 154[66], 161[66], 168[66], 233
Jahnke, M., 288[107], 308
James, D. R., [83], 307
Jänchen, M., 126[223], 128[196], 130[223], 173[196], 201[223], 238, 239
Janik, B., 252[89], 267[89], 307
Jansen, E. F., 83[54], 88[103], 106, 108

Janssen, F. W., 187[207,208], 239
Jarrell, J. A., 332[16], 351
Jazaraman, S., 253[91], 277[91], 307
Jenkins, R. A., 66[77,78], 76
Jenning, E., 174[176], 237
Jensen, I. M., 128[181], 180 [181], 237
Jensen, M. A., 214[289], 220 [289], 242
Jerchtel, D., 53[51], 75
Jespersen, N. D., 35[23], 74
Johansson, A., 134[57], 232, 285 [103], 292[123], 294[123], 308, 309
Johansson, G., 126[96,188,335], 130[233], 131[170], 188[335], 189[188], 204[233], 234, 237, , 238, 240, 244, 245[45], 249 [3], 250[3,18,25], 251[30, 101], 259[25], 260[30], 262 [18,30], 282[101], 304-306, 308
Johnson, J., 128[205], 179[205], 238
Johnson, J. A., 27[15], 73
Johnson, P., 127[184], 185[185], 220[185], 238
Jones, D. O., 49[50], 75
Jong, T. H., 130[221], 199[221], 239
Jonsson, S., 292[128], 309
Jordan, J., 286[104], 308
Joseph, J. P., 131[82], 169[82], 233
Joseph, M. D., 252[63], 272[63], 306
Julliard, J. H., 91[144], 92[144, 153], 93[154], 110, 128[135], 177[135], 235
Juneau, M., 49[49], 75

Kadish, A. H., 113[1], 230
Kadziauskiene, K., 214[287], 215 [287], 221[287], 222[287], 224 [287], 225[287], 242
Kameno, Y., 114[7], 231
Kamerro, J., 128[119], 197[119], 235
Kamin, R. A., 129[217], 172[217], 239

Kamo, N., 129[212], 195[212], 239
Kamoun, P., [20], 300[165], 302 [165], 311, 351
Kaplan, N., 17[7], 73, 104[135], 109
Kaplan, N. O., 25[12], 73
Kapoulas, A., 250[9], 304
Karube, I., 104[160], 110, 120[36, 37], 126[244], 128[122], 131 [167,244], 180[167], 196[122], 197[122], 203[244], 211[279], 214[279,282-286,290,292,300], 215[279,290,296,299,304], 219 [292,304], 221[279,290], 222 [282], 223[283], 224[279,299], 225[284-286,296], 232, 235, 237, 240-242, 250[17], 251[19], 262 [17,19], 304
Karube, J., 130[160], 131[164], 237
Kasiprzak, D., 252[63], 272[63], 306
Katchalski, E., 83[42,51,58,59], 88[102], 89[110-114], 90[110, 114], 95[58,121-123,125], 96 [111], 106, 108, 109, 253[80], 276[80], 307
Kato, S., 230[329], 243
Katz, S. A., 116[16,17], 231
Kawashima, T., 131[172], 192[172], 237
Kawauchi, Y., 130[231], 131[231], 203[231], 240
Kay, G., 84[65], 85[85], 88[65], 107
Kedem, O., 83[59], 106
Kennedy, J. F., 103[128], 109
Kerwin, R. M., 187[208], 239
Keyes, M., 250[50], 257[50], 306
Keyes, M. H., 77[9,10], 104[7,10], 104, 128[141], 208[273], 210 [273], 236, 241, 251[29], 259 [29], 305
Kiang, C., [14], 130[156], 166 [156], 231, 236
Kiang, C. H., 325[17], 351
Kiang, H., 130[78,79], 166[78], 167[79], 233
Kiang, S. W., 325[8], 329[9,10], 351
Kimura, K., 209[267,268], 241

Kirch, P., 293[144], *310*
Kissel, T. R., 145[70], *233*
Kissov, A. A., 77[8], 104[8], *104*
Kjellen, K., 130[162,163], 190 [162,163], 215[162], *237*
Klatt, L. N., 252[72], 283[72], 284[72], *307*
Klenke, H.-J., 195[148], *236*
Kmetec, E., 128[205], 179[205], *238*
Knoblock, E. C., 65[70], *76*
Knowles, J. R., 88[107], *108*
Knox, J. M., 276[84], *307*
Kobos, R., 126[86,87], 202[86, 87], 211[280], *234, 241*
Kobos, R. K., 196[215], 212[302, 309,310], 214[309,310], 215 [301,302,312], 220[310], 223 [301], 224[302], *239, 242, 243*
Kochsieck, K., 195[148], *236*
Koono, K., 113[2], *230*
Körmendy, L., 46[40], *74*
Korosi, A., 128[129], 129[129], 172[129], 193[129], 209[129], *235*
Kost, N., 300[150], 303[150], *310*
Kovach, P. M., 127[332], 186 [332], *244*
Koyama, M., 129[180], 177[180], *237*
Kramer, D., 209[263], *241*, 245 [1], *304*, 343[28], 344[28], *352*
Kramer, D. N., 31[20], 63[62], 64[63-65], 67[82,83], *73, 75, 76*, 80[21,22], *105*, 128[114], *235*, 251[21], 263[21], 300 [160], 304[160], *305, 311*
Krisam, G., 292[130], 293[144], *309, 310*
Kronwall, G., 292[128], *309*
Krueger, J. A., 214[311], *243*
Kuan, J. C. W., 329[6], *351*
Kuan, J. W., 321[19], 324[23, 24], 325[8], 329[9-11], *351, 352*
Kuan, S., 66[81], *76*, 117[22], 127[98], 128[124], 130[156], 166[156], 200[124], 226[337],

[Kuan, S.], 227[338], *231, 234-236, 244*
Kuan, S. S., 127[110], 128[117], 130[78,79], 131[58], 136[58], 166[78], 167[79], 168[58], 186 [110], 196[117], 197[117], *233-235*, 324[23], 325[8,17], 329[6, 9-11], 330[3], *351, 352*
Kubo, T., 214[282,283], 222[282], 223[283], *241*
Kuhn, R., 53[51], *75*
Kulys, J., 120[33], 127[214], 128 [214], 129[214], 130[225], 195 [214], 202[225], 214[287], 215 [287], 221[287], 222[287], 224 [287], 225[287], *232, 239, 242*
Kulys, J. J., 104[133],   , 129 [198,213], 173[198], 195[213], *238, 239*
Kumagai, H., 44[35], *74*
Kumar, A., 113[1], 127[112], 128 [120], 186[112], 197[120], 209 [270], *230, 234, 235, 241*
Kunz, H. J., 250[12], 254[12], *304*
Kuriyama, S., 214[293], 218[293], *242*
Kusling, J. F., 130[158], 199 [158], *236*
Kyzlink, J., 130[161], 190[161], *237*

LaDue, J., 39[27], *74*
Landgraff, L., 251[90], 282[90], *307*
Landriscina, C., 66[75], *76*
Larsen, N., 208[252], 209[252], *240*
Larsson, P., 126[188], 189[188], 215[333], 221[333], *238, 244*
Lau, H., 321[5], 323[22], *351*
Lau, H. K., 321[19], *351*
Laurent, J., 177[140], 178[140], 206[140], *236*
Lawny, F., 129[153], 130[153], 131[153], 180[153], *236*
Ledingham, W. M., 253[98], 280 [98], *308*
Lee, Y., 300[148], 303[148], *310*
Leh, M., 120[35], *232*
Leon, L. P., 251[54], 252[54,94],

[Leon, L. P.], 253[54,86], 267 [54,94], 280[86], 281[54], *306, 307*
Leung, F. Y., 252[89], 267[89], *307*
Levin, D., 88[92], *108*
Levin, H. W., 249[10], 250[10], *304*
Levin, Y., 88[102], 89[111,112], 96[111], *108*
Liberti, A., 127[60], 136[60], 142[60], 145[60], 146[60], 147 [60], 148[60], 149[60], 150 [60], 152[60], 153[60], 154 [60], 162[60], 197[60], *233*
Lilly, M. D., 85[85], *107*
Lindahl, L., 215[333], 221[333], *244*
Lindau, G., 83[44], *106*
Linde, H., 31[21], *73*
Lineweaver, H., 21[10], *73*
Liu, C., 129[189], 194[189], *239*
Liu, C. C., 129[202,203], 130 [190], 176[202], 177[203], 191 [130], *238*
Llenado, R., 116[20], 126[95], 127[67,68], 131[168], 142[67, 68], 143[68], 150[67,68], 162 [67,68], 182[95], 198[67,68], 209[261,262], *231, 233, 234, 237, 241*
Llenado, R. A., 117[25], *231*
Lofrumento, E., 66[75], *76*
Lojda, Z., 81[34], *105*
Lopiekes, D. V., 85[71], *107*
Lostia, O., 251[92], 252[92], 253[92], *307*
Lovett, S., 208[254], *240*, 251 [32], 261[32], *305*
Lubrano, G. J., 122[44], 125 [50], 126[50,94,134], 128[54, 69,134], 129[69], 133[50,54], 136[50,54], 139[44], 143[54], 148[54], 150[54], 151[50,54], 157[50,54,59,69], 161[54,69], 162[54,69], 163[50,54,59], 171 [54,69,134], 183[50], 187[94], 192[50,54,69,94], *232-234*
Lundberg, J., 134[57], *232*, 285 [103], 292[123], 294[123], *308, 309*

Lundstrom, I., 210[274], *241*
Luppa, D., 208[249], *240*
Luttrell, G. H., 251[24], 259[24], *305*
Lynn, M., 126[84], 202[84], *233*
Lyons, C., 121[42], 124[42], 129 [42], 170[42], *232*

Mabbott, G. A., 66[80], *76*
MacDonald, A., 245[2], 246[2], 252[59], 253[82], 266[59], 276 [82], *304, 306, 307*
Macholan, L., 126[277], 127[184], 130[161,210], 185[184], 190[161, 210], 209[264], *237, 238, 239, 241*
MacIntosh, F. C., 82[37], *105*
Mahenc, J., 129[145,211,276], 193[145,211], 195[150,151], *236, 239, 241*
Maheni, J., 195[150,151], *236*
Makino, Y., 113[2], *230*
Malinauskas, A., 120[33], 127[214], 128[214], 129[214], 130[225], 195 [214], 202[225], *232, 239*
Maloy, J., 145[71], 147[71], 149 [71], 152[71], 205[238], *233, 240*, 300[155], 301[155], *310*
Mandenius, C. F., 292[127], 293 [140], 296[140], 333[25], *309, 310, 352*
Manecke, G., 84[61], *106*
Manoylov, S. E., 84[66], 88[66], *107*
Mansson, M., 126[188], 189[188], *238*
Marconi, W., 251[20,23,92,95], 252[92,95], 253[92,95], 259 [23], 264[23], 273[95], 276 [95], 292[132], *305, 307, 308, 309*
Mardashev, S. R., 81[29], *105*
Marini, M. A., 286[105], 288[114], *308*
Mark, H. B., 66[73], *76*
Martin, C. J., 286[105], 288[114], *308*
Mascini, M., 127[60,100], 131 [169], 136[60], 142[60], 145

[Mascini, M.], [60], 146[60], 147[60], 148[60], 149[60], 150 [60], 152[60], 153[60], 154 [60], 162[60], 170[169], 183 [100], 197[60], 212[303], 214 [303], 215[303], 217[303], 219 [303], *231, 233, 234, 237, 242*
Mason, M. K., 332[16], *351*
Mason, S. G., 82[37], *105*
Mason, W., 208[248], *240*
Matsumoto, K., 127[113], 191 [113], 214[300], *235, 242*
Matsunaga, T., 120[36,37], 214 [284-286], 215[299,304], 219 [304], 224[299], 225[284,285], *232, 241, 242*
Matsuoka, H., 229[325], *243*
Mattiasson, B., 85[84], *107,* 115 [10], 128[234], 129[155], 134 [57], 193[155], 203[232], 204 [234], 215[333], 221[333], 228 [317], *231, 232, 236, 240, 243, 244,* 285[103], 290[117], 291 [117,131], 292[117-120,123,128, 129,133], 293[117,118,120,133, 135-138,140,145], 294[117,120, 123], 295[120,135], 296[120, 129,138,140], 297[118,120,138], 298[120,133], 299[133], *308-310*
Mauerer, P. H., 85[87], *107*
May, S., 251[90], 282[90], *307*
Mazzola, G., 252[100], 281[100], *308*
McCaughey, C., 31[22], *73*
McDonald, A., 85[74], *107*
McLaren, A. D., 81[27], *105*
McLaren, J. V., 85[70], *107*
McMurtey, K., 115[12], *231*
McQueen, R., 125[48], *232*
Meinenhofer, H., 84[62], *106*
Mell, L., 145[71], 147[71], 149 [71], 152[71], 205[238], *233, 240*
Melrose, G. J. H., 88[89], 95 [124], *107, 109*
Menten, M., 19[9], *73*
Messing, R. A., 250[13], 254 [13], *304*
Meyerhoff, M., 207[255], 208 [255], 209[255,269], 215[295], 226[295], 227[339], 228[318], *240-244*

Meyerhoff, M. E., 127[332], 186 [332], 214[313], 215[312], 217 [313], 220[313], *243, 244*
Meyerson, L., 115[12], *231*
Michael, L., 209[271], *241*
Michaelis, L., 19[9], *73*
Michelson, A. M., 85[72], *107*
Milano, M., 60[59], *75*
Millar, B. S., 27[15], *73*
Miller, J. N., 300[162], *311*
Miller, R. C., 252[67], 267[67], *306*
Mindt, W., 128[131,132], 173[131, 132], 195[131,132], *235*
Mitsuda, S., 214[284,285], 225 [284,285], *241, 242*
Mizutani, F., 129[340], 194[340], *244*
Mohrbacher, R. J., 252[61], 253 [61], 269[61], 277[61], 278 [61], *306*
Molin, N., 104[156], *110*
Momoi, H., 44[34,35], *74*
Montalvo, J., 116[18], 209[272], 210[272], *231, 241*
Montalvo, J. G., 131[53,63,74,80, 171], 133[53], 136[53,63], 139 [53], 145[53,63], 152[53], 153 [53], 154[53], 155[53], 157[74], 160[53,63], 162[53,63], 163[53], 167[63,80], 168[53,74], 170[53, 63], 205[53], *232, 233*
Moore, M., 119[31], 128[336], 197 [336], *231, 244*
Mor, J.-R., 129[201], 176[201], *238*
Mori, T., 83[49], *106*
Moriarty, B., 215[294], 220[294], *242*
Morioka, A., 229[322,324,325], *243*
Morishita, M., 129[200], 176[200], *238*
Morisi, F., 251[20,23,92,95], 252 [92,95], 253[92,95], 259[23], 264[23], 273[95], 276[95], *305, 307, 308*
Morris, D., 250[8], 256[8], 257 [8], 266[8], *304*
Morris, D. L., 252[66,69], *306*
Mosbach, K., 77[2], 79[2], 85[84], 104[2], *104, 107,* 123[46], 124

[Mosbach, K.], [46], 126[188],
127[105,192], 128[46], 129
[192], 130[46,62,192], 131
[46], 134[57], 136[62], 162
[46], 163[46], 169[62], 172
[46], 184[105,192], 189[188],
194[192], 199[46,62], *232, 233,
234, 238,* 285[102,103], 287
[113], 288[102,108,113], 289
[113], 290[117], 291[113,117,
131], 292[117,118,120,121,123-
127,129], 293[117,118,120,124,
125,135,136,138,141,145], 294
[117,120,123,125], 295[120,
121,125,135,141], 296[120,124-
126,129,138], 297[118,120,138],
298[120], *308-310,* 333[25], *352*
Mottola, H., 114[5], *230*
Mrochek, J. E., 251[53], *306*
Mukerji, A. K., 123[47], 130[47],
151[47], 156[47], 162[47], 163
[47], 198[47], *232*
Müller, K., 128[196], 173[196],
238
Munnecke, D., 293[136], *309*
Murachi, T., 77[13], *104,* 252
[93], 253[93], 270[93], 280
[93], *307*
Murador, E., 252[56,61], 253[61],
269[61], 277[61], 278[56,61],
279[56], 280[56], *306*
Musha, S., 126[209], 129[209],
188[209], *239*

Nachlas, M., 54[54], *75*
Nagamura, Y., 229[325], *243*
Nagasawa, Y., 229[323,326], *243*
Nagda, N. L., 129[202], 176[202],
238
Nagy, G., 129[76,101], 131[58,
65], 136[58,65], 160[65], 161
[76], 162[76], 163[76], 168
[58,65], 170[65], 171[76], 181
[101], 185[101], *233, 234*
Nakamo, K., 209[267,268], *241*
Nakano, N., 114[7], 128[119],
197[119], *231, 235*
Nakayama, K., 114[8], 128[121],
197[121], *231, 235*

Nanjo, M., 126[55,56,93], 129[56,
93], 130[77], 131[55,165], 133
[55,56], 157[55,56], 161[55],
162[55], 163[55], 164[77], 174
[56], 183[56], 187[93], 192[55],
204[165], *232, 233, 234, 237*
Narayanan, S., 252[94], 267[94],
307
Nelson, J. M., 77[1], *104*
Neujahr, H., 128[116,187], 130
[162,163,187], 189[187], 190
[162,163], 215[162], 222[162],
235, 237, 238
Neujahr, H. Y., 128[115], *235*
Newirth, T. L., 253[74], 280[74],
307
Ngo, T., 127[107], 186[107], 208
[245], *234, 240,* 247[48], *306*
Ngo, T. T., 104[131], *109,* 249
[4,5], 250[4,5], *304*
Nikolaev, A., 81[29], *105*
Nikolayev, A. Y., 83[48], *106*
Nikolelis, D., 129[143], 202[143],
236
Nikolelis, D. P., 114[5], *230*
Nilsson, H., 123[46], 124[46], 128
[46,234], 130[46,62,220], 131
[46], 136[62], 162[46], 163[46],
169[62], 172[46], 199[46,62,
220], 204[234], 228[317], *232,
233, 239, 240, 243,* 293[138],
296[138], 297[138], *310*
Nilsson, L. G., 251[101], 282[101],
308
Noma, A., 114[8], *231*
Nonna, A., 128[121], 197[121], *235*
Noy, G. A., 77[14], *105*

Obana, H., 214[290,292], 215[290],
219[292], 221[290], *242*
O'Driscoll, K. F., 250[9], *304*
Ogata, K., 88[106], *108*
Ögren, L., 126[96], 130[233], 131
[170], 204[233], *234, 237, 240,*
249[3], 250[3,18,60], 251[30,
101], 252[60], 260[30], 262[18,
30], 272[60], 282[101], *304-
306, 308*
Oimomi, M., 17[8], *73*
Okada, S., 252[93], 253[93], 270
[93], 280[93], *307*

Okano, T., 228[319], *243*
Okinaka, S., 44[35], *74*
Okumura, H., 229[323], *243*
Olgren, L., 130[233], 131[170], 204[233], *237, 240*
Olliff, C. J., 130[159], 200 [159], *237*
Ollis, D. F., 205[240], *240*
O'Loughlin, P. D., 252[166], *311*
Olsen, A. C., 253[76], *307*
Olsen, L., 208[248], *240*
Olson, A. C., 83[54], 88[103], *106, 108*
Olson, C., 64[67], *75*, 120[32], 126[92], 129[142], 174[142], 188[92], *231, 234, 236*
Olsson, B., 128[234], 204[234], *240*
Onitiri, A. C., 253[97], 274 [97], *308*
Onyezili, F. N., 253[97], 274 [97], *308*
Osajima, Y., 127[113], 191[113], *235*
Osaki, T., 129[200], 176[200], *238*
Osann, G., 10[3], 52[3], *73*
Osawa, H., 128[218], 172[218], 207[218], 208[218], *239*
Ottesen, M., 88[106], *108*
Ozawa, H., 88[96], *108*

Palleschi, G., 118[26], 127[100], 183[100], *231, 234*
Papariello, G. J., 123[47], 130 [47,158], 151[47], 156[47], 162[47], 163[47], 198[47], 199 [158], *232, 236,* 251[24], 259 [24], *305*
Papastathopoulos, D., 118[27], 126[88,89], 200[88,89], 202 [88,89], *231, 234*
Papastathopoulos, D. S., 129 [143], 202[143], *236*
Pardue, H., 49[50], 60[59], 65 [68,69], *75*
Pascoe, E., 81[33], *105*
Patel, A., 81[32], *105*
Patel, A. B., 85[73], 88[101], *107, 108*
Patel, R. P., 85[71], 88[98], *107, 108*

Paul, F., 247[39], 248[39], 251 [39], 257[39], *305*
Pauling, L., 5[2], *73*
Pavan, B., 252[62], 253[62], 267 [62], 269[62], 270[62], 271[62], 277[62], 280[62], *306*
Pecht, M., 88[92,102], 89[112], *108*
Pederson, H., 206[242], *240*
Pegon, Y., 104[162], *111*
Pei, P. T.-S., 286[104], *308*
Pennington, S. N., 85[73], 88[101], *107, 108*, 287[106], *308*
Pesliakiene, M. V., 129[198], 173 [198], *238*
Peterson, G., 81[27], *105*
Peterson, J. W., 332[16], *351*
Pfeiffer, D., 104[157], *110*, 126 [90,223], 128[196], 130[223], 173[196], 201[90,223], *234, 238, 239*
Phillips, A., 208[253], *240*
Pin, A., 252[62], 253[62], 267 [62], 269[62], 270[62], 271[62], 277[62], 280[62], *306*
Pitcher, W., 104[164], *111*
Pittalis, F., 251[92], 252[92], 253[92], *307*
Plass, M., 28[18], *73*
Plowman, K. M., 23[11], *73*
Poltorak, O. M., 83[45,46], *106*
Porath, J., 85[79,80], 88[97,109], 89[80], *107, 108*
Potezny, N., 252[166], *311*
Potter, L. T., 68[87], *76*
Price, S., 85[71], 88[98], *107, 108*
Pring, B., [83], *307*
Prosperi, G., 251[92], 252[92], 253[92], *307*
Purdy, W. C., 65[70,71], 66[76], *76*
Pye, E., 253[74], 280[74], *307*
Pyon, H. Y., 212[302], 215[302], 224[302], *242*

Quiocho, F. A., 88[105], *108*

Racine, P., 128[131,132], 173[131, 132], 195[131,132,148], *235, 236*
Ramsey, T. A., 196[215], *239*

Razumas, V., 120[33], *232*
Read, D., 62[60], *75*
Rechnitz, G., 104[161], *110*,
116[20,21], 118[27], 126[86-
89,95], 127[103], 128[123,
181], 129[229,230], 131[168],
180[168], 182[95], 200[88,89,
123], 202[86-89], 203[229,
230], 204[237], 207[255], 208
[255], 209[255,261,262,269],
211[281], 212[230,303], 214
[230,293,303], 215[230,295,
297,303,305], 217[230,303],
218[293], 219[230,303], 220
[305], 223[297], 226[295], 227
[339], 228[318], 230[327,331],
231, 234, 235, 237, 239-244
Rechnitz, G. A., 116[17], 117
[25], 126[226], 127[67,68],
142[67,68], 143[68], 150[67,
68], 162[67,68], 198[67,68],
202[226], 212[309,310], 214
[289,291,309,310,313,315], 215
[312,316], 217[291,313], 218
[315], 220[289,310,313], *231,
233, 239, 242, 243*
Reed, D. L., 68[86], *76*
Reed, J., 62[60], *75*
Rees, A. W., 84[63], *106*
Reisel, E., 253[80], 276[80],
307
Rey, A., 121[38,39], *232*
Rhodius, R., 83[44], *106*
Rice, D. J., 215[301], 223[301],
242
Richards, F. M., 83[57], 88[105,
107], *106, 108*
Richardson, T., 127[109], 184
[109], *234*
Richterich, R., 44[33,36], *74*
Ridgway, T. H., 66[73], *76*
Riechel, T. L., 126[226], 202
[226], 212[309], 214[309], 215
[297,312], 223[297], *239, 242,
243*
Riecke, E., 293[136], *309*
Riemann, W., 250[27], 256[27],
305
Riendeau, C. J., 252[61], 253
[61], 269[61], 277[61], 278
[61], *306*

Rietz, B., 313[1], 326[1,12,13],
351
Riordam, J. F., 85[86], *107*
Riseman, J. H., 214[311], *243*
Risinger, L., 251[101], 282[101],
308
Rocks, B. F., 300[161,162], *311*
Rofe, A. M., 252[166], *311*
Rohm, T., 63[61], *75*
Romette, J. L., 129[204], 177
[204], *238*
Rony, P. R., 83[41], *106*
Ross, J. W., 214[311], *243*
Rossi, E., 44[33,36], *74*
Rosso, C., 252[56], 278[56], 279
[56], 280[56], *306*
Roy, A., 45[39], *74*
Roy, A. B., 27[16], *73*
Ruelius, H. W., 187[207,208], *239*
Rusling, J. F., 251[24], 259[24],
305
Ruzicka, J., 169[81], *233*

Sablin, P., 215[333], 221[333],
244
Sabotka, H., 88[91], *108*
Sack, R., 208[258], *240*
Sack, R. A., 49[46], *75*
Sadar, M. H., 77[11,12], 104[11,
12], *104*
Sadar, S., 125[48], *232*
Sailer, D., 250[27], 256[27], *305*
Saini, R., 91[142], *110*
Salak, J., 88[93], *108*
Salleh, A. B., 253[98], 280[98],
308
Samalius, A. S., 129[198], 173
[198], *238*
Sanfridsson, B., 292[118], 293
[118], 297[118], *309*
Sansur, M., 251[54], 252[54], 253
[54], 267[54], 281[54], *306*
Sasaki, K., 129[340], 194[340],
244
Sato, Y., 129[180], 177[180], *237*
Satoh, I., 104[160], *110*, 126
[244], 128[122], 131[164,167,
244], 180[167], 196[122], 197
[122], 203[244], 211[279], 214
[279,300], 215[279], 221[279],
224[279], *235, 237, 240, 241,*

[Satoh, I.], *242*, 250[17], 251
  [19], 262[17,19], 293[141],
  295[141], *304, 310*
Satoh, J., 126[91], 188[91], *234*
Sawai, M., 229[323,326], *243*
Sayers, C., 288[109], 293[142],
  294[142], *308, 310*
Schäl, W., 250[27], 256[27], *305*
Schanel, L., 130[210], 190[210],
  239
Schapira, G., 44[32,37], *74*
Schapira, R., 44[32], *74*
Schejter, A., 88[104], *108*
Scheller, F., 104[157], *110*, 126
  [90,223], 128[196], 129[197],
  130[223], 173[196,197], 201
  [90,223], 214[208], 215[306],
  222[288,306], *234, 238, 239,
  242*
Schery, H., 252[68], 273[68],
  306
Schick, H. F., 84[64], *107*
Schifreen, R., 288[116], *309*
Schifreen, R. S., 77[17], *105*,
  292[134], 294[134], *309*
Schiller, J. G., 129[203], 130
  [190], 177[203], 191[190], *238*
Schindler, J. G., 131[216], 168
  [216], *239*, 250[27], 256[27],
  305
Schmidt, H. L., 292[130], 293
  [144], *309, 310*
Schlapfer, P., 128[131,132], 173
  [131,132], 195[131,132], *235*
Schleifer, A., 130[158], 199
  [158], *236*
Schmidt, E., 71[88], *76*
Schmidt, F., 71[88], *76*
Schormuller, J., 46[43], 47[44-
  45], *74*
Schram, E., 300[164], 302[164],
  311
Schultz, R., 129[133], *235*
Schütz, E., 26[13], *73*
Schütz, J., 26[14], *73*
Scott, M., 288[108], 292[124],
  293[124], 296[124], *308, 309*
Seary, R., 44[31], *74*
Seegopaul, P., 115[11,166], *231,
  237*

Sehon, A. H., 83[53], *106*
Seijo, H., 214[300], *242*
Seitz, W. R., 300[154], 301[154],
  310
Seligman, A., 54[54], *75*
Semersky, F. G., 250[14], 254[14],
  304
Serrir, D. R., 252[72], 283[72],
  284[72], *307*
Seyer, J., 128[196], 173[196],
  238
Shargool, P., 208[245], *240*
Shearer, C. M., 123[47], 130[47],
  151[47], 156[47], 162[47], 163
  [47], 198[47], *232*
Shields, A., 288[107], *308*
Shimura, Y., 129[340], 194[340],
  244
Shinbo, T., 17[8], *73*, 129[212],
  195[212], *239*
Shinohara, R., 229[325], *243*
Shu, F., 116[19], 117[19], 125
  [49], 127[49,104], 129[104], 131
  [49], 150[72], 163[49], 169[49],
  181[49], 182[104], 184[104],
  231-234
Shu, F. R., 129[278], 172[278],
  241
Shults, M. C., 129[276], *241*
Shuto, S., 229[323], *243*
Siegler, K., 129[197], 173[197],
  238
Silman, H., 89[110], 90[110], *108*
Silman, I. M., 83[42,58,59], 89
  [113], 95[58,123], *106, 108*
Simon, R., 65[69], *75*
Simon, R. K., 65[71], *76*
Singer, S. J., 84[64], *106*
Sir Ram, J., 85[87], *107*
Sjoholm, I., 300[157], 302[157],
  310
Skogberg, D., 127[109], 184[109],
  234
Smith, J. B., 253[86], 280[86],
  307
Smith, L., 46[42], *74*
Smith, M., 126[92], 188[92], *234*
Smith, R., 116[18], *231*
Smith, R. M., 131[171], *237*
Smith, R. S., 77[16], 97[16], *105*

Snyder, L. R., 251[54], 252[54], 253[54,86], 267[54], 280[86], 281[54], *306, 307*

Solsky, R. L., 230[331], *244*

Somers, P. J., 85[70], 104[158], *107, 110*

Sonobe, N., 126[91], 188[91], *234*

Spassov, G., 209[260], *241*

Spotorno, G., 251[92], 252[92], 253[92], *307*

Stakbro, W., 328[18], *351*

Stasny, M., 250[12], 254[12], *304*

Stauffer, J. F., 57[57], *75*

Sternberg, R., 128[136,193], 177 [136,138,139,140], 178[138, 140,193], 179[139], 206[139, 140], *235, 236, 238*

Stiffel, D. N., 250[14], 254 [14], *304*

Stillhart, H., 44[36], *74*

Stotz, E., 11[5], 15[5], 33[5], 50[5], *73*

Stowell, E., 31[22], *73*

Stubley, E., 119[30], *231*

Sudo, T., 229[326], *243*

Sugita, H., 44[34], *74*

Sugiura, M., 129[212], 195[212], *239*

Sulciman, A., 333[26], *352*

Sumi, Y., 251[52], 276[52], 283 [52], *306*

Sundaram, P. V., 251[55], 252 [55,65,70], 253[70,75,78,81, 87,91,99], 277[55,85,91], 278 [87], 281[78,99], 283[70], *306-308*

Sung, H. Y., 127[110], 186[110], *234*

Surinov, B. P., 84[66], 88[66], *107*

Suzuki, H., 215[296], 225[296], *242*

Suzuki, S., 104[160], *110*, 120 [36,37], 126[91,244], 128[122], 129[180], 130[160], 131[164, 167,244], 177[180], 180[167], 188[91], 196[122], 197[122], 203[244], 211[279], 214[279, 282-286,290,292,300], 215[279,

[Suzuki, S.], 290,296,298,299, 304], 219[292,304], 221[279, 290], 222[282], 223[283], 224 [279,298,299], 225[284-286,296], 229[322,324,325], 230[328-330], *232, 234, 235, 237, 240-242*, 250 [17], 251[19], 262[17,19], 292 [122], *304, 309*

Svendsen, I., 88[106], *108*

Svensson, A., 292[126,129], 296 [126,129], *309*

Svensson, K., 292[128], *309*

Svirmickas, G. S., 129[213], 195 [213], *239*

Szabo, A., 114[4], 116[4], *230*

Szabo, S., 114[4], 116[4], *230*

Tabachnick, M., 88[91], *108*

Tabata, M., 252[93], 253[93], 270 [93], 280[93], *307*

Takahashi, K., 17[8], *73*, 126[91], 188[91], *234*

Tannenbaum, S. R., 288[107], *308*

Tarp, M., 131[73], 150[73], 156 [73], 157[73], 169[73], *233*

Taylor, P., 128[205], 179[205], *238*

Teraoka, N., 215[304], 219[304], *242*

Thevenot, D. R., 128[136,193], 177[136,138,140], 178[138,140, 193], 206[140], *235, 236, 238*

Thomas, D., 104[155], *110*, 126 [225], 127[102,182,183], 129 [153,204], 130[153], 131[153], 174[128], 177[204], 180[153], 184[182], 185[183], 228[320, 321], *234-239, 243*

Thomas, L., 129[147], 208[246,247, 250], *236, 240*

Thomas, L. C., 66[79], *76*

Thompson, H., 116[21], 128[123], 200[123], *231, 235*

Thompson, M., 119[30], *231*

Toftgard, R., 300[156], *310*

Tokarsky, J., 130[157], 196[157], *236*

Tominaga, N., 131[172], 192[172], *237*

Toriyama, M., 119[29], *231*

Torstensson, A., 126[188,335],

[Torstensson, A.], 188[335], 189[188], *238, 244,* 250[25], *305*

Tosa, T., 83[49], *106*

Toul, Z., 126[277], *241*

Toyokura, Y., 44[34,35], *74*

Tran-Minh, C., 116[15], 126[199, 227], 127[199], 128[174,175], 129[199], 131[199], 167[174, 175], 170[199], 174[199], 184 [199], 202[227], 204[174,175], *231, 237-239,* 292[143], 293 [143], 294[143], *310*

Treichel, I., 214[314], 218[314], *243*

Triplett, R. B., 88[99,100], *108*

Tsou, K., 54[54], *75*

Tsubomura, H., 229[323,326], *243*

Tsuchida, T., 128[178,341], 175 [178], 200[341], *237, 244*

Tyson, B., 64[64], *75*

Udenfriend, S., 68[85], *76*

Ugarova, N. N., 300[149-152], 303[149-152], *310*

Umbreit, W. W., 57[57], *75*

Updike, S., 214[314], 218[314], *243*

Updike, S. J., 80[23-25], *105,* 121[43], 124[43], 129[43,276], 136[61], 170[43], *232, 233, 241,* 249[6], 250[6], 252[57], 254[6], 265[57], *304, 306*

Usdin, V., 79[20], 97[20], *105*

Ushikubo, S., 17[8], *73*

Vallee, B. L., 85[86], *107*

Vallin, D., 292[143], 293[143], 294[143], *310*

Vallon, J. J., 104[162], *111*

Van, W., 104[164], *111*

Van der Ploeg, M., 81[33,34], *105*

Van Duijn, P., 81[33,34], *105*

Vanuxem, D., 129[146], 193[146], 195[152], *236*

Varga, J., 128[116], *235*

Vasileva, T., 300[151], 303[151], *310*

Vasta, B., 79[20], 97[20], *105*

Vaughn, A., 316[2], 326[2], *351*

Velosy, G., 114[4], 116[4], *230*

Venkatasubramanian, K., 91[142], 103[126], *109, 110*

Vieth, W. R., 91[142,143], 103 [126], *109, 110*

Vodrazka, Z., 88[93], *108*

Voleski, B., 250[37], 257[37], *305*

Vonderschmitt, D. J., 252[64], 264[64], *306*

Von Storp, L. H., 117[23,24], 118 [23], 129[76], 161[76], 162[76], 163[76], 171[76], 208[24], *231, 233*

Vorobeva, E. S., 83[45,46], *106*

Vretbald, P., 85[79], *107*

Wada, M., 215[298], 224[298], *242*

Wallace, T., 120[34,35], *232*

Walsh, K. A., 83[55], *106*

Walters, R., 127[185], 185[185], 215[294], 220[185,294], *238, 242*

Wan, J., 81[31], *105*

Wang, C. H., 68[86], *76*

Wang, J., 204[235,236], *240,* 264 [42,43], *305*

Warburg, O., 10[4], 50[4], 57[56], *73, 75*

Ward, P., 252[89], 267[89], *307*

Wasilewski, J. C., 286[104], *308*

Wasserman, R., 253[87], 277[85], 278[87], *307*

Watanabe, T., 214[300], *242*

Watanabe, Y., 292[122], *309*

Watson, B., 77[9], 104[9], *104,* 250[14], 251[29], 254[14], 259 [29], *304, 305*

Wawro, R., 127[103], *234*

Weaver, J. C., 288[107,110], 293 [110], 294[110], *308,* 332[16], *351*

Wechs, L., 128[118], 197[118], *235*

Weeks, L., 114[6], *230*

Weetall, H. H., 85[67,68], 90[115-119], 91[116-119], 93[120], 94 [120,165], 95[120], 104[120, 164], *107, 109, 111,* 252[73], 253[73], 283[73], 285[73], *307*

Weibel, M. K., 249[7,10], 250

[Weibel, M. K.], [7,10,15], 251
[33], 254[7,33], 255[7], 257
[15], 261[33], *304, 305*
Weise, H., 129[197], 173[197],
238
Weitzmann, R., 208[259], *240*
Werner, M., 104[134], *109*, 252
[61], 253[61], 269[61], 277
[61], 278[61], *306*
Wheller, K. P., 85[77], *107*
White, C., 103[128], *109*
White, W., 127[108], 184[108],
212[108], *234*
Whittaker, M., 40[26], 41[26],
74
Whittam, R., 85[77], *107*
Wieland, T., 81[26], *105*
Williams, D. C., 300[154], 301
[154], *310*
Williams, D. L., 128[129], 129
[129], 172[129], 193[129], 209
[129], *235*
Williams, R. T., 130[159], 200
[159], *237*
Wilson, G. S., 129[217,278], 172
[217,278], *239, 241*
Winartasaputra, H., 66[81], *76*
Wingard, L. B., 129[202,203],
176[202], 177[203], *238*
Winquist, F., 293[138], 296
[138], 297[138], *310*
Wold, F. J., 83[60], *160*
Wolfson, S., 130[157], 196[157],
236
Wolfson, S. K., 127[106], 129
[203], 177[203], 184[106],
234, 238
Wollenberger, U., 214[288], 215
[306], 222[288,306], *242*
Woodbridge, J., 45[39], *74*
Worsfold, P., 119[30], *231*

Wright, J. M., 130[159], 200[159],
237
Wroblewski, F., 16[6], 39[27], 50
[6], *73, 74*

Yacynych, A., 126[206], 128[127],
174[127], 182[206], 183[206],
235, 238
Yagi, T., 209[267,268], *241*
Yamada, H., 126[244], 131[244],
203[244], *240*
Yamada, K., 127[113], 191[113],
235
Yamamoto, N., 229[323,326], *243*
Yao, S., 130[157], 196[157], *236*
Yao, S. J., 127[106], 129[203],
177[203], 184[106], *234, 238*
Yao, T., 126[209], 129[209], 188
[209], *239*
Yasuda, T., 214[282,283,290,292],
215[290,296], 219[292], 221
[290], 222[282], 223[283], 225
[296], *241, 242*
Yatsimirskii, K. B., 39[24], *73*
Yaverbaum, S., 104[164], *111*
Yoda, K., 128[178,341], 175[178],
200[341], *237*
Yoshino, F., 128[218], 172[218],
207[218], 208[218], *239*
Yourtree, D. M., 251[51], 264
[51], 293[139], *306, 310*
Yuan, C., 226[337], 227[338], *244*

Zaborsky, O., 103[130], *109*
Zahn, H., 84[62], *106*
Zimmer, M., 49[48], *75*
Zimmerman, H., 42[29,30], 45[38],
74
Zimmerman, R., 327[15], *351*
Zimmerman, R. L., 325[4], *351*

Subject Index

Absorption curve,
  NAD and NADH, 59
  NADP and NADPH, 59
Acetate, electrode for, 126,
    159, 214, 216, 224
Acetate kinase, 366
Acetylcholine, electrode for,
    126, 202
S-Acetyl coenzyme A synthetase,
    366
Acetylesterase, 366
α-N-Acetyl-galactosaminidase,
    366
N-Acetyl β-D-glucosaminidase,
    366
β-N-Acetyl hexosaminidase, 366
Acetyl β-methyl choline, elec-
    trode for, 126
N-Acetylneuraminic acid aldo-
    lase, 367
Acetylserine sulfhydrylase,
    electrode for, 208
Acetyl trypsin, 367
Acid phosphatase (see Phospha-
    tase, acid)
Aconitase, 367
Activation energy, 32, 94
Activator, effect on enzyme,
    27, 28
Activity, enzyme, definition
    of, 14, 15
Actomyosin, 367
Acylase, 361, 367
Acyl coenzyme a oxidase, 367
Acyl coenzyme a synthetase, 367
Adenosine, electrode for, 126,
    201, 214, 218

Adenosine deaminase, 367
Adenosine 5' monophosphate, elec-
    trode for, 126, 201, 202
Adenosine phosphates, structures
    of, 6
Adenosine-5'-phosphosulfate
    kinase, 367
Adenosine 5'-triphosphatase, 367
Adenosine 5'-triphosphate (ATP):
    absorption and fluorescence, 59
    electrode for, 126, 201
    enzyme reactor for, 292, 295,
        299, 300, 302, 303
Adenosine-5'-triphosphate
    sulfurylase, 368
5-Adenosyl-L-homocysteine hydro-
    lase, 368
5'-Adenylic acid deaminase, 368
Adsorption techniques, immobiliza-
    tion, 83
Agarase, 368
L-Alanine aminotransferase, 368
    SSSF assay, 312, 326, 327
L-Alanine dehydrogenase, 368
Albumin, assay of, 292
Alcohol:
    assay with crystals, 333
    electrode for, 121, 159, 183,
        186-191, 195, 214, 221, 222
    enzyme reactor for, 250, 251,
        258, 259, 282, 296
    instruments for assay, 334
Alcohol dehydrogenase, 361, 368
Alcohol oxidase:
    electrodes for, 208
    source of, 368
Aldehyde dehydrogenase, 368

Aldehydes, electrodes for, 126
Aldolase, 361, 369
Alkaline phosphatase (*see* Phosphatase, alkaline)
Amine oxidase:
  electrode for, 208
  source of, 369
Amines, electrodes for, 126
Amino acid arylamidase, 369
D-Amino acid oxidase, 369
Amino acids:
  properties of, 2-4
  spectral characteristics, 59
  structure of, 2-4
D-Amino acids, electrodes for, 126, 159, 182
L-Amino acid oxidase, 369
L-Amino acids:
  enzyme electrodes for, 116-118, 125, 126, 158, 174, 181, 183, 184, 208
  enzyme reactors for, 249, 251
  instruments for assay, 334, 350
Aminoacyl t-RNA synthetase, 369
δ-Amino levulinate dehydratase, 370
Aminopeptidase, 370
Ammonia:
  determination of, 60, 61
  enzyme electrode for, 214, 216, 223
Amperometry, use in enzymatic analysis, 64, 65
Amygdalin:
  characteristics of electrode, 142, 143
  enzyme electrode for, 127, 159, 198
α-Amylase:
  enzyme electrode for, 113, 207, 208, 210
  enzyme reactor for, 250, 257
  source of, 361, 370
β-Amylase, 370
γ-Amylase, 370
Amyloglucosidase, 361, 371
Androgens, electrode for, 222
Androsterone, assay of, 300, 303
Antibodies, electrode for, 226, 230

Antidiuretic hormone, electrode for, 214, 216, 220
Antigens, electrode for, 226-229
Apotryptophanase, 371
Apyrase, 371
Arginase:
  electrode for, 207, 208
  source of, 371
L-Arginine:
  enzyme electrode for, 127, 184, 185
  enzyme reactor for, 249, 250
Arginine decarboxylase, 371
Arginine kinase, 371
Argininosuccinate lyase, 371
Aromatase, 371
Arylamine transferase, 371
Aryl sulfatase, 371
Ascorbate oxidase, 372
Ascorbic acid:
  enzyme electrode for, 127, 191, 192
  enzyme reactor for, 292
L-Asparaginase:
  electrode for, 208
  source of, 372
L-Asparagine:
  enzyme electrode for, 127, 159
  enzyme reactor for, 249, 250, 251, 273, 276
Aspartase:
  source of, 372
  specificity of, 9, 10
L-Aspartate:
  enzyme electrode for, 212, 214, 216, 220
  enzyme reactor for, 251, 276, 283
L-Aspartate aminotransferase:
  assay with SSSF, 312, 326, 327
  enzyme reactor for, 252
  source of, 372
L-Aspartate carbamoyltransferase, 372
Aspartate decarboxylase, 372
Automation:
  multiple-point method of analysis, 51, 52
  two-point methods, 51, 52
Autozyme tubes, 281

Benzoyl arginine ethyl ester,
    assay of, 292, 294, 295
Berthelot reaction, 61
Bifunctional reagents, 83, 84
Biological fluids, examination
    of, 41
BOD, electrode for, 214, 216,
    224, 225
Brilliant cresyl blue, 55
Bromelain, 372, 361
Buffer:
    effect on enzyme reactions,
        35, 38
    types, 36, 37
Butyrylthiocholine, electrodes
    for, 128

Calculase, 372
CAM, 263, 341, 345, 346
Carbamate kinase, 373
Carbohydrates:
    enzyme electrode for, 113,
        121, 124, 128, 158, 170-
        178, 180, 183, 214, 216,
        220, 221
    enzyme reactors for, 249, 265-
        275, 296, 297
Carbonic anhydrase, 373
Carboxy esterase, 373
Carboxypeptidase, 361, 373, 374
Carla Erba instrument, 333, 334
Carnitine acetyltransferase,
    374
Catalase, 361, 374
Catechol, electrode for, 128,
    189, 190
Catechol-O-methyltransferase,
    374
Cathepsin, 374
Cellobiose, enzyme reactor for,
    292
Cellulase, 374
Cephalosporin:
    electrode for, 214, 216
    enzyme reactor for, 292, 298
Cephalosporinase, 375
Ceramide trihexosidase, 375
Ceruloplasmin, 375
Chitinase, 375
Chloramphenicol acetyltrans-
    ferase, 375

Chloroperoxidase, 375
Cholesterol:
    assay with SSSF, 312, 324
    commercial instrument for, 333,
        334, 337
    determination of, 114
    enzyme electrode for, 114, 118,
        120, 128, 159, 196, 197,
        214, 216, 222
    enzyme reactor for, 251, 282,
        292, 294, 295
    enzyme stirrer for, 312, 330,
        331
    esters, 128, 196, 197, 292
Cholesterol esterase, 375
Cholesterol oxidase, 375
Choline acetyltransferase, 375
Choline kinase, 375
Choline oxidase, 375
Cholinesterases:
    assay with SSSF, 312, 325
    electrode for, 117, 208
    immobilized, 89
    source of, 361, 376
Choloylglycine hydrolase, 376
Chondroitinase, 376
Chondro-4-sulfatase, 376
Chondro-6-sulfatase, 376
Chromopapain, 376
Chymotrypsins:
    electrode for, 208
    source of, 362, 376, 377
α-Chymotrypsinogen A, 377
Citrate lyase, 377
Citrate synthase:
    electrode for, 208
    source of, 377
Citrulline, assay of, 277
Classification of enzymes, 11-17,
    353-360
Clinibond, 267, 277-280
Clostripain, 377
Coenzyme A:
    structure of, 6
    use of, 9
Coenzymes, 5-9
    spectral characteristics, 59
    stability of, 71
    structure of, 6-8
Collagen, membranes for immobili-
    zation, 91-93

Collagenases, 377
Columns, enzyme, 245-304
Commercial instruments, 333-350
Conductimetry, methods, 121
Copper:
  enzyme electrode for, 128, 204
  enzyme reactor for, 293, 295,
    296
Covalent binding of enzymes, 83-
    93
Creatinase, 377
Creatine:
  assay with SSSF, 312, 331,
    332
  electrode for, 128
  enzyme reactor for, 251
Creatine kinase:
  enzyme reactor for, 251
  isoenzymes, 16
  source of, 362, 378
Creatininase, 378
Creatinine:
  commercial instruments for,
    350
  enzyme electrode for, 118,
    128, 200, 201
  enzyme reactor for, 252, 267,
    277-280, 292, 294
Cyanide, enzyme reactor for, 292
Cyclic nucleotide phospho-
    diesterase, electrode for,
    207, 208
Cysteine, electrode for, 214
Cytidine phosphate, structure
    of, 6
Cytochrome C peroxidase, 378
Cytochrome C reductase, 378
Cytochrome oxidase, 378

Deoxyribonuclease, 362, 378
Deoxyribonucleic acid ligase,
    379
Deoxyribonucleic acid polymer-
    ase, 379
Dextranase, 362, 379
Diagnostic value of enzymes,
    39-45
Dialysis membrane, 152, 153
Diamine oxidase, 379
  to study enzyme reactions,
    113-121

Diamines, electrodes for, 126
Dianisidine, 55
Diaphorase, 379
Diastase, 380
Dichloroindophenol, 54
Dihydrofolate reductase, 380
Dihydroorotate dehydrogenase, 380
Dihydropteridine reductase, 380
Dihydropyrimidinase, 380
Diphosphopyridine nucleotide
    (see Nicotinamide adenine
    dinucleotide)
Dip sticks, 331, 332
Direct injection enthalpimetry,
    286
DNA nucleotidylexotransferase,
    380

Elastase, 380
Electrochemical detectors, for
    reactors, 248-251, 253-265
Electrochemical methods, 62-66
Electrodes:
  enzymes, 112-230
  substrate, 206-210
Emulsin, 386
Endoglycosidase, 381
Endonuclease, 381
Endopeptidase, 381
Enolase, 381
Enteropeptidase, 381
Enzymatic acalysis:
  basis of, 1-10
  in beverages, 46
  in botany and agriculture, 47
  experimental techniques, 47-69
  in food processing, 45, 46
  measurement techniques, 56-67
  in medicine, 39-46
  in microbiology, 47
  principles of, 9-24
  specificity of, 9, 10
Enzyme activities:
  in blood diseases, 40
  in bone diseases, 44
  for diagnosis and therapy, 39-
    46
  for differential diagnosis, 40,
    41
  electrodes for, 206-210
  ethnic correlation, 46

[Enzyme activities]
in fermentation, 45, 46
in food processing, 45, 46
in heart diseases, 40, 41
myocardial infarct, 42-44
in liver diseases
acute hepatitis, 40, 42, 43
carcinoma, 40
chronic hepatitis and
cirrhosis, 40, 44
obstructive jaundice, 40-44
vital hepatitis, 42, 44
in muscle diseases, 40, 44, 45
in pancreas diseases, 40
in serum, 40
in tears, 40
in urine, 40
Enzymatic activity:
definition of, 14, 15
measurement of, 56-67
specificity of, 9, 10
Enzyme classification, 11-17,
353-360
Enzyme coils, 349, 350
Enzyme electrodes, 112-230
construction of, 125, 132-135
performance characteristics,
137-164
selectivity and interferences,
157-164
Enzyme kinetics, 19-23
theory of, 204-206
Enzyme layers, 311-327
Enzyme pipette, 277
Enzyme reactions:
effect of activators, 27, 28
effect of buffers, 35-38
effect of enzyme, 26, 27
effect of inhibitors, 28-31,
163, 164
effect of ionic strength, 38,
39
effect of substrate, 20, 21,
24, 25
effect of temperature, 31-33
factors affecting, 24-35
pH dependence, 34
properties of, 17-24
Enzyme reactors, 245-304
electrochemical detectors,
248-251, 253-265

[Enzyme reactors]
luminescence detectors, 298-304
thermal detectors, 285-298
UV-visible detectors, 251, 252,
265-285
Enzyme stirrer, 307-331
Enzymes:
basic properties, 1-12
classification of, 11-17, 353-
360
as reagents, 1, 2, 10, 11, 39-45
specificity of, 9, 10, 157-164
stability of, 69-71
structure, 2-6
Enzyme thermistor, 286-291
Equilibrium method, for assay,
47, 48
Erepsin, 381
Esperase, 381
Estradiol dehydrogenase, 381
Ethanol:
assay with crystal, 333
assay with enzyme stirrer, 312,
329
enzyme electrode for, 128, 186,
187, 191, 214, 216, 219,
221, 222
enzyme reactor for, 250, 251,
258, 259, 282, 292, 296
Exonuclease, 381
Experimental techniques, 47-69

Fatty acid thioesterase, 381
Ferricyanide, 55
Fibrinolysin, 381
Ficin, 362, 382
Figlu enzymes, 382
Fixed concentration method, 49, 50
Fixed time method, 51, 52
Flavin adenine dinucleotide (FAD):
absorption and fluorescence, 59
structure of, 7
use of, 9
Flavin mononucleotide (FMN):
absorption and fluorescence, 59
structure of, 7
Flow injection analysis, 272
Fluorescence, solid surface
analysis, 311-327
Fluoride ion, electrode for, 128,
204

Fluorometry:
  in analysis, 66-68
  solid surface analysis, 311-
    327
Formaldehyde, assay with
  crystal, 333
Formaldehyde dehydrogenase, 382
Formate dehydrogenase, 382
Formic acid, electrode for, 126,
  159, 214, 216, 224
Formyltetrahydrofolate synthe-
  tase, 382
β-Fructofuranosidase, 382, 383
D-Fructose, electrode for, 214,
  221
Fructose-1,6-diphosphatase, 382
D-Fructose-6-phosphate kinase,
  382
L-Fucose dehydrogenase, 383
α-L-Fucosidase, 383
Fumurase, 383

GABase, 383
Galactokinase, 383
D-Galactose:
  assay with dip stick, 312,
    331, 332
  commercial instruments for,
    333, 334, 337, 338
  enzyme electrodes for, 115,
    128, 179, 180
  enzyme reactor for, 250, 252,
    257, 273, 292
β-Galactose dehydrogenase, 383
Galactose oxidase:
  source of, 383
  use in assay, 11
D-Galactose-1-phosphate uridyl
    transferase, 383, 384
α-Galactosidase, 384
β-Galactosidase, 384
Galactosyltransferase, 384
Gas membrane electrodes, 115-
  117
Gelatinase, 384
Gentamicin, enzyme reactor for,
  292
Gentamicin acetyltransferase,
  384
Gestrogens, electrode for, 222
Glass, immobilization of
    enzymes on, 90, 91

Glucanase, 384
D-Gluconate, enzyme electrode for,
  128, 180
Gluconate kinase, 384
Gluconolactonase, electrode for,
  209
Glucosaminidase, 384
D-Glucose:
  assay with dip stick, 312, 331,
    332
  assay with SSSF, 312, 325
  enzyme electrode for, 113, 121,
    124, 128, 158, 170-178, 180,
    183, 214, 216, 220, 221,
    249, 250
  enzyme reactor for, 252-257, 265-
    275, 291, 292, 294, 298-301
  enzyme stirrer for, 312, 329,
    330
  instrument for, 333-338, 341,
    342, 348-350
Glucose dehydrogenase, 384, 385
Glucose oxidase:
  electrode for, 209
  source, 362, 385
  stability of, 115, 117, 121,
    123, 124, 128
  use in assay of glucose, 10, 11,
    115, 117, 121, 123, 124, 203
D-Glucose-6-phosphate, enzyme
    reactor for, 300, 302
Glucose-6-phosphatase, 385
Glucose-6-phosphate dehydrogenase,
  362, 363, 385
Glucose phosphate isomerase:
  assay of, 312
  source of, 385
α-Glucosidase, 385
β-Glucosidase:
  characteristics and commercial
    source, 117, 362, 386
  electrode for, 209
β-Glucuronidase:
  electrode for, 209
  source of, 362, 386
L-Glutamate, enzyme electrode for,
  127, 159, 184, 194, 195,
  214, 216, 218, 219
L-Glutamate decarboxylase, 386
L-Glutamate dehydrogenase, 386
Glutamate oxaloacetate trans-
    aminase (GOT) (see
    Aspartate aminotransferase)

Glutamate pyruvate transaminase
   (GPT) (*see* Alanine gluta-
   mate transaminase)
Glutaminase:
   electrode for, 209
   source of, 387
L-Glutamine, enzyme electrode
   for, 129, 159, 182, 203,
   212, 214, 217-219
L-Glutamine decarboxylase, 387
L-Glutamine synthetase, 387
Glutamyl transpeptidase:
   assay with SSSF, 312, 326
   source of, 387
Glutathione reductase, 387
Glutathione transferase, 387
Glyceraldehyde-3-phosphate
   dehydrogenase, 362, 388
Glycerate dehydrogenase, 388
Glycerol, enzyme reactor for,
   281
Glycerol dehydrogenase, 388
Glycerol kinase, 388
Glycerol-1-phosphate dehydro-
   genase, 363, 388
Glycerol-3-phosphate dehydro-
   genase, 388, 389
Glycosidases, 389
Glycogen synthase, 389
Glycollate oxidase, 389
Glyoxalase, 389
Glyoxylate reductase, 389
Guanase:
   characteristics of, 202
   enzyme reactor for, 250, 264
   source of, 389
Guanine, enzyme electrode for,
   129, 202
Guanosine, structure of, 7
Guanylate kinase, 389

Handling of biochemical
   reagents, 69-71
Heart diseases, detection of,
   40, 41
Helicoid reactors, 245, 247,
   248, 257, 274, 275
Hemicellulase, 389
Hesperidinase, 389
Hexokinase:
   assay of activity, 312
   source of, 362, 363, 390
Histidinase, 390

L-Histidine, enzyme electrode for,
   127, 185, 214, 220
Histidine decarboxylase, 390
Hyaluronidase, 390
Hydrogenase, electrode for, 209
3-Hydroxy CoA dehydrogenase, 390
Hydroxybutyrate dehydrogenase:
   assay by SSSF, 312
   source of, 390
Hydroxydecanoyl thioester
   dehydrase, 391
Hydroxy mandelonitrile lyase, 391
Hydroxyprostaglandin dehydrogenase,
   391
Hydroxysteroid dehydrogenases, 391
Hydroxysteroid oxidase, 391
Hydroxysteroids, enzyme reactor
   for, 252, 281, 282

IgG, electrode for, 228, 229
IEM, 341, 345, 346
Immobilized enzymes:
   advantages of, 78
   basic applications of, 103, 104
   commercial availability, 361-365
   enzyme electrodes, 112-230
      construction of, 125, 132-137
      interferences, 157-164
      performance characteristics,
         137-164
      types of electrodes, 126-131,
         164-206
      using whole cells, 211-216
   enzyme reactors, 245-304
   enzyme stirrer, 327-331
   immunoelectrodes, 226-230
   immunoreactors, 297-299
   methods of preparation, 78-93
      experimental procedures, 98-
         103
   properties of, 93-98
   solid surface fluorescence
      analysis, 311-327
   substrate electrodes, 206-210
Immobilized substrates, 206-210
Immunoelectrodes, 226-230
Immunoreactors, 297-299
Inhibitors:
   commercial instruments for assay
      of, 341, 343-346
   competitive, 29, 30
   enzyme reactor for, 262, 263,
      283, 284, 293, 295

[Inhibitors]
  irreversible, 29-31
  noncompetitive, 30
  reversible, 28, 29
Inorganic pyrophosphatase, 391
Inosine, structure of, 7
Inositol dehydrogenase, 391
INPETTE, 277
Insecticides:
  commercial instruments for,
    341, 343-346
  enzyme reactors for, 251, 263,
    293, 300, 304
Insulin:
  enzyme electrode for, 221
  enzyme reactor for, 293
Interferences, in enzyme elec-
    trodes, 157, 160-164
Invertase, 391
Ionic strength, effect on
    enzymes, 38, 39
Ion selective electrodes, 62, 63
Isocitrate dehydrogenase, 392
Isoenzymes:
  definition, 16, 17
  electrophoretic separation, 16
Isoleucine RNA synthetase, 392
Isomerases, classification of,
    14

Kallikrein, 392
Katal, 16
Ketoglutarate dehydrogenase, 392
Kimble Analyzer, 333, 334, 342
Kinetics:
  chemical reactions, 17, 18
  enzyme reactions, 19, 48-52
KODAK Clinical Analyzer, 332

Lactase (see β-Glucosidase)
D-Lactate dehydrogenase, 392
L-Lactate:
  enzyme electrode for, 129,
    159, 188, 191, 193-196,
    214, 220, 221
  enzyme reactor for, 252, 283,
    293, 295, 300, 302
  instrument for assay of, 333,
    334, 347
L-Lactate dehydrogenase:
  assay by SSSF, 312, 327
  characteristics of, 120, 129

[L-Lactate dehydrogenase]
  enzyme electrode for, 209
  in heart diseases, 40, 42, 43
  isoenzymes, assay of, 16, 226-
    228
  in liver diseases, 40, 42, 43
  source of, 363, 392, 393
Lactose:
  enzyme electrode for, 129, 180,
    181
  enzyme reactor for, 250, 252,
    257, 274, 293, 296
  instruments for, 333, 334, 337,
    338
Lag phase, 23, 24
Lag time, 23
Laminarinase, 393
Lecithin:
  enzyme electrode for, 129
  enzyme reactor for, 250, 262
Leeds and Northrup Analyzer, 333,
    334, 342
L-Leucine:
  enzyme electrode for, 182-184
  enzyme reactor for, 249, 250
Leucine aminopeptidase:
  in liver diseases, 40, 42, 43
  source of, 370
Ligases, classification of, 14
Lipase, 393, 394
Lipids, enzyme reactor for, 251,
    262
Lipoprotein lipases, 394
Lipoxidase, 394
Liver, enzyme activity patterns
    in, 42-44
Luciferase, 394
Luminescence, detectors for
    reactors, 298-304
Lyases, classification of, 13, 14
L-Lysine:
  enzyme electrode for, 127, 183-
    185
  instrument for assay of, 350
L-Lysine decarboxylase, 394
Lysozyme:
  electrode for, 209, 214, 226
  source of, 394

Macerase, 395
L-Malate, enzyme electrode for,
    129, 191

L-Malate dehydrogenase:
  electrode for, 209
  source of, 363, 395
L-Maltose:
  enzyme electrodes for, 130,
      178, 180, 181
  enzyme reactors for, 251, 252
  instrument for, 333, 334, 338
Mandelonitrile lyase, 395
Mannosidase, 395
Manometry, 57
Mass spectrometry, 332, 333
McIlvaine buffer, 36
Measuring techniques, 56-59
  electrochemical, 62-66
  fluorescence, 66-68
  spectrophotometry, 57-62
Membrane reactors, 245, 247,
      248, 257, 274, 275
Mercury, assay by inhibition,
      130, 204, 262, 263, 293,
      295, 296
L-Methionine:
  enzyme electrode for, 127,
      182-184, 186
  instrument for, 350
Methylene blue, 54
Methylene tetrahydrofolate
      dehydrogenase, 395
Michaelis constants:
  determination of, 21, 22
  of immobilized enzymes, 95,
      97
Michaelis-Menten equation, 19-22
Microbes, electrode for, 214
Microbial electrodes, 211-226
Microencapsulation of enzymes,
      82, 83
Molar enthalpies, 286
Molecularity of reactions, 17
Monoamine oxidase, characteris-
      tics and commercial
      products, 115, 369
Monoamines, electrodes for, 126
Multiple-point method, 51, 52
Mutarotase, 395
Mycozyme, 395
Mylase, 395
Myocardial infarct, 42, 43
Myokinase, 396
Myosin, 396

Naringinase, 396
Nessler's reagent, 61
Neuraminidase, 396, 397
Nicotinamide adenine dinucleotide
      (NAD):
  absorption curve, 59
  enzyme electrode for, 130, 201,
      214, 223, 224
  fluorescence, 59
  stability, 71
  structure, 7
  use of, 9
NAD kinase, 396
NAD nucleotidase, 396
NAD pyrophosphorylase, 396
Nicotinamide adenine dinucleo-
      tide, reduced form (NADH):
  absorption curve, 59
  assay by mass spectrometry, 312
  enzyme electrode for, 130, 202,
      214, 224
  enzyme reactor for, 300, 302,
      303
  fluorescence, 59
  stability, 71
  structure of, 7
NADH peroxidase, 396
Nicotinamide adenine dinucleotide
      phosphate (NADP):
  absorption and fluorescence, 59
  stability of, 71
  structure of, 8
  use of, 9
Nicotinamide adenine dinucleotide
      phosphate, reduced form
      (NADPH):
  absorption and fluorescence of,
      59
  enzyme reactor for, 300, 302
  stability of, 71
  structure of, 8
Nicotinic acid, enzyme electrode
      for, 214, 216, 223, 224
Ninhydrin reaction, 61
Nitrate:
  assay with SSSF, 312, 325
  enzyme electrode for, 130, 159,
      165-167, 214, 216, 223
  enzyme reactor for, 251, 252,
      264, 283
Nitrate reductase, 397

Nitrilotriacetic acid, enzyme
    electrode for, 212, 214,
    224
Nitrite:
  enzyme electrode for, 130,
    159, 165, 167
  enzyme reactor for, 251, 264
Nuclease, 397
Nucleoside-5'-diphosphate kinase,
    397
Nucleoside monophosphate kinase,
    397
Nucleoside phosphorylase, 397
3'-Nucleotidase, 397
5'-Nucleotidase:
  electrode for, 207, 209
  source of, 397
Nucleotide pyrophosphatase, 397
Nylon enzyme tubes, 245-304
Nystatin, electrode for, 214,
    216

Octopine dehydrogenase, 397
Organophosphorus insecticides:
  enzyme reactors for, 251, 263,
    293, 300, 304
  instrument for assay, 341,
    343-346
Ornithine carbamytransferase,
    398
OTHERS, 246-304
Owens-Illinois Analyzer, 333,
    334, 342
Oxalate:
  enzyme electrode for, 130, 196
  enzyme reactor for, 252, 293
Oxalate decarboxylase, 398
Oxalate oxidase, 398
Oxidation-reduction methods,
    119, 120
Oxidoreductases, classification
    of, 12
Oxygen electrode:
  basics, 63
  use in measurement of enzyme
    reactions, 113-115
Oxynitrilase, 398

Pancreatin, 398
Papain:
  immobilization of, 89, 90
  source of, 363, 398

Pectinase, 398, 399
Pectinesterase, 399
Peletier device, 287, 288
Penicillin:
  enzyme electrode for, 130, 159,
    198-200
  enzyme reactor for, 251, 252,
    259, 273, 293, 294, 296-298
Penicillinase, 399
Pentosanase, 399
Pepsin, 363, 399
Pepsinogen, 399
Peptidase, 399
Peroxidase:
  source of, 363, 399
  use in assays, 10
Peroxide:
  enzyme electrodes for, 130
  enzyme reactors for, 293, 294,
    298-300
Pesticides, enzyme reactors for,
    251, 263, 293, 300, 304
pH:
  effect on electrode response,
    150, 151
  effect on enzyme reactions, 34
  effect on immobilized enzymes,
    94, 96, 97
Phenazine methosulfate, 55
Phenols:
  enzyme electrode for, 130, 190,
    191, 214, 216, 222
  enzyme reactors for, 293
L-Phenylalanine, enzyme electrode
    for, 117, 127, 182, 184,
    185, 214, 216, 218
Phenylalanine ammonia lyase, 399,
    400
Phenylalanine decarboxylase, 400
Phenylalanine hydroxylase, 400
Phenylethanolamine N-methyltrans-
    ferase, 400
Phosphatase, acid:
  assay with SSSF, 326
  isoenzymes, 16
  source of, 400
Phosphatase, alkaline:
  assay of with SSSF, 326
  electrode for, 119, 120, 209
  isoenzymes, 16
  in liver diseases, 40-42
  source of, 363, 400, 401

Phosphate, inorganic:
  enzyme electrode for, 130,
    159, 164, 165
  enzyme reactor for, 252, 283-
    285
Phosphodiesterases, 401
Phosphoenolpyruvate, enzyme
    reactor for, 253, 282
Phosphoenolpyruvate carboxylase,
    402
Phosphoglucomutase, 402
Phosphogluconate dehydrogenase,
    402
Phosphoglucose isomerase, 402
Phosphoglycerate dehydrogenase,
    402
Phosphoglycerate kinase, 402
Phosphoglycerate mutase, 402
Phospholipase, 403
Phosphomannose isomerase, 403
Phosphoribosyl pyrophosphate
    amidotransferase, 403
Phosphoribulo kinase, 403
Phosphorylases, 404
Phosphorylase kinase, 404
Phosphotransacetylase, 404
Physical entrapment of enzymes,
    79-82
Phytase, 404
Piezoelectric crystal detector,
    333
Plasminogen, 404
Polarimetric methods, 62
Polynucleotide kinase, 404
Polynucleotide phosphorylase,
    404
Polynucleotide synthetase, 404
Polyol dehydrogenase, 405
Polyphenol oxidase, 405
Potentiometry, use in measure-
    ment of enzyme reactions,
    63, 64
Preincubation, 23, 24
Prolidase, 405
Prolinase, 405
Pronase, 364
Prostaglandin dehydrogenase,
    405
Prostaglandin ketoreductase,
    405
Prostaglandin reductase, 405
Prostaglandin synthase, 405

Proteases:
  electrode for, 209
  source of, 364, 405, 406
Proteinase, 364, 406
Protein kinase, 406
Proteolytic enzyme, 406
Pullulanase, 406
Purine nucleoside phosphorylase,
    406
Pyridoxyl-5-phosphate, structure
    of, 8
Pyroglutamate aminopeptidase, 406
Pyrophosphatase, 406
Pyruvate:
  enzyme electrode for, 130, 214,
    221, 224
  enzyme reactor for, 253, 282,
    283, 300, 302
Pyruvate decarboxylase, 406
Pyruvate kinase, 364, 406, 407

Radiochemical methods, 68, 69
Reaction kinetics, 19-22
Receptor-destroying enzyme, 407
Renin, 407
Resazurin, visualization of UV
    reactions, 55
Response time, of enzyme elec-
    trodes, 145-154
Restriction endonucleases, 407,
    408, 409
Reverse transcriptase, 409
Rhodanase:
  characteristics of, 117
  electrode for, 209
  source of, 409
Ribonucleases, 364, 409, 410
Ribonucleic acid ligase, 410
Ribonucleic acid polymerase, 411
t-RNA synthetase, 411
Ribonucleic acid transfer enzyme,
    411
Ribonucleoside diphosphate reduc-
    tase, 411
Ribonucleoside triphosphate
    reductase, 411
Ribonucleotide reductase, 411
D-Ribulose-1,5-diphosphate
    carboxylase, 411
D-Ribulose-5-phosphate-3-
    epimerase, 411
Roche Analyzer, 334, 347

Saccharopine dehydrogenase, NAD,
    lysine forming, 411
Secretin, 411
Semisolid surface fluorescence
    (SSSF), 311-327
Semicarbazide, electrode for,
    131
Serine, electrode for, 214
Setric Analyzer, 333, 334, 347
Silicone rubber pads, 311-327
    advantages of, 319, 320
    assay with, 321-327
    experimental, 314-319
Silver, assay of using enzyme
    reactor, 293
Slope methods, 49, 50
Solea-Tacussel electrode instru-
    ment, 334, 348, 349
Solid surface fluorescence, 311-
    327
Solid membrane electrodes, 117-
    119
Sorbitol dehydrogenase, 411, 412
Sphingomyelinase, 412
Specific activity, 15
Spectrophotometric methods, for
    assay of enzyme reactions,
    57-62
Stability:
    of biochemical reagents, 69-71
    of coenzymes, 71
    of enzymes in serum, 69-71
    of enzyme electrodes, 137-144
Starch:
    enzyme electrode for, 179, 201
    enzyme reactor for, 251, 253
Steroids, enzyme electrode for,
    214
Stirrers, enzyme, 327-331
Streptodornase, 412
Streptokinase, 412
Substrate inhibition, 25
Succinate, enzyme electrode for,
    131, 159
Succinic thiokinase, 412
Sucrose:
    commercial instrument for,
        333, 334
    determination of, 34
    enzyme electrode for, 131,
        179-181, 214, 216, 221

[Sucrose]
    enzyme reactor for, 251, 253,
        273, 274, 293, 296, 298
Sulfate:
    enzyme electrode for, 131, 159,
        165, 166
    enzyme reactor for, 253, 283,
        284
Superoxide dismutase, 412
Syphylis, electrode for, 230

Tacussel analyzer, 334, 337, 338,
    348, 349
Taka diastase, 412
Tautomerase, 412
Technicon Instrument, 333, 334,
    342
Temperature dependence:
    of enzyme electrodes, 151
    of enzyme reactions, 31-33
    of immobilized enzymes, 95, 96,
        98
Terminal deoxynucleotidyltrans-
    ferase, 412
Testosterone, enzyme reactor for,
    253, 300, 305
Tetrazolium salts, use in visualiz-
    ing UV reactions, 53-55
Thermal detectors, for enzyme
    reactors, 285-298
Thiamine pyrophosphate, structure
    of, 8
Thiophorase, 412
Thiosulfate, electrode for, 131
Thrombin, 412
Thrombokinase, 413
Thymidylate synthetase, 413
Thyroid peroxidase, 413
Transaldolase, 413
Transformylase, 413
Transfructolyase, 413
Transglucolyase, 413
Transhydrogenase, 413
Transketolase, 413
Trapping reagents, 53
Triglycerides, enzyme reactors
    for, 257, 281, 293, 295
Triosephosphate isomerase, 413
Tripeptide synthetase, 314
Tris buffer, 37

Trypsin:
  immobilized, 89
  source of, 364, 413
Trypsin inhibitors, 364, 365,
    414
Trypsinogen, 414
L-Tryptophan:
  enzyme electrode for, 184
  enzyme reactor for, 253, 275
Tryptophanase, 414
Tyramine, electrode for, 131
Tyrosinase (see Polyphenol
    oxidase)
L-Tyrosine:
  enzyme electrode for, 117,
    125, 127, 158, 184-186
  enzyme reactor for, 293, 298
  instrument for, 350
L-Tyrosine decarboxylase, 414

Ultraviolet methods, visualiza-
    tion of, 53-55
Universal Sensors, 334, 350
Units of enzyme activity, 14-16
Urea:
  assay with dip stick, 312, 332
  assay by mass spectrometry,
    312, 332, 333
  assay by SSSF, 312, 321
  commercial instruments for,
    333, 334, 338, 339, 340,
    350
  enzyme electrode for, 116,
    131, 158, 167-170, 174
    interferences in, 160, 161
  enzyme reactors for, 251, 253,
    259-262, 266, 267, 268,
    269, 270-273, 276, 277,
    293, 294
  enzyme stirrer for, 312, 328
Urease:
  effect of pH and buffer on,
    38
  electrode for, 116, 209, 210
  isolation of, 1
  source of, 365, 414, 415
  stability of, 139-142
  units of activity, 15

Uric acid:
  assay by SSSF, 312, 321
  commercial instruments for, 333,
    334, 337, 350
  enzyme electrode for, 131, 159,
    192, 193
  enzyme reactor for, 251, 253,
    254, 261, 267, 269, 271,
    278, 280, 293, 300, 302
Uricase, 365, 415
Uridine-5'-diphosphogalactose-4-
    epimerase, 415
Uridinediphosphoglucose dehydro-
    genase, 415
Uridinediphosphoglucose pyrophos-
    phorylase, 415
Uridine phosphate, structure of,
    8
Uridyl transferase (see Galactose-
    1-phosphate uridyl trans-
    ferase)
Urokinase, 415
Urophrinogen-1-synthetase, 416
UV-visible detectors, for enzyme
    reactors, 251, 252, 265-285

Variable time method, 50, 51
Visual color tests, eip stick,
    331, 332
Visualization of UV reactions,
    53-56
Vitamin B, electrode for, 214,
    216

Washtime, of enzyme electrode,
    156, 157
Wheat germ lysate, 416
Whole cell electrodes, 211-226

Xanthine oxidase, 416
L-Xylulose reductase, 416
Xymolean, 416

Yellowsprings Instruments, 333,
    334, 342

Zinc, assay by enzyme reactors,
    251